CONTENT AND MODALITY

Content and Modality

Themes from the Philosophy of Robert Stalnaker

Edited by
JUDITH THOMSON
and
ALEX BYRNE

CLARENDON PRESS · OXFORD

OXFORD
UNIVERSITY PRESS

Great Clarendon Street, Oxford OX2 6DP

Oxford University Press is a department of the University of Oxford.
It furthers the University's objective of excellence in research, scholarship,
and education by publishing worldwide in

Oxford New York

Auckland Cape Town Dar es Salaam Hong Kong Karachi
Kuala Lumpur Madrid Melbourne Mexico City Nairobi
New Delhi Shanghai Taipei Toronto

With offices in

Argentina Austria Brazil Chile Czech Republic France Greece
Guatemala Hungary Italy Japan Poland Portugal Singapore
South Korea Switzerland Thailand Turkey Ukraine Vietnam

Oxford is a registered trade mark of Oxford University Press
in the UK and in certain other countries

Published in the United States
by Oxford University Press Inc., New York

First published 2006

British Library Cataloguing in Publication Data

Data available

Library of Congress Cataloging in Publication Data

Content and modality : themes from the philosophy of Robert Stalnaker / edited by
Judith Thomson and Alex Byrne.
p. cm.
Includes index.
ISBN-13: 978–0–19–928280–7 (alk. paper)
ISBN-10: 0–19–928280–3 (alk. paper)
1. Modality (Logic) 2. Logic. 3. Language and languages—Philosophy. 4.
Stalnaker, Robert. I. Thomson, Judith. II. Byrne, Alex.
BC199.M6C65 2006
191—dc22 2006020132

Typeset by Laserwords Private Limited, Chennai, India
Printed in Great Britain
on acid-free paper by
Biddles Ltd., King's Lynn, Norfolk

ISBN 0–19–928280–3 978–0–19–928280–7

1 3 5 7 9 10 8 6 4 2

Contents

Preface vii

1. Actors and Zombies 1
   Daniel Stoljar
2. The Frege–Schlick View 18
   Sydney Shoemaker
3. Character Before Content 34
   Paul M. Pietroski
4. Idiolects 61
   Richard G. Heck, Jr.
5. There Are Many Things 93
   Vann McGee
6. Stalnaker on the Interaction of Modality with Quantification and Identity 123
   Timothy Williamson
7. Conditional-Assertion Theories of Conditionals 148
   William G. Lycan
8. Non-Catastrophic Presupposition Failure 164
   Stephen Yablo
9. The Story of 'Fred' 191
   Frank Jackson
10. Stalnaker and Indexical Belief 204
    John Perry
11. Understanding Assertion 222
    Scott Soames
12. Responses 251
    Robert Stalnaker

Publications by Robert Stalnaker 296
List of Contributors 301
Index 303

Preface

Robert Stalnaker is a major presence in contemporary philosophy. His contributions over the years to philosophy of language, philosophy of mind, and metaphysics have had, and continue to have, a profound impact on work in all of those areas.

In philosophy of language, Bob's work has ranged widely over conditionals, presupposition, context, assertion, indexicals and belief attribution. His possible-worlds semantics for counterfactuals, his treatment of presupposition as a pragmatic phenomenon, and his account of assertions as effecting changes in the conversational context, are now staples of philosophy of language—and of semantics as studied in linguistics departments.

A central preoccupation of Bob's work in philosophy of mind has been "the problem of intentionality"—roughly, the problem of explaining how mental states, words, pictures, and so forth, can represent things. Three ideas underpin his approach to the problem. First, that the direction of explanation runs from thought to language. (Thus he rejects what he calls the "linguistic picture," which takes language as the fundamental vehicle of representation.) Second, that intentionality is to be explained in terms of causal and counterfactual relations. Third, that propositions are individuated by their possible-worlds truth conditions. His approach to the problem of intentionality fits neatly with externalism about mental content, of which he has been one of the major defenders. These ideas and their consequences have been at the heart of contemporary debates in philosophy of mind, and Bob has been a pivotal figure in all of them.

Although Bob has always had a deep interest in metaphysics, his general attitude toward it is cautious and mildly skeptical. This shows itself in particular in his defense, in opposition to David Lewis, of a moderate (actualist) realism about possible worlds. For Bob, possible-worlds talk is primarily a useful tool: for instance, he adopts a version of Lewis's counterpart theory in order to defuse puzzles about so-called "contingent identity". Influential themes present throughout Bob's writings on metaphysics are the importance of distinguishing semantic and logical questions from metaphysical ones, an emphasis on formal frameworks as offering ways of clarifying metaphysical questions, and an abiding suspicion of conceptual analysis and the a priori.

Bob joined the Department of Linguistics and Philosophy at MIT in 1988, and he has been a major figure in its activities ever since. His classes and seminars are attended by linguists as well as philosophers. (His class on modal logic is particularly popular, despite the fact that students emerge from it in a state of exhaustion.) He is an active participant in our Proseminar, which is required of our first-year graduate students, and he regularly teaches classes and seminars in philosophy of language, philosophy of mind, epistemology, and metaphysics. He supervises Ph.D. theses in all of those areas, and students express the greatest admiration for the depth and sensitivity of his contributions to their work. He is endlessly willing to give time to his students and colleagues. And while his standards are very high, and his criticism can utterly devastate a cherished idea, that is only Part I; Part II is always an effort to see how one's project might be emended and improved. His kindness has been deeply appreciated by generations of students, and by all of us who have lived and worked with him over the years.

The breadth and importance of Bob's contributions to philosophy is demonstrated in the essays that his friends and colleagues have given us for publication in this volume. Bob's "Responses" is itself a substantial and significant essay, in which he replies to their comments and criticism, and indicates where he currently stands on the issues they raise. We are proud to have had the opportunity of publishing this volume in his honor; we are pleased to be able to speak for the philosophy community at large in presenting it to him—with admiration and affection.

Judith Thomson
Alex Byrne

1

Actors and Zombies*

Daniel Stoljar

1. Much of contemporary philosophy of mind is dominated by the intersection of three topics: physicalism, the conceivability argument, and the necessary a posteriori. I will be concerned here (i) to describe (what I take to be) the consensus view of how these topics intersect; (ii) to explain why I think this account is mistaken; and (iii) to briefly sketch an alternative.

2. The first of our trio, physicalism, is the thesis that, not necessarily but as a matter of fact, everything is physical. This thesis stands in need of clarification. For one thing, we need to be told what it is to be physical. This is a difficult and somewhat neglected question, but I want to set it aside. A rough and ready understanding will do for present purposes. Another aspect of the thesis requiring clarification is the sense in which it pertains to everything. There are a number of proposals about how to explain this, but here it is sufficient to identify physicalism as the thesis that the physical truths entail (in the sense of necessitate) *all* the truths, and so all the psychological truths. If this is physicalism, and if it is true, then it is contingent, i.e. true not necessarily but as a matter of fact. For it is contingent *which* truths are the physical and psychological truths at any given world. If the physical truths concern only extension in space and time, and the psychological truths concern ectoplasm, the physical truths will not entail the psychological. On the other hand, if the physical truths are as multifarious and complex as those that (we assume) obtain in our world, and if the psychological truths concern experiences more or less as they are construed by folk psychology, physicalism might be true.

Physicalism is if true contingent, but there is nevertheless a necessary truth lurking in the shadows that is important for our purposes to bring out. For suppose that as a matter of fact physicalism is true, and thus the physical truths do entail the psychological truths. Then there must be a statement S which summarizes the complete physical truths including the physical laws and principles that obtain in our world, i.e. the truths that in fact obtain; likewise there must be a statement S* which summarizes all the psychological truths. Now consider the

* I am very grateful for comments from Ben Blumson, Jonathan Dancy, Tyler Doggett, and Andy Egan.

truth-functional conditional formed from these, 'if S then S*' and call this 'the psychophysical conditional'. If physicalism is true, the psychophysical conditional is necessarily true. The reason is that S necessitates S* and it is *not* contingent which truths S summarizes even if it is contingent that the truths it summarizes are the complete physical truths of our world; mutatis mutandis for 'S*'. To put it differently, the expressions 'the physical truths' and 'the psychological truths' may compatibly with their meaning be associated with different truths at different worlds; not so for 'S' and 'S*'. Hence, if physicalism is true, and if physicalism is the thesis that the physical truths entail the psychological truths, the psychophysical conditional is necessary.

3. Our second topic is the conceivability argument. The first premise of this argument is that it is conceivable that the psychophysical conditional is false; that is, it is conceivable that there is a situation in which the antecedent of the conditional is true and the consequent false. The second premise is that if this is conceivable then it is genuinely (i.e. metaphysically) possible. However, if it is genuinely possible that the psychophysical conditional is false, that conditional is, at best, contingent. But as we have seen, if physicalism is true, the conditional is necessary. Hence, if the premises of the conceivability argument are true, physicalism is false.

Why is it conceivable that there is a situation in which the antecedent of the psychophysical conditional is true and the consequent false? The usual way to develop this point is to consider the idea of a zombie, where, as Robert Stalnaker (2002, 239) puts it, zombies are "creatures that are physically exactly like ordinary people, but have no phenomenal consciousness. A zombie world is a world physically exactly like ours, but with no phenomenal consciousness at all. The sun shines in such worlds, but the lights are out in the minds of the unfortunate creatures who live in them". The idea of a zombie in turn prompts a particular implementation of the conceivability argument. As Stalnaker (p. 239) says: ". . . it is conceivable, or conceptually possible, that there be zombies. From this it is inferred that zombies, and zombie worlds, are metaphysically possible", and from this in turn it is inferred that physicalism is false.

The idea of a zombie makes the conceivability argument less abstract than it might otherwise be, but it also raises problems. Sydney Shoemaker (e.g., 1999), for example, argues that the idea of a zombie is incoherent, and so not conceivable, in view of the fact that there are constitutive connections between experience and beliefs about experience. What Shoemaker says may well be right, but it would be mistaken to go on to suppose (and in fact Shoemaker does not suppose) that considerations of this sort will undermine the conceivability argument. For these considerations attack at best an example. They do not attack the underlying argument. In this respect, the situation is akin to Putnam's famous (1981) attack on skepticism, in which it is argued that the causal theory of reference undermines various brain-in-a-vat examples. What Putnam says might (*might*) be right, but it will not undermine skepticism *tout court*, for the skeptic may mount his argument on the basis of a different example (cf. Campbell 2002). The same point applies to those suggestions that emphasize the constitutive connections between experience and belief.

There is also a more general concern about the conceivability argument, whatever precisely the example is that lies in the background. This is that the notions in terms of which it is stated are notoriously unclear. The concern is serious, but I doubt those who discuss the conceivability argument against physicalism are under any special obligation to allay it; and indeed this fact will be important in what follows. For the conceivability argument we are concerned with is in important respects analogous to arguments that are used and accepted throughout philosophy, and in philosophy of mind in particular. For example, consider a very different argument of Putnam's (1965): the perfect actor objection to (philosophical) behaviorism. Perfect actors are people that behave actually and potentially exactly like ordinary people but have quite different phenomenal states. It seems conceivable, and so possible, that there are such people. And, if this is possible, behaviorism is false, for behaviorism entails that behavioral truths entail the psychological truths. It is standard practice in philosophy of mind to assume that this sort of argument is successful—a standard practice I assume is perfectly legitimate. But it is bad form to use a method of argument against theories you don't like, and then turn hypercritical when the same method is deployed against theories you do.

4. Turning now to our third topic, a truth is a priori—to put it roughly—just in case (fully) understanding it is sufficient for knowing that it is true; and a truth is necessary just in case it is true in all possible worlds. Traditionally, it was assumed that these two features are co-extensive: all and only priori truths are necessary. But what Kripke (1980) and others showed is that it is possible to have a truth that is both necessary and a posteriori. (It was also argued, more controversially, that it is possible to have truths that are contingent and a priori; but we will set aside this idea in what follows.) One of Kripke's examples is the identity statement 'heat = molecular motion'. This statement, he says, is true at all possible worlds (or at any rate is true at all possible worlds at which heat exists); and yet it is also a posteriori in the sense that mere understanding it does not entail knowing that it is true. Of course, every example is controversial in some sense, and this one is no different. But it simplifies matters greatly if we assume in what follows that Kripke is right on this point and that 'heat = molecular motion' is a necessary a posteriori truth. At any rate, that will be my procedure.

5. So far we have introduced our three topics; it remains to introduce the consensus view about them. The consensus view has two parts. The first points to the possibility of a version of physicalism I will call ***a posteriori physicalism***. We have seen that if physicalism is true, the psychophysical conditional is necessary. But now let us ask: is the psychophysical conditional a priori or a posteriori? The answer to this is not determined by any assumption we have made so far. Physicalism itself is contingent, and presumably too it is a posteriori. But it does not

follow that if physicalism is true, the psychophysical conditional is a posteriori. After all, the *modal* status of physicalism might diverge from that of the psychophysical conditional; why should the same not be true of its *epistemic* status? On the other hand, while our assumptions do not entail anything about the epistemic status of the conditional, they *do* make salient the possibility that it is a necessary a posteriori truth, and as such exhibits the same combination of modal and epistemic features that is exhibited by statements such as 'heat is motion of molecules'. Those who assert that this is the case are a posteriori physicalists; those who assert this is not, i.e., that the psychophysical conditional is a priori, are *a priori physicalists.*

So the first part of the consensus view is a posteriori physicalism; the second is the suggestion that the a posteriori physicalist *is*, while the a priori physicalist *is not*, in a position to answer the conceivability argument. The claim here is not simply that if a posteriori physicalism is true, the argument can be answered *somehow*. That point is obvious; the conceivability argument is an argument *against* physicalism, so the truth of physicalism entails it can be answered somehow. The claim of the consensus view is rather that, in explaining how exactly the argument goes wrong (assuming it does) one must draw on the distinctive claim of a posteriori physicalism, i.e. the claim that the psychophysical conditional is a posteriori. Of course different proponents of the consensus view may have different views about *just how* to respond to the argument in the light of this claim. But what is distinctive of the view, or at least the second part of the view, is the assertion that it is uniquely the a posteriori physicalist who can answer the argument. Turning this around, what is distinctive of the view is that physicalists must meet the conceivability argument by becoming a posteriori physicalists.

6. Is the consensus view correct? I don't think so. I don't disagree with the first part of the view, i.e. the claim that the psychophysical conditional is a posteriori. But I disagree that this fact, if it is a fact, bears on the conceivability argument. So my disagreement is with the second part.

Since my criticism focuses on the second part of the consensus view, it is different from a well-known criticism of the view that focuses on the first, due mainly to Jackson (1998) and Chalmers (1996). This criticism says that it is mistaken to suppose that the psychophysical conditional is a posteriori in the first place, or at any rate it is mistaken to suppose this if physicalism is true. According to proponents of this criticism, there are premises in philosophy of language (and perhaps epistemology) from which it follows that if physicalism is true, the psychophysical conditional is (not merely necessary but) a priori. Clearly in this case the question of what to say about the second part of the consensus view is moot. If the psychophysical conditional is *not* a posteriori, it cannot be this fact about the conditional that answers the conceivability argument. On the other hand, if the psychophysical conditional *is* a posteriori, it is presumably an open question whether this fact about it answers the conceivability argument.

The problem with the well-known criticism is that the premises from philosophy of language (and perhaps epistemology) from which it proceeds are extremely controversial. What is at issue here is what Stalnaker has called in a number of places (e.g. 2002, 208) 'the generalized Kaplan paradigm'. Stalnaker himself rejects the generalized Kaplan paradigm; others defend it. My own view is that the matter is unclear. Take the highly complicated, nuanced and sophisticated version of the description theory advanced by, for example, Jackson (1998)—this is one version of what Stalnaker means by the generalized Kaplan paradigm; and now take the highly complicated, nuanced and sophisticated version of the anti-description theory advanced by, for example, Stalnaker. How is one to decide between them? I don't deny the issue might in principle be settled; it is rather that I myself don't see any clear way to settle it. So I will not engage this issue in what follows. Rather I will assume, as against the generalized Kaplan paradigm, that a posteriori physicalism is possible, and in fact is true. I think the consensus view is mistaken even given that assumption.

I have said that I want to set aside the well-known criticism. But it bears emphasis that the debate surrounding this criticism contributes greatly to the consensus position being the consensus position, and in fact this is my excuse for using the label. The reason is that this debate encourages the thought that *if* the psychophysical conditional were a posteriori, this *would* have a major impact on the conceivability argument. In fact, both sides in the debate about the first part of the consensus view seem to proceed under the assumption that this last conditional claim is true. So, while many philosophers think outright that the a posteriori nature of the psychophysical conditional answers the conceivability argument, many more philosophers agree that it *would* answer the argument *if* it were a posteriori.

7. I have distinguished two parts of the consensus view and said I have no quarrel with the first. What then is my quarrel with the second? The basic criticism can be stated very simply. The conceivability argument is an argument to the conclusion that the psychophysical conditional is not necessary. But, shorn to essentials, the response on behalf of the a posteriori physicalist is this: the psychophysical conditional is necessary *and* a posteriori. But now I ask you to forget your prejudices and look afresh at this answer. How can this alone possibly constitute a persuasive response? In general, if I have an argument from a set of premises Q1, Q2 . . . QN, to a conclusion P, it is not a persuasive response to me to simply assert not-P. How then can it possibility be a persuasive response to me to assert not-P *and* R for any R apparently unrelated to the premises? The assertion that the psychophysical conditional is necessary *and* a posteriori is on its face no more of a response to the conceivability argument than the outright assertion that the conditional is necessary; or that it is necessary and is really very interesting; or that it is necessary

and by the way that's a lovely shirt you're wearing. On the face of it, the consensus view is a spectacular non-sequitur.

Perhaps this way of putting matters makes my criticism of the consensus view sound a bit sophistical. So let me put things slightly differently. Nowhere in the conceivability argument is there any explicit mention of the a posteriori. Strictly and literally what we have been told is something about conceivability, and then something else about possibility. So it is a mystery—at least it is a mystery to *me*—how the notion of the a posteriori is supposed to enter the picture. At the very least we require a story in which the connection of the a posteriori to conceivability is explained. Unless such a story is produced, we have no answer here to the conceivability argument.

8. No doubt proponents of the consensus view are at this point bursting to tell me the story. I will consider some proposals in a minute. But first I want to point out that the criticism I have just made of a posteriori physicalism—that, at least on the surface, it does not answer the conceivability argument—is closely related to a similar point made by Kripke in *Naming and Necessity*, at any rate as I understand him. (In what follows I will state Kripke's point in my own terms rather than his.)

The way in which the matter comes up for Kripke is via a comparison of a conceivability argument about experiences (his example is pain) with a conceivability argument about secondary qualities (his example is heat). We have seen that zombies are people who are physically just like us but who lack phenomenal consciousness. But imagine now a type of physical object physically just like the pokers that exist in our world but which uniformly lack heat; call them zpokers. Offhand, it looks conceivable, and so possible, that there be zpokers. But then heat, or at any rate heat in pokers, must be something over and above the physical, i.e. must be something over and above motion of molecules. In short, there is a conceivability argument about heat—call it CA (heat)—that parallels the one we have been considering—call it CA (pain).

Now the line of thought suggested by the comparison between CA (pain) and CA (heat) may be summarized as follows. First premise: CA (heat) is unsound—after all, we *know*, or at any rate have assumed, that 'heat is motion of molecules' is necessary and a posteriori; so an argument to the conclusion that it is not necessary must be mistaken. Second premise: CA (heat) is analogous to CA (pain). Conclusion: CA (pain) is unsound too. Moreover, the reason that this line of thought is important for us is that it naturally suggests that the second part of the consensus view is true. After all, the most salient philosophical fact about CA (heat) is that it involves a necessary a posteriori truth, i.e. 'heat is motion of molecules'. Moreover, it is natural to assume that it is *this* fact that explains the failure of CA (heat). More generally, if we arrange things so that the connection between pain and the physical is in all respects like the connection between heat and the physical, we would have an answer to the conceivability argument against physicalism; in short, the second part of the consensus view is true.

But, as is of course well known, Kripke rejects this line of thought, on the ground that there is no relevant analogy between the two arguments. In the case of heat, we may distinguish heat itself from sensations thereof. And this distinction permits us to deny that it is conceivable there be zpokers. What is conceivable instead is that there be pokers that produce no sensations of heat; but this is a different matter. In the case of pain, however, there is no distinction between pain and sensations of pain. At least in the intended sense, pain just is a sensation of pain, and thus there is no possibility of producing a response to the argument that turns on a 'distinction' between them; there is none. (To be sure, there may be another sense in which pain is something in your toe. But this does not affect the substance of the issue. Kripke could have made his point by contrasting heat and sensations of heat directly.)

What is the relation between Kripke's discussion and our own? Well, we started from the question: what is the connection between the fact (assuming it to be fact) that the psychophysical conditional is necessary and a posteriori, on the one hand, and the conceivability argument on the other? We also noted that it is at least unobvious how this question is to be answered. Kripke's discussion can be usefully thought of as starting in the same place. It is just that he goes on to consider and dismiss a suggestion about how the connection might be explained. In short, Kripke's discussion is further evidence that our basic criticism of the consensus view is correct.

9. Unless there is some way to connect a posteriority with the conceivability argument, the consensus view is a non-sequitur. Kripke in effect discusses one way in which this connection might be explained, but the suggestion runs aground on the difference between heat and pain. But of course, even if this *particular* suggestion is unsuccessful, it scarcely follows that *nothing* similar is. So I want next to examine a related suggestion due to Stalnaker. Stalnaker makes the suggestion I want to focus on through the voice of a character he calls Anne; but I will take the liberty in what follows of assuming that the position is his. Of course, whoever in fact holds the position, it is important and needs to be discussed.

10. Stalnaker begins by considering a philosopher Thales who asserts that water is, not a compound like H_2O, but some sort of basic element. Stalnaker himself refers to this element, following Putnam, as 'XYZ', but I will call it 'Thalium'. Surely it is an empirical fact that the stuff we call 'water', the stuff we use to fill bathtubs and water the garden, is H_2O rather than Thalium. Similarly, surely it is an empirical fact that we live in an H_2O world rather than a Thalium world. This suggests that, properly understood, the word 'water' is, as Stalnaker puts it, "theoretically innocent" (p. 247). In using it, we refer to something, but we don't prejudice its nature. To put the point slightly differently, the fact that 'water' refers to H_2O is to be explained, not merely by the way in which we use the word, but by the way in which we are embedded in our environment. If we lived in a world that Thales thinks is the actual world, and we used the word rather as we use it actually, our word would in that case have referred to Thalium.

Now just as it is an empirical fact that we live in the H_2O world rather than the Thalium world, Stalnaker says, it is an empirical fact that we live in a materialist world rather than a dualist world. And this suggests that properly understood words such as 'experience', 'pain' and so on are theoretically innocent too. In using them, we refer to something without prejudicing its nature. The fact—assuming it to be a fact—that 'pain' refers to some neural or physical condition is to be explained, not merely by the way in which we use the word, but by the way in which we are embedded in our environment. If we lived in a world that the dualist thinks is the actual world, we would use the word rather as we use it actually, but our word 'pain' would in that case have referred to a non-physical property.

These considerations prompt an account of what has gone wrong in the conceivability argument that is different from, but related to, the suggestion considered by Kripke. In effect, the suggestion considered by Kripke was that, in advancing a conceivability argument about heat, we are confusing the conceivability of (1) with that of (2):

(1) There is molecular motion in the poker but no heat in the poker.
(2) There is molecular motion in the poker but nothing in it causing heat sensations.

Or, if a similar argument were to be advanced by Thales against the hypothesis that water is H_2O—call such an argument CA (water)—the suggestion considered by Kripke would be that Thales is confusing the conceivability of (3) with that of (4):

(3) There is H_2O in the bathtub but no water in the bathtub.
(4) There is H_2O in the bathtub but nothing in it causing perceptions as of water.

However, Kripke went on to say, these points are no help at all in the case of the conceivability argument against physicalism, i.e., CA (pain). For here the parallel suggestion would be that a proponent of the argument is confusing the conceivability of (5) with that of (6):

(5) There are people physically like us but which lack pain.
(6) There are people physically like us but which lack states that cause sensations of pain.

And this parallel suggestion fails, Kripke argues, since there is no way to make sense of the idea that (5) has been confused with (6).

Stalnaker's alternate proposal is that in mounting CA (water), Thales is confusing (3) not with (4) but with:

> (7) In a Thalium world considered as actual, there is H_2O in the bathtub but no water in the bathtub.

Moreover, this point *does* have application to CA (pain). For it is now available to us to say similarly that here we are confusing (5) not with (6) but with:

> (8) In a dualist world considered as actual, there are c-fibers firing in me but I am not in pain.

The phrase 'world considered as actual' is due to an important paper by Davies and Humberstone (1982), and has a technical meaning within two-dimensional modal logic, a topic to which Stalnaker has made seminal contributions. The details of these ideas are difficult, but I think there is no harm in the present context to interpret what is intended as follows:

> (7*) There is H_2O in the bathtub and there is no water-as-Thales-understands-water in the bathtub (i.e., there is no Thalium in the bathtub).
>
> (8*) There are c-fibers firing in me and I am in not in pain-as-the-dualist-understands-pain.

On this interpretation, Stalnaker's proposal is that in CA (water) we confuse (3) with (7*) and in CA (pain) we confuse (5) with (8*). And the significance of this suggestion is that both (7*) and (8*) is in the context unobjectionable. It is not impossible that what Thales says is true, so it is not impossible that there is Thalium in the bathtub. But this does not undermine the hypothesis that water is H_2O. Similarly, it is not impossible that what the dualist says is true, so it is not impossible that there are c-fibers firing in me and I am not in pain-as-the-dualist-understands-it. But this does not undermine the hypothesis that physicalism is true.

11. Stalnaker's suggestion is ingenious, but I have two objections. To see the first, consider again the perfect actor argument against behaviorism. We have seen that this argument proceeds from the premises, first, that it is conceivable that there are perfect actors, i.e., people psychologically distinct from us but behaviorally identical, and second, that what is conceivable is possible. The conclusion of the argument is that behaviorism is false, for behaviorism entails that behavioral truths entail the psychological truths. As I have said, I take it to be quite obvious that this argument is successful, and that what we have here is a good argument against behaviorism.

But unfortunately Stalnaker is in no position to say this. For there is no reason at all why the behaviorist might not respond to these arguments in precisely the way that he recommends we respond to the conceivability argument. In particular, there is nothing in Stalnaker's account to prevent a behaviorist from responding as follows. "The perfect actor argument fails because it confuses pain with pain-as-the-anti-behaviorist-understands-it. Everyone agrees that pain understood *that* way could come apart from behavior, but if you assume that you have begged the question against me. The question is whether pain as we ordinarily conceive of it can come apart from behavior, and this the argument does not show." I take it that there is something seriously wrong with the idea that a behaviorist might respond to the perfect actor argument in this way, and so there is likewise something seriously wrong with Stalnaker's proposal.

One might reply by pointing out that there are many *other* reasons to resist behaviorism—empirical reasons, say. True enough, but irrelevant: I am not denying that there might be other arguments against behaviorism; of course there are. Nor am I saying that Stalnaker's position commits him to behaviorism; of course it doesn't. What I am saying is that Stalnaker's response to the conceivability argument has the bad consequence that a good argument against behaviorism turns out to be a bad argument. His response provides the materials to respond to Putnam's perfect actor argument; but since we *know* that the latter argument is a good one, there must be something mistaken about his response.

Alternatively, one might reply by gritting one's teeth. Stalnaker has prescribed a drug to rid us of the conceivability argument. The drug has a side effect, but perhaps this is something we should learn to tolerate, a bad consequence outweighed by good. I think this response forgets just how plausible the perfect actor argument is as a refutation of behaviorism. In the standard philosophy of mind class you begin with dualism and show that it is implausible, and then you turn to behaviorism and show that it is implausible, and then you move onto other things. But how did you persuade the students behaviorism is implausible? At least a large part of this case is provided by the perfect actor argument (and similar conceivability arguments such as Block's (1981) blockhead argument). These arguments are completely compelling to undergraduates, and I think the reason for that is that they *are* completely compelling. So casting the consequence of Stalnaker's proposal that I have pointed out as tolerable is not an option.

12. In any case, there is a further reason why gritting one's teeth is no response to the problem about perfect actors. This is that it is plausible to suppose that the technique for defeating the conceivability argument that Stalnaker advances would defeat *any* conceivability argument at all, or at least any conceivability argument of the sort we are considering.

To illustrate, take any two distinct truths A and B. Suppose someone argues that it is conceivable that A is true and B is not, and concludes that it is possible that A is true and B is not, and that in consequence the truth of B is something 'over and above' the truth of A. Someone who adopted Stalnaker's strategy as

I understand it (and put in schematic form) might respond as follows: "Distinguish B from B-as-understood-as-over-and-above-A; for short, distinguish B from over-and-above-B. When you claim that it is conceivable that A is true and B is not, all that is genuinely conceivable is that A is true and over-and-above-B is not. But from this nothing follows: everyone agrees that it is possible that A is true and over-and-above B is not." The problem for Stalnaker is that, if this strategy worked, it could be used against any conceivability argument of this form. So either no conceivability argument like this is sound, or the strategy is unsound. I assume that some conceivability arguments are sound; for example I assume that the perfect actor argument is sound. So the strategy is mistaken.

Stalnaker's defense of the a posteriori physicalism runs into a problem that in my view is endemic to many contemporary attempts to respond to the conceivability argument: it overgenerates. As we have noted, the conceivability argument against physicalism is in structure identical to arguments that are used throughout philosophy. This fact suggests the following condition of adequacy on any candidate response to that argument: if you think you have isolated a factor that constitutes the mistake in the conceivability argument against physicalism, check to see if that factor is present in parallel arguments you accept; if so, consign your proposal to the flames. The problem for Stalnaker, I am suggesting, is that his proposal fails to meet this condition of adequacy. (For parallel criticisms of other contemporary attempts to respond to the conceivability argument, see Stoljar in press-a, and in press-b.)

13. I said earlier that I had two objections to Stalnaker's account. The first, which we have just been discussing, is that it mistakenly gives the behaviorist the materials to respond to Putnam's perfect actor objection, a point that generalizes to other conceivability arguments as well. The second is that what Stalnaker says has nothing to do with the epistemic status of the psychophysical conditional. For suppose—perhaps impossible—that a priori physicalism is right and the psychophysical conditional is necessary and a priori. Of course a priori physicalists face the conceivability argument too. How are they to respond? There is nothing to prevent them from arguing as Stalnaker does, or—what I assume to be the same thing—as his proxy Anne does. Anne is a B-type materialist, or what I am calling here an a posteriori physicalist, and what she says about the conceivability argument she is perfectly entitled to say. Still, there is no reason why an A-type or a priori physicalist might not say the same. In fact some a priori physicalists *do* say the same or at least very similar things. One is David Braddon Mitchell (see 2003; see also Hawthorne 1997).

So Stalnaker's strategy for responding to the conceivability argument is open to both versions of physicalism. How serious is this as a criticism of the strategy? In one sense it's not serious at all. Its availability to both positions does not render the view implausible. Indeed, in this respect, Stalnaker's proposal is similar to the one discussed in connection with Kripke. Kripke suggested that the way around the CA (heat) was to distinguish heat from heat sensations and then pointed out that such a response is unavailable to someone seeking a response to CA (pain). What Kripke says is plausible, but the epistemic status of physicalism plays no role in it. Suppose I were an a priori physicalist, not only about pain but about heat as well. I would *still* need an answer to both the CA (heat) and to CA (pain). And if what Kripke says is right, I would have an answer to CA (heat) but would have no answer to CA (pain).

So in one sense it is no criticism of what Stalnaker says that it is available to both versions of physicalism. On the other hand, it is very natural, on reading of Anne's intervention into the debate about the conceivability argument, to suppose that it is somehow her being an *a posteriori* physicalist that permits her to make the response that she does. After all, the *only* thing we know about Anne is that she is an a posteriori physicalist: "Don't look for a real-world analogue for this character", we are told, "at least not one with this name" (284). If what I have been saying is right, Anne's being a certain kind of physicalist is irrelevant: her being an a posteriori physicalist is one thing, and her advancing the strategy she does is quite another.

Furthermore, the observation that Stalnaker's proposal is available to both the a priori and the a posteriori physicalist lends additional weight to our criticism of the consensus view. The a posteriori physicalist obviously has to say *something* to the proponent of the conceivability argument; *every* physicalist has to say something to the proponent of the conceivability argument. But when we look in detail at what Stalnaker suggests qua defender of a posteriori physicalism, we find that what is being said has nothing to do with the physicalism in question being of the a posteriori variety. So, contrary to the consensus view, the fact that distinguishes the a posteriori physicalist from other sorts of physicalist is not the fact that answers the argument.

14. I have suggested that the claim that the psychophysical conditional is necessary and a posteriori by itself does nothing to answer the conceivability argument, and that two initially promising suggestions (one discussed by Kripke, one advanced by Stalnaker) about how to develop a posteriori physicalism lead nowhere. At this point you might object that I have simply been dense.

"Surely," you might say, "the construction 'it is conceivable that p' *just means* 'it is not a priori that not-p'. And, since 'it is not a priori that p' just means 'it is a posteriori that p,' the connection you are looking for is very short indeed. In particular 'it is conceivable that not-p' is logically equivalent to 'it is not a priori that not not-p', and by the definition of the a posteriori, and double negation elimination, this in turn is equivalent to 'it is a posteriori that p'."

I think, as against this, that there is no point denying that 'it is conceivable that p' has a reading according to which it means 'it is not a priori that not p'. The notion of conceivability can be legitimately spelled out in a number of different ways; this is one of those ways. But the idea that, in the *specific* context of the conceivability argument, this is what 'it is conceivable that p' means is quite another matter. When Putnam tells us about perfect actors, I don't think he means to be saying merely that it is not a priori false that there are perfect actors. I think he means to be saying that a certain case appears to be possible or (if this is different) is imaginable. On the other hand, talk of what seems to be possible or of what is imaginable is prima facie different from talk of what is or is not a priori.

It might be replied that while this is true prima facie, it is not true all things considered, and in particular, 'it is imaginable that p'—to focus on this notion for the moment—itself just means 'it is not a priori that not p'. However, I think this last equivalence is decidedly implausible (cf. Yablo 1993). For consider any of the standard examples of necessary a posteriori truth—say, 'water is H_2O'. It is clear that it is a posteriori that water is H_2O and so of course it is not a priori that water is H_2O. Is it likewise imaginable that water is *not* H_2O? I think not. As Kripke argued, it is not at all clear that we can imagine water not being H_2O. Of course we can imagine related things. For example, we can imagine water not producing perceptions as of water; perhaps also we can imagine water that isn't Thalium. But none of this is strictly speaking imagining that water is not H_2O. More generally, therefore, the idea that 'it is imaginable that p' just means 'it is not a priori that not p' is open to counterexample. More generally still, this very shortest way to connect the notion of the necessary a posteriori with the notion of conceivability, and so defend the consensus view, is implausible.

15. I noted earlier that, while the particular method that Kripke discusses for connecting the topic of the necessary a posteriori with the topic of the conceivability argument breaks down, nothing we have said *proves* that no proposal along these lines could work. Obviously, it remains true that nothing has been proved. Still, I think our previous reflections make very plausible the hypothesis that there is in fact no connection here, and hence the second part of the consensus view is mistaken.

More generally, there would appear to be two topics: first, the necessary a posteriori and associated matters; second, what if anything has gone wrong in the conceivability argument and associated matters. The interesting suggestion of the consensus view is that these two topics are intimately connected. However, in light of what we have said, a more plausible view is that they are not connected in any obvious way.

Of course, that leaves us with at least two daunting projects. One is to fit the necessary a posteriori into a smooth picture of our thought and talk and the way in which that thought and talk relates to the world. This is something I have already indicated I will not do, for the simple reason I have no idea how. The other is to say something sensible about where and how the conceivability argument goes wrong (assuming it does). This too is a long story, but I think here I have something to say. I will devote the final sections of the paper to very short account of what this is.

16. Summarizing his interpretation of Kripke's achievement, Stalnaker (1997, 168) writes:

The positive case for the theses that Kripke defends is not novel philosophical insight and argument, but naïve common-sense. The philosophical work is done by diagnosing equivocations in the philosophical arguments for theses that conflict with naïve common-sense, by making the distinctions that remove the obstacles to believing what it seems intuitively most natural to believe.

Viewed from a sufficiently high level of abstraction, something similar is true in the case of the conceivability argument. There is a response to the conceivability argument that is intuitively very natural to believe. The case for this response is, if not naïve common-sense, then at least scientifically and historically informed common-sense. And the work in defending this response is mainly in identifying and undermining the philosophical reasons for dismissing or ignoring it.

What then is the response to the argument that is intuitively so natural to believe? The natural response is—wait for it!—that we are missing a piece of the puzzle; that is, we are ignorant of a type of truth or fact which (a) is either physical or entailed by the physical; and (b) is itself relevant to the nature of experience. To say this is not to say that we will remain forever ignorant of this type of truth, nor that our ignorance must concern basic physics—it may concern a fact that supervenes on basic physics but which is nevertheless not psychological. (Remember there are *many* such facts.) The positive case for this response can certainly be made, but it largely consists in reminding ourselves of our epistemic position. It is an obvious empirical fact that we are ignorant of the nature of consciousness—there is no reason why a response to the conceivability argument may not draw on that fact along with anyone else. It is also true that historically we have been in similar situations before (cf. Stoljar 2005, 2006). These facts provide good, but not demonstrative, evidence that this is our situation here too.

How does the hypothesis of ignorance answer the argument? Well, consider the claim that there are people like us in all physical respects but who lack phenomenal consciousness. The phrase 'all physical respects' contains a quantifier, and so we may ask about its domain, and so about the interpretation of the central claim of the conceivability argument. Suppose the domain is construed broadly, so as to include absolutely all respects; in particular, so it includes respects relevant to experience but of which we are ignorant. (The hypothesis of ignorance in effect says there are such respects.) Then the conceivability claim would put pressure on physicalism, but it is doubtful that we can genuinely conceive of the relevant situation. How am I supposed to conceive various respects about which I have no knowledge? On the other hand, suppose the quantifier is construed narrowly, to include only those respects or types of respects of which we are not ignorant. Then the conceivability claim is plausible, but it will not put any pressure on physicalism. For the physicalist will be on good ground responding that the possibility claim at issue only seems possible because it is driven by a conceivability claim that does not take all relevant respects into account.

Not only does the proposal answer the argument, it does so in a way that speaks to the concerns that emerged in the course our previous discussion. In effect, there were two such concerns. First, the proposal leaves it open whether the psychophysical conditional is a priori or not, so in that sense we are not being offered a version of the consensus view. Second, the proposal satisfies the condition of adequacy on any response to the conceivability argument that we formulated when thinking about Stalnaker's proposal. According to this condition of adequacy, any proposal about where the mistake is in the conceivability argument must be checked against conceivability arguments we accept. In the context of our discussion, this condition of adequacy resolved itself into the following question: does the epistemic response have the effect of granting to the behaviorist the materials to respond to the perfect actor objection? But the answer to this question is 'no', and the reason is that there is a major discrepancy in the way in which a behaviorist appeals to behavioral truths, on the one hand, and the way in which the physicalist appeals to physical truths on the other. Behavioral truths are, and are intended to be by the behaviorist, truths that we can be established on the basis of direct perception: behavioral dispositions, or any rate their manifestations, are supposed to be available to perception. That was the basic rationale of the behaviorist program. And it is very plausible that no truth of that *sort* will be of any help in thinking about the perfect actor objection to behaviorism; a fortiori, no unknown truth of that sort will be of any help in thinking about behaviorism. On the other hand, physical truths meet no such epistemological condition; in fact, it is far from obvious that they meet any positive condition at all apart from being non-experiential. Hence there is room here for an ignorance-based or epistemic response to the conceivability argument.

17. So in briefest outline is the epistemic response to the conceivability argument. Why have so many missed it? No doubt part of the story is our tendency to discount our own ignorance. But another, and perhaps ultimately more interesting, reason derives from a powerful view of what philosophical problems are and what contributions to them should be.

A statement of the view I have in mind can be found in the famous passage from the *Investigations* in which Wittgenstein says:

> We must do away with all explanation, and description alone must take its place. And this description gets its power of illumination—i.e. its purpose—from the philosophical problems. These are, of course, not empirical problems; they are solved rather by looking into the workings of our language, and that in such a way as to make us recognize those workings: in despite of an urge to misunderstand them. The problems are solved, not by giving new information, but by arranging what we have always known. Philosophy is a battle against the bewitchment of our intelligence by means of language. (1954, 47)

The most famous line in this passage is probably the last one, but for me the penultimate one is the most important. At least in philosophy of mind, this idea about philosophical problems is remarkably influential and persistent, much more influential and persistent than the Wittgensteinian apparatus within which it first appeared—so, at any rate, it seems to me. Frank Jackson (1998), to take one modern example, says that what he calls serious metaphysics is "discriminatory at the same time as being complete or complete with respect to some subject matter" (p. 5). Similarly, John Perry (2001) describes his approach by saying that it "won't be physiological or neurological, nor even . . . .very phenomenological. [It] will be logical, semantical and philosophical" (p. 118). As I read things, the suggestion implicit in Perry's remark, and explicit in Jackson's, is that in an important sense all the relevant empirical facts are in; we just need a way to think through those facts. Wittgenstein, Jackson and Perry are remarkably different in other respects, but on this matter they speak with a single voice, or so it seems to me. All three are united in the idea that solving the problem presented by the conceivability argument does not involve any new information; it involves rather rethinking the information already in our possession. On the other hand, this idea precisely is in conflict with informed common-sense, for, when confronting the conceivability argument, the view of informed common-sense is precisely that new information is required.

18. I have my own views about how to respond to this conflict, but this is not the place to pursue them. I certainly don't mean in these sketchy remarks to recommend a blanket rejection of this account of what philosophical problems consist in. For one thing, it is quite clear that *some* philosophical problems *do* conform to this general description. But the idea that philosophical problems *as a class* do, and that the problems represented by the conceivability argument do in particular, seems to me to be something of a dogma. One consequence of dropping the dogma is that a more particularist approach to philosophical problems comes into view—perhaps there is *nothing* much to say *in general* about what a philosophical problem is like. But another more immediate effect is the removal of one of the main impediments to informed common-sense when it comes to the conceivability argument against physicalism.

REFERENCES

Braddon-Mitchell, David 2003 'Qualia and Analytic Conditionals', *Journal of Philosophy* 100:111–35.

Block, N. 1981. 'Psychologism and Behaviorism', *Philosophical Review,* 90, 5–43.

Campbell, J. 2002. 'Berkeley's Puzzle', in John Hawthorne and Tamar Szabó Gendler (eds.) *Conceivability and Possibility* (Oxford: Oxford University Press).

Chalmers, D. 1996. *The Conscious Mind* (Oxford: Oxford University Press).

Davies, M, and Humberstone, L. 1980. 'Two Notions of Necessity', *Philosophical Studies,* 38, 1–30.

Hawthorne. J. 2002. 'Advice for Physicalists', *Philosophical Studies,* 109, 17–52.

Jackson, F. 1998. *From Metaphysics to Ethics* (Oxford: Oxford University Press).

Perry. J. 2001. *Knowledge Possibility and Consciousness* (MIT Press).
Putnam, H. 1965. 'Brains and Behaviour', in R. J. Butler (ed.), *Analytical Philosophy, Vol.2,* (Blackwell).
—— 1981. *Reason Truth and History.* (Cambridge: Cambridge University Press).
Shoemaker 1999. 'On David Chalmers' *The Conscious Mind, Philosophy and Phenomenological Research*, 59, 439–44.
Stalnaker 1997. 'Reference and Necessity', in Crispin Wright and Bob Hale (eds.), *Blackwell Companion to the Philosophy of Language.* Blackwell; reprinted in Stalnaker 2003*b*; references to the reprinted version.
—— 2002. 'What is it like to be zombie?', in John Hawthorne and Tamar Szabó Gendler (eds.) 2002. *Conceivability and Possibility* (Oxford: Oxford University Press); reprinted in Stalnaker 2003*b*; references to the reprinted version.
—— 2003. 'Conceptual Truth and Metaphysical Necessity'. In Stalnaker 2003b.
—— 2003*b*. *Ways a World Might Be: Metaphysical and Anti-Metaphysical Essays* (Oxford: Oxford University Press).
Stoljar, D. (2005). 'Physicalism and Phenomenal Concepts', *Mind and Language.*
—— (2006). *Ignorance and Imagination: On the Epistemic Origin of the Problem of Consciousness* (New York: Oxford University Press).
Wittgenstein, L. 1954. *Philosophical Investigations* (London: Macmillan).
Yablo, S. 1993. 'Is Conceivability a Guide to Possibility?', *Philosophy and Phenomenological Research*, vol. 53, I, 1–42.

2

The Frege–Schlick View

Sydney Shoemaker

What I have called the Frege–Schlick view is the view that the relations of qualitative similarity, identity, and difference are well defined only for the intrasubjective case—so, for example, my experience of red is neither qualitatively like yours nor, as the inverted spectrum hypothesis would have it, qualitatively like your experience of some other color.[1] This view is suggested by things Frege says in "The Thought," and by things Schlick says in "Positivism and Realism."[2] In the text of my "The Inverted Spectrum" (hereafter my 1981) I mentioned this view as one compatible with the functionalist account of qualitative similarity I offered there, but set it aside as unacceptably at variance with common sense. But in a postscript to the 1984 reprinting of that paper I raised a difficulty for the view—what Robert Stalnaker refers to as "Shoemaker's paradox"—and suggested that avoiding the problem might require acceptance of the Frege–Schlick view. I remained reluctant to accept the Frege–Schlick view, and eventually came up with a solution to "Shoemaker's paradox" that, so I thought, removes the apparent reason for accepting that view—this was presented in my 1996. Stalnaker has argued in support of the Frege–Schlick view, maintaining that my 1984 view was right, and that my subsequent attempt to avoid it was unsuccessful.[3] I will attempt here to answer his arguments—and so to defend my 1981 self and my 1996 self against my 1984 self and Stalnaker.

Stalnaker's critique of my views is so sympathetic and generous that it seems almost churlish to attempt to reply to it. But I think that the issues here are

[1] Shoemaker 1981 and 1984.

[2] Frege declares to be "unanswerable", indeed really nonsensical, the question "does my companion see the green leaf as red, or does he see the red berry as green, or does he see both as one colour with which I am not acquainted at all," and says that "when the word 'red' does not state a property of things but is supposed to characterize sense-impressions belonging to my consciousness, it is only applicable within the sphere of my consciousness" (Frege 1956, 299). Schlick says that "The proposition that two experiences of different subjects not only occupy the same place in the order of a system but are, in addition, qualitatively similar has no meaning for us. Note well, it is not false but meaningless: we have no idea what it means" (Schlick 1959, 299). In his earlier review of Husserl's *Philosophie der Arithmetik* Frege asserts only that "nobody even knows how far his image (say) of red agrees with somebody else's" (Frege 1952, 79), without concluding that such comparisons are "nonsensical"; so perhaps he did not then accept the Frege–Schlick view.

[3] Stalnaker 2003.

important enough to warrant an airing. The Frege–Schlick view has been little discussed, and Stalnaker's paper is the only extended defense of it that I know of. If it is correct, this has important implications for issues about phenomenal consciousness that are very much at the center of current discussion—issues about the phenomenal character, the "what it is like," of sensations and perceptual experiences. Recent discussion often pits those who accept representationist views about phenomenal character against those who believe that the nature of phenomenal character is such as to allow for the possibility of spectrum inversion, i.e., of cases in which the representational character of experiences and their phenomenal character vary independently of one another, and are differently associated in different sorts of perceivers. If the Frege–Schlick view is correct, the positions on both sides of this debate may rest on misconceptions.

In my 1984 I thought that I might have to give up what I called the "standard view of qualia." This view says that the phenomenal character of experiences is determined by intrinsic properties of them, qualia, that these are the primary relata of the relations of qualitative similarity and difference, and that these relations are internally related to the qualia, in the sense that fixing the qualia of experiences fixes the relations of qualitative similarity and difference amongst them. I thought, for reasons to be considered later on, that, given certain plausible assumptions, combining the standard view of qualia with a functionalist account of qualitative similarity and difference leads to a contradiction. I was (and am) firmly committed to the functionalist account of qualitative similarity and difference, so I thought that the standard view of qualia might have to go—and I took it that giving it up would amount to accepting the Frege–Schlick view. Stalnaker's version of the Frege–Schlick view combines this rejection of the standard view of qualia and acceptance of the functionalist view of qualitative similarity.

It is in fact not clear that the Frege–Schlick view requires rejection of the standard view of qualia. It could be held—and this may have been Frege's view—that the qualitative character of experiences is determined by what I will call "private properties", properties that can be instantiated only in the experiences of a single person. These, it could be held, are intrinsic properties of experiences and the primary relata of the relations of qualitative similarity and difference, and are such that these relations are internal to them. This would imply, of course, that experiences of different persons are qualitatively different; but they would not be different in the way experiences of red are different from experiences of green. Each person's qualia would be "alien" relative to every other person's qualia.[4]

[4] It might be questioned whether the private properties view should count as a version of the Frege–Schlick view. If the experiences of different people instantiate different qualia, there will be qualitative differences amongst them, even though there are no qualitative similarities. But on the version of the Frege–Schlick view Stalnaker defends experiences of different people will presumably

But the private properties version of the Frege–Schlick view seems to be incompatible with physicalism. Assuming physicalism, it would seem that if there are qualia as conceived by the standard view they must be physically realized. It would be highly implausible to hold that the physical properties that realize them are ones that cannot be shared by experiences of different persons—and if the physical realizers can be shared, so can the qualia. So my focus here will be on versions of the Frege–Schlick view that reject the standard view of qualia. But it will be useful to have the private properties view as an object for comparison.

Stalnaker holds that his version of the Frege–Schlick view, while rejecting the standard view of qualia, does not amount to eliminativism about qualia. It allows that there are qualia, but takes them to be relational properties of experiences rather than intrinsic ones. What we must now consider is what sort of relational properties they might be.

There is one conception of qualia as relational that is not available to Stalnaker. It has been suggested in a recent paper by David Hilbert and Mark Kalderon, and also in a recent book by Austen Clark, that the qualitative character of color experiences is determined by their position in the subject's color experience space, i.e., by their similarities and difference from other experiences in the repertoire of the subject.[5] This is a view that rules out the possibility of a symmetrical color experience space, and so the possibility of undetectable spectrum inversion. It certainly construes qualitative character as relational. But on this view, if different subjects have identically structured color experience spaces, which is certainly possible, then when their color experiences occupy corresponding positions in their spaces they will be qualitatively identical. And this of course is incompatible with the Frege–Schlick view.

In a number of places I have argued for the possibility of spectrum inversion by arguing first that *intra*subjective spectrum inversion is possible, and then arguing from the possibility of that to the possibility of *inter*subjective inversion—in my 1996 I call this the "intra-inner argument." The argument from the possibility of intrasubjective inversion to the possibility of intersubjective inversion involves imagining a case in which during a certain interval one subject undergoes intrasubjective inversion and another does not, where both before and after the inversion the two subjects are alike in what color discriminations they can make and what color similarity and difference relations they perceive. The claim is that either before the inversion or after it the experiences of the two subjects when viewing the same colors must be different in their qualitative character. This of course assumes that qualitative similarity and difference are well defined

have different "relational qualia," and in that sense will differ qualitatively. What will not be true on either version is that there is a single color quality space, every position in which is related to every other position by a chain of qualitative similarities, such that qualia instantiated in the experiences of different people are all located within this space. Experiences of different people will not differ in the way my experiences of red differ from my experiences of green.

[5] Hilbert and Kalderon 2000 and Clark 2000.

for the intersubjective case, and assumes the falsity of the Frege–Schlick view. But in addition to rejecting this step in the intra-inter argument, Stalnaker questions my argument for the possibility of intrasubjective inversion, and he appears to think that the Frege–Schlick view should reject the possibility of intrasubjective inversion. There certainly seems no reason why a proponent of the private properties version of the Frege–Schlick view should reject this possibility, and it is not obvious to me why it should be rejected by a proponent of the version that rejects the standard view of qualia. But considering Stalnaker's objection to my argument will be a useful way of approaching the question of what sort of relational properties qualia might be, on the Frege–Schlick view.

One way of arguing for the possibility of intrasubjective spectrum inversion is just to imagine the case in which someone reports that overnight the appearance of everything has changed—red things look the way green things used to look and vice versa, and likewise for other pairs of colors. This is open to the objection that what accounts for the person's report might be not a change in the ways things appear but a change in how the person remembers the ways things appeared in the past. To finesse this objection, I have imagined the inversion occurring by stages.[6] In the first stage the subject reports that things of most colors look the way they looked earlier, but things of a few shades of colors look the way their complementaries looked earlier. The truth of his report would require a verifiable change in his discriminatory abilities, and if this change occurred it would be ruled out that the person's report was due simply to a change in how he remembers his past experiences. We then imagine a series of further changes of the same magnitude that add up to a total inversion. (I imagine that each change is followed by a "semantic accommodation" resulting in the person's being disposed to apply color words to the same colors as before, despite the differences in appearance.)

Stalnaker challenges this argument with the following analogy. Assume, for the sake of argument, a purely relational theory of space. This would rule out the possibility of everything in the world moving ten feet in a certain direction. Someone might claim that we can imagine such a change occurring by stages—first one group of things moves ten feet in a certain direction, then another group, then another, and so on, till in the end everything has moved. Stalnaker says, rightly, that imagining this does not amount to imagining that the entire universe has moved. What we have at each stage is certain things moving *relative to the rest of the things there are*. At each stage the reference class, the class of things relative to which something moves, is somewhat different. There is no basis here for the conclusion that everything has changed position in any absolute sense.

Location, on the relational theory, is a property something has only relative to other things. And the suggestion might be that qualitative character is likewise a property an experience has only relative to other experiences. Just as the reference

[6] See my 1981*b* and 1996.

class changes at each stage in the spatial case, so (it is suggested) it changes at each stage in the series of color inversions. So there is no basis for the claim that at the end of the latter series, when the color quality space of the subject has the same structure it did before the series of inversions began, the phenomenal character of experiences of red is different from what it was earlier.

But on examination the analogy breaks down. The relativity of location goes with a certain sort of relativity of distance—and distance here is supposed to be the analogue of qualitative similarity. But (on the relational theory of space Stalnaker is working with) it is only *diachronic* distance—distance between something at one time and something (perhaps the same thing) at a different time—that is relative. Something at time t2 is at a distance from where it was at time t1 if its synchronic distances from some group of other things at t2 are different from its synchronic distances from those same things at t1, where the synchronic distances of those things to one another are (at least for the most part) the same at t2 as they were at t1. Synchronic distance is not in the same way relative. But there is no difference in status between synchronic qualitative similarity and diachronic qualitative similarity that corresponds to this. In particular, the similarity (or difference) of an experience at time t1 and an experience at time t2 does not constitutively involve the synchronic similarity and difference relations between the first experience and other experiences at t1 and between the second experience and other experiences at t2. If I am in a darkened room, and you flash two colored lights one after the other, I can say whether the experiences of the lights are similar or different without being in a position to compare either with other experiences had at the same time—and there seems no reason to think that this diachronic similarity or difference is in any way constituted by synchronic similarities and differences.

I do not think that Stalnaker intends his spatial analogy to carry a great deal of weight by itself. I have gone into it for two reasons. First, I don't think that it amounts to an effective challenge to the inversion by stages argument for the possibility of total intrasubjective spectrum inversion. Second, and more importantly, I think that consideration of the analogy shows that the way location is relative cannot be used to explain how qualitative character can be relational. So it remains unclear in what way, compatibly with the Frege–Schlick view, qualia could be relational.

Stalnaker follows this analogy with one that compares qualitative similarity and difference with relations between utilities as conceived by Von Neumann–Morgenstern utility theory. This is not used to question the argument for the possibility of intrasubjective inversion, but simply to present a case in which it is intuitively plausible to say that we have similarity and difference relations that are well defined only for the intrasubjective case. That it certainly does. I prefer coffee to tea, and we can recast this as the statement that my liking for coffee is greater than my liking for tea; but the criteria for the truth of this give no basis for comparing my liking for coffee with your liking for tea. But I do not find that this analogy clarifies for me in what way qualia are supposed to be relational properties of experiences. Drinking coffee and drinking tea occupy different positions in my preference structure, and I suppose that this gives my likings for these activities different relational properties. One might compare this to the fact that red and green occupy different positions in my color quality space, and that gives the experience of red and the experience of green different relational properties. But I don't think this gives us what we want.

What qualia are supposed to characterize are token experiences—here, token experiences of red and green. Now the claim that qualia are relational cannot mean simply that in virtue of having the qualia they do experiences stand in relations of similarity and difference; it is, of course, part of the standard view of qualia as intrinsic properties that the qualia of experiences ground the qualitative similarity and difference relations amongst experiences. What the standard view holds, however, is that the relations of similarity and difference amongst qualia, and the experiences that have them, are *internal* relations. The view that qualia are relational must hold that qualitative similarities and differences amongst experiences are *external* relations, and that having a (relational) quale is just a matter of standing in certain of these relations to other experiences. But to say that these are external relations should mean that experiences that are intrinsically the same can differ in what relations of this sort they stand in. What will be the intrinsic properties of experiences, if not their qualia?

One answer would be: their physical properties. To make sense of this we need some way of saying what the physical properties of an experience are. Assuming physicalism, it would seem that they must be properties that realize the functional role that makes an event or state an experience. Can we make sense of the idea that two experiences might have the same functional role, and realize it physically in exactly the same way, and yet differ in their relations of qualitative similarity and difference to other experiences? I don't see how. The functional role of an experience will include the ways in which its co-occurrence with other experiences will lead to judgments of similarity or difference, abilities to discriminate, abilities to recognize, etc. So it will include its qualitative similarity and difference relations to other experiences. What count as the physical properties of a token experience will be determined by such a functional role. It seems, then, that if two token experiences share exactly the same physical properties they must be functionally identical, and must share exactly the same qualitative similarity relations to other experiences—so they must be qualitatively identical and have the same qualia. So I do not see how we can get any handle on the notion of qualia as non-intrinsic, relational properties of experiences.

Let's return to my liking for coffee and my liking for tea. My preferences might change so that I come to like tea more than coffee—so it would then be true that my liking for tea is greater than my liking for coffee. This is perhaps analogous to the case in which I undergo a partial inversion, such that red things look to me the way green things used to look, and vice versa—or, as we might put it, my experience of red becomes just like my experience of green used to be, and vice versa. What the latter comes to is that after the change my token experiences of red are qualitatively like my token experiences of green before the change. Can we say in the former case that there are token likings, and that before the change my token likings for coffee were greater than my token likings for tea, whereas after the change my token likings for tea are greater than my token likings for coffee? This seems questionable at best. It is not at all clear that this case gives us mental particulars that stand in relations that can only hold intrasubjectively; and if it doesn't, it doesn't give us relational properties of particulars that illustrate what sort of properties relational qualia would be. What we have in this case seems to be a triadic relation between coffee (or the drinking of it), tea (or the drinking of it), and a person—or perhaps a four-term relation having a moment of time as an additional term. Perhaps someone will suggest that we have something analogous in the case of color experience comparisons, e.g., a relation having as terms some colors, an observer, and a time. This would certainly disallow intersubjective experience comparisons. But it would also seem to amount to eliminativism about qualia rather than a relational view of them.

Stalnaker says that it is not his aim to defend representationism, but says he is "sympathetic to the general strategy of trying to explain qualitative content in terms of representational content, and skeptical about the coherence of thought experiments such as the inverted spectrum that attempt to pull them apart" (p. 222). I share the sympathy expressed here, but not the skepticism. I distinguish standard representationism, which holds that the phenomenal character of experiences is determined by what objective properties (colors, shapes, etc.) they represent, and a version of representationism which says that the phenomenal character of experiences is an aspect of their representational content that is in a certain way more subjective. On the latter version the phenomenal character of experiences consists in their representation of aspects of objective properties, "qualitative characters," that are individuated by how they affect experience, given certain conditions.[7] I reject standard representationism, but accept the second version.

One of my complaints about standard representationism is that it cannot do justice to certain phenomena of color constancy—the fact that something in shadow can look different from something not in shadow without looking to have a different color, and the fact that something in shadow and something of a lighter shade of color can look the same without looking to have the same color. It is important that such cases do not involve any misperception or perceptual illusion. The version of representationism I favor can handle this nicely. We can

[7] See my 2006. In an earlier version of the view I held that the phenomenal character consists in what I first called "phenomenal properties" and later called "appearance properties," where these are properties things have in virtue of producing or being disposed to produce experiences of certain sorts. See my 1994 and my 2000.

say that in the one case different instantiations of the same color present different qualitative characters, while in the other instantiations of different colors present the same qualitative characters. On this account of the matter, the same color will have a number of different qualitative characters, different ones of which it presents under different viewing conditions. And once we allow this, we should be open to the idea that what qualitative character a color presents depends not only on viewing conditions but on the nature of the perceptual system of the observer. And that will give us the possibility of spectrum inversion.

Now this account would not be available to a proponent of the Frege–Schlick view, except on the private properties version of it. For qualitative characters of colors are individuated by what sorts of experiences they produce, and this requires that experiences have intrinsic qualia that define the relevant sorts. At any rate, I have not found any conception of non-intrinsic qualia that will do the job.

How will the Frege–Schlick view handle the color constancy phenomena I have mentioned? Well, it of course embraces a functionalist account of qualitative similarity and difference as relations amongst experiences, so it can say what it is for the experiences in the one case (of things of the same color, one in shadow and the other not) to be different and for the experiences in the other case (of things of different colors, one in shadow and the other not) to be the same. But it cannot say that these samenesses and differences in the experiences are samenesses and differences in their representational content. It cannot, it seems to me, honor G. E. Moore's transparency intuition. So in an important way, it is less friendly to representationism than the account I have offered. Stalnaker concludes his paper with the sentence "It does seem to be a common-sense view to think that the qualitative character of experience has something essential to do with the ways things appear to us to be" (p. 238). I agree, but this seems to me to go against the Frege–Schlick view (or at any rate its compatibility with common sense) rather than in favor of it.

It is time to confront "Shoemaker's paradox." I shall present Stalnaker's version of the argument, since it is more reader friendly than my own. In Case A Alice is capable of being in either of two qualitative states, as different as red and green, which are physically realized by physical states Px and Py. In Case B Bertha is like Alice except that she has a backup system. Call her primary system **α** and her backup system **β**. When the **β** system is active the possible qualitative state realizers are Pz and Pw. If Bertha goes from Px (with **α** active) to Pz (with **β** active) she notices no difference—the tomato looks the same. Likewise if she goes from Py (with **α** active) to Pw (with **β** active). In Case C Clara has only one visual system, but it is the **β** system rather than the **α** system. In case D Dorothy has the two visual systems **α** and **β**, but they are connected differently than in case A; if Dorothy was in Px (with **α** active) and goes into Pz (with **β** active) she reports that the tomato looks different, while if she goes from Px to Pw she reports that it looks the same. We label the qualitative states with the person and the physical realization state; so, for example, Q(A,x) is the quale experienced by Alice when in state Px.

The following qualia identities seem reasonable:

1. Q(A,x) = Q(B,x).
2. Q(C,z) = Q(B,z)
3. Q(D,z) = Q(C,z)
4. Q(D,x) = Q(A,x)
5. Q(B,x) = Q(B,z).

In support of 1, and implicitly of 2–4, Stalnaker says that "It does not seem reasonable to suppose that the presence of an *inactive* backup system could affect the qualitative character of the experience" (p. 234), and appeals to the fact that Alice and Bertha (in 2–4, Clara and Bertha, Dorothy and Clara, Dorothy and Alice) are in the same physical state at the time of the experience. In support of 5 is the fact that there is no introspectable difference for Bertha between being in the Px state and being in the Py state.

It follows from 1–5 that Q(D,x) = Q(D,z). But this is incompatible with the fact that for Dorothy there is an introspectable difference, as great as that between red and green, between being in states Px and Pz.

Stalnaker says that the source of the problem is that "Shoemaker's account of qualia relies on two different criteria of identity for qualitative properties . . . The functional theory provides a criterion for intrapersonal identities in terms of discriminatory capacities, and identity of physical realization properties provides a criterion for interpersonal qualia identities. One cannot assume that the two equivalence relations can coherently be put together . . ." (p. 235). I think it is a bit misleading to speak of two different *criteria* here. The claim that qualitative properties having the same realizers are identical is not a separate criterion of qualitative identity, but just an analytic consequence of the notion of realization—the instantiation of a realizer is sufficient for the instantiation of the property realized, and properties having the same sets of (possible) realizers cannot help but be identical. What the argument just given calls into question is not the compatibility of two different criteria of qualitative identity, but the view that qualitative identity (and similarity and difference) holds in virtue of intrinsic properties of the relata that are physically realizable, i.e., the conjunction of what I earlier called the "standard view of qualia" and physicalism.

But I think it is important when speaking of realization to distinguish *core realizers* and *total realizers*.[8] In the above example, Px, Py, Pz, and Pw are core realizers of qualia. The total realizers would bring in the way the systems are connected and, what Stalnaker emphasizes, the memory systems of the subjects. Let's say that in Bertha the two systems are connected in way C, while in Dorothy they are connected in way C*, and let these modes of connection include the memory

[8] For this distinction, see my 1981.

mechanisms. Then we can represent the realizers of Bertha's qualia as Px&C, Py&C, Pz&C, and Pw&C, and we can represent the realizers of Dorothy's as Px&C*, Py&C*, Pz&C*, and Pw&C*. Letting X be the property of having only one system, we can represent the total realizers of Alices's qualia as Px&X and Py&X, and those of Clara's qualia as Pz&X, and Pw&X.

The core realization designators in 1–5 should be replaced by the corresponding total realization designators. And once this is done the truth of these identities is not so obvious. The first becomes "Q(A,x&X) = Q(B,x&C," and no longer can it be said that "Alice and Bertha are in the same state, physically, at the time of the experience" (p. 234). Whether or not Px&X and Px&C are realizers of the same quale, they are not the same realizer of it.

My response to the argument in my 1996 was to argue that we can deny that 1–4 (modified so that they speak of total realizers) all hold without calling into question the functionalist view of qualitative similarity and difference. I envision the case in which someone changes from being like Alice, having only the system $\boldsymbol{\alpha}$, to being like Bertha, having both systems $\boldsymbol{\alpha}$ and $\boldsymbol{\beta}$, joined in way C, and has an experience with Px&X before the change and an experience with Px&C after it. We can suppose that the person reports at the later time that her Px&C experience is just like she remembers her Px&X experience being, and that in general her behavior is as the functionalist account says it should be if the experiences are qualitatively the same. But as I say, all of this could occur in a case in which a person's memory system is tampered with, and the functionalist account should not say that in such a case we would have qualitative identity. Adding system $\boldsymbol{\beta}$ to system $\boldsymbol{\alpha}$ in a creature will of course require changing the creature's memory system, and it will change it in different ways depending on whether the systems are combined in way C or in way C*. And I claim that this undermines the case, based on the apparent satisfaction of the functionalist criterion, for saying that the later Px&C experience is qualitatively the same as the earlier Px&X experience. In effect, what we have is a case of memory tampering.

Stalnaker allows that this response avoids the contradiction, but questions whether it saves the common-sense view. He points out that according to it, "later changes in a person's perceptual and memory system, even differences in unrealized counterfactual possibilities, can affect the qualitative character of one's experience" (p. 236). This is ruled out by his claim that the presence of an inactive backup system could not affect the qualitative character of the experiences (the presence of the backup system would bring with it the counterfactual possibilities, and its being inactive would mean that these possibilities are unrealized). It has the consequence that if Bertha has a flexible brain, which has a backup system that will take over if the part of the brain that realizes a certain quale is damaged, while Alice has a less flexible brain, and would lose the capacity for that kind of qualitative experience if such damage occurred, then even if the core realizations are the same, and the brain damage never occurs, their qualia are different. What can seem even more damning, it has the consequence that if Alice once had a flexible brain like Bertha's, but the backup system atrophied as she aged, the qualitative character of her experience changes with the change in her brain, though the change is inaccessible to introspection.

I want to focus on Stalnaker's claim that it is not reasonable to suppose that the presence of an inactive backup system could affect the qualitative character of the experience. To test this, consider a case I present in my 1996, and that Stalnaker discusses. To put it his way, "Suppose we have a person—call her Ellen—with the same α and β but in whom they are connected first in one way (as in Bertha), then in the other (as in Dorothy)" (p. 237). Supposing the backup systems are *not* inactive, we get a kind of incoherence. During her Bertha-like period, Ellen's Px and Pz experiences will be qualitatively alike, while during her Dorothy-like period they will be qualitatively different. But during her Dorothy-like period she will remember her earlier Px experiences as being like her more recent Px experiences, and her earlier Pz experiences as being like her more recent Pz experiences, though she will also remember her earlier Px and Pz experiences as being qualitatively alike and will remember (and introspect) her recent and current Px and Pz experiences as being qualitatively different. It should be noted that during her Bertha-like period she will be unable to distinguish the things perceived by means of Px experiences from the things perceived by means of Pz experiences, while during her Dorothy-like period she will distinguish these things with ease. Plainly it cannot be the case both that her Px experiences were qualitatively the same throughout and that her Pz experiences were qualitatively the same throughout, given that the Px and Pz experiences were qualitatively the same during the Bertha-like period and qualitatively different during the Dorothy-like period.

But now consider two different versions of the case. In one of them the β system was inactive throughout, while in the other the α system was inactive throughout. Stalnaker's claim would suggest that in the first version Ellen's Px experiences are qualitatively the same throughout the period, both while she is Bertha-like and while she is Dorothy-like, while in the second version her Pz experiences are qualitatively the same throughout the period. And this might seem to be supported by the fact that, as we can suppose, Ellen's memory reports and behavior are as they should be, given the functionalist view of qualitative similarity, if these relations of qualitative similarity hold. Her situation is like Bertha's in the case where she never suffers brain damage and so never has her backup system activated.

I suggest, however, that this verdict about these versions of the case is called into question by the version of the example in which both systems are active (intermittently—both cannot be active at the same time) during both the Bertha-like period and the Dorothy-like period. There it certainly is not true that Ellen's memory reports and behavior are as they should be, given the functionalist view of qualitative similarity and difference, if the Px experiences are all qualitatively alike and the Pz experiences are all qualitatively alike. Yet we can suppose that the sequences of Px and Pz experiences in this case are exactly like the sequences of Px and Pz experiences in the versions of the case in which one or the other of the systems was inactive throughout—the members of these sequences can be put in a one–one correspondence such that corresponding members of these sequences have exactly similar causes and occur in exactly similar circumstances. How can it be that in one such sequence all of the Px (Pz) experiences are qualitatively alike while in the other they are not?

During the period when Ellen was Bertha-like, and the **β** system was inactive, it seems right to say that her Px experiences were qualitatively alike. But this is not just because of how she behaved and what she reported; it must also be in part because of how she was disposed to behave and report, and so because of what counterfactuals are true of her. And one relevant counterfactual is this: if the Px experiences had occurred just when they did, caused by just the things that did cause them, but the backup system had sometimes been active during this period, she still would have behaved as the functionalist criterion of qualitative similarity says she should behave if the Px experiences during that interval were all qualitatively alike. But during the longer period in which Ellen was first Bertha-like and then Dorothy-like, and the **β** system was inactive, that counterfactual would not be true, as is shown by the version of the case in which both systems were intermittently active throughout the longer period. It seems to me that the functional criterion of qualitative sameness and difference should be understood as being satisfied only when a counterfactual conditional of that sort holds. The failure of the relevant counterfactual to hold is an indication that we have something like memory tampering.

This shows, I think, that "differences in unrealized counterfactual possibilities" *can* affect the qualitative character of experiences; more specifically, that the presence of an inactive backup system can have a bearing on this. But our case (that of Ellen) in which the relevant counterfactual failed to hold was one in which there was a change in the way the systems were connected. There seems to be no reason to say that the counterfactual fails in the case where Alice's backup system atrophied; supposing that her backup system was never activated while it was viable, there is no reason to think that had it been activated the functionalist case for taking her earlier (pre-atrophy) Px experiences and her later (post-atrophy) Px experiences as qualitatively alike would have been undermined.

It is certainly plausible to think that in the atrophied Alice case the earlier and later Px experiences are the same, despite the fact that their qualia have different total realizers—let these be Px&C and Px&X. In my 1996 I discussed such a case (presented to me by Stalnaker in correspondence), and offered a way of respecting the intuition that this is so. My general strategy is to use the functional account of qualitative similarity to determine what states count as qualia realizers, and to determine what qualia realizers are realizers of the same quale, and then take the latter facts to be the basis of intersubjective comparisons of qualitative character. And I suggested that while Px&C and Px&X cannot be instantiated in atrophied

Alice (or anyone) at the same time, they can count as realizers of the same quale if their instantiations in the same individual at different times yield intrasubjective qualitative identity as defined by the functionalist account, where this includes the requirement that there not be a change in memory mechanism of the sort there is in the case of Ellen, i.e., one involving a change in the way systems α and β are connected. I took this to be so in the case of atrophied Alice. But this of course takes me from the frying pan to the fire. (I think that Stalnaker probably noticed this, but generously refrained from mentioning it.) I cannot, it would seem, allow that atrophied Alice's earlier and later Px experiences are qualitatively the same without accepting the identities 1–4. And I can't accept those along with 5 without being stuck with a contradiction. So to hold onto my view, I must reject the plausible intuition that there is no change in the qualitative character of Alice's Px experiences when her backup system atrophies. And I have not the reason I have in the case of Ellen for saying that the change in Alice involves memory tampering.

But there is a principled reason, other than the desire to avoid the contradiction implied by (1)–(5), for denying that Alice's pre-atrophy Px experiences and her post-atrophy Px experiences are qualitatively alike. If one accepts that qualia are physically realized, and also accepts that qualia are multiply realizable in a way that permits different instantiations of the same quale in a single subject to be grounded in instantiations of different realizers, then one must hold that there is a mechanism that implements the transition from an instantiation of one of the realizers to the instantiation of a different one in a way that yields the functionally appropriate effects. The mechanism will be such as to make the transition functionally equivalent, in psychologically relevant ways, to there being successive instantiations of the same quale realizer. This mechanism—call it the change-of-realizer mechanism—will of course include a memory mechanism which is such that instantiations of the different realizers have the same effects on the subject's subsequent memories. The change-of-realizer mechanism will have to be part of each of the total realizers of the quale; without it they will not be realizers of the same quale, and there is nothing that could make one of them rather than any of the others a realizer of *that* quale. But this imposes the requirement that any case in which there is a change in how a quale is realized must be a case in which the different total realizers share the same change-of-realizer mechanism. In cases where this requirement is violated it may appear that the functionalist criterion of qualitative similarity is satisfied, but it is not in fact satisfied. This is so in the version of the Ellen case in which one of her systems was inactive throughout, and there we had an independent reason for thinking that the appearance is misleading. I think it is also so in the case of atrophied Alice.

Stalnaker takes the case of Ellen to support the Frege–Schlick view. He says that "the purely relational account of qualia that grounds the Frege–Schlick view not only rejects interpersonal qualia identities, but also claims that intrapersonal comparisons across time may hold only relative to perhaps arbitrary assumptions about the accuracy of qualitative memories" (p. 237), and he sees the case of Ellen as supporting this. He ends his discussion of this case by saying that Ellen must be misremembering *something* about the qualitative character of her past experiences, but that there need be no fact of the matter about which experiences she is misremembering.

This is not the only place where Stalnaker says that there can be no fact of the matter about whether experiences of the same person at different times are qualitatively identical or similar. He claims this also in the case where a person wears inverting glasses that initially make things look upside down. After a while the person accommodates; but when he removes the glasses things again look upside down for a while. Concerning two hypotheses about the qualitative similarities and differences that hold amongst the person's experiences while this is going on he says "I don't think that even naïve common sense supports the judgment that there must be an answer to the question of which of these hypotheses is correct" (p. 231). And he says the same about a case that involves the rewiring of the nervous system so as to change (at least temporarily) the felt location of pains.

Certainly the Frege–Schlick view can allow for such indeterminacies. But one can allow for them without accepting the Frege–Schlick view. Holding that qualitative similarity and difference can hold interpersonally does not commit one to holding that there is always a determinate fact of the matter whether experiences belonging to different persons are qualitatively alike. On a view like mine, such indeterminacies would stem from indeterminacies as to whether different physical properties are realizers of the same quale—this would yield in the first instance indeterminacies as to whether experiences of the same person at different times instantiating these properties are qualitatively alike, but it would also yield indeterminacies as to whether instantiations of these properties in experiences of different persons are qualitatively alike. It can be somewhat indeterminate what the functional role of a property is, what causal features go into it, and it can for that reason be indeterminate whether two properties both play that functional role—or whether the causal features of the realized property are a subset of the causal features of the putative realizers. And this sort of indeterminacy can equally affect intrasubjective and intersubjective comparisons. The indeterminacy might involve memory mechanisms. What is indeterminate might be whether a physical difference between memory mechanisms disqualifies properties whose instantiation involves the physically different mechanisms from being total realizers of the same quale. So it might be indeterminate whether we have a case of "memory tampering." But this will bear equally on the intrasubjective and the intersubjective case. It is of course compatible with there being such cases of indeterminacy that there are plenty of cases where there is no indeterminacy, and that this is as much true in the intersubjective case as in the intrasubjective case. It is compatible with it that if you and I are functionally and physically alike, what it is like for you to see red, or to feel pain, is like what it is for me to see red, or to feel pain.

Holding a view that combines the functionalist account of qualitative similarity and difference with the claim that qualitative similarity and difference can hold intersubjectively has a cost. The cost is having to hold that it is false or at least not determinately true that atrophied Alice's earlier Px experiences, had when her backup system was viable but inactive, are qualitatively the same as her later Px experiences, had after the backup system atrophied. That, I allow, is prima facie counterintuitive (perhaps a bit less so if we say "not determinately true" rather than "false"). But whether this cost means that we should reject the view depends on what the cost is of holding the alternative view, namely the Frege–Schlick view. And I think that cost is far greater.

The main cost of holding the Frege–Schlick view is that it involves abandoning the intuition that it can be the case, and (most of us would think) almost certainly is the case, that when different persons are physically and functionally alike, what it is like for them to see red, smell lilacs, feel pain, etc. is the same or similar. Stalnaker says that the main reason I resist the Frege–Schlick view is that I think that "no one can plausibly deny the coherence of the hypothesis that there has been an intrapersonal spectrum inversion" (p. 223). But that isn't the main reason; the main reason is the compelling character of the intuition just expressed. The closest the Frege–Schlick view can come to respecting that intuition is to allow that physically and functionally similar persons have, in similar circumstances, experiences similar in objective representational content, e.g., in what colors they represent things as having. And that isn't close enough. One doesn't have to invoke the possibility of spectrum inversion in order to see that the "what it is like" of experiences is not determined by objective representational content—color constancy phenomena of the sort mentioned earlier are sufficient to show that. While Stalnaker thinks that the Frege–Schlick view does a better job than the standard view of qualia of respecting the connection between qualitative character and representational content, I argued earlier that the reverse is true. What I have called the private properties version of the Frege–Schlick view seems unacceptable because it is committed to the denial that qualia can be physically realized, and so seems incompatible with physicalism. And as I have argued, it is not easy to see how the Frege–Schick view can abandon the private properties view without quining qualia altogether. Neither of Stalnaker's analogies—that involving the relational theory of space, and that involving the Von Neumann–Morgenstern utility theory—seems to provide a model that makes sense of the notion of "relational qualia."

For these reasons I remain unconvinced that the Frege–Schlick view provides a viable alternative to the account I have offered. But I wish I had something more satisfying to say about the case of atrophied Alice.[9]

[9] Thanks to Carl Ginet for comments on an earlier version.

REFERENCES

Clark, A. 2000: *A Theory of Sentience* (Oxford: Oxford University Press).

Frege, G. 1952: "Illustrative Extracts from Frege's Review of Husserl's *Philosophie der Arithmetik*," in P. Geach and M. Black (eds.), *Translations from the Philosophical Writings of Gottlob Frege* (Oxford: Basil Blackwell).

—— 1956: "The Thought, A Logical Inquiry," *Mind*, 65.

Hilbert, D., and Kalderon, M. 2002: "Color and the Inverted Spectrum," in S. Davis (ed.), *Color Perception: Philosophical, Psychological, Artistic and Computational Perspectives* (Oxford: Oxford University Press).

Shoemaker, S. 1981*a*: "Some Varieties of Functionalism," *Philosophical Topics*, 12, I, 83–118.

—— 1981*b*: "The Inverted Spectrum," *The Journal of Philosophy*, 74:7, 357–81.

—— 1984: Postscript to reprinting of "The Inverted Spectrum" in Shoemaker, *Identity, Cause and Mind* (Cambridge: Cambridge University Press).

—— 1994: "Phenomenal Character," *Nous*, 28, 21–38.

—— 1996: "Intrasubjective/intersubjective," in Shoemaker, *The First Person Perspective and Other Essays* (Cambridge: Cambridge University Press).

—— 2000: "Introspection and Phenomenal Character," *Philosophical Topics*, 28:2, 247–273.

—— 2006: "On the Ways Things Appear," in T. Gendler and J. Hawthorne (eds.), *Perceptual Experience* (Oxford: Oxford University Press).

Stalnaker, R. 2003: "Comparing Qualia Across Persons," in Stalnaker, *Ways a World Might Be* (Oxford: Clarendon Press). Originally published in *Philosophical Topics*, 26, 1 & 2, 1999.

3

Character Before Content

Paul M. Pietroski

Speakers can use sentences to make assertions. Theorists who reflect on this truism often say that sentences have *linguistic meanings*, and that assertions have *propositional contents*. But how are meanings related to contents? Are meanings less dependent on the environment? Are contents more independent of language? These are large questions, which must be understood partly in terms of the phenomena that lead theorists to use words like 'meaning' and 'content', sometimes in nonstandard ways. Opportunities for terminological confusion thus abound when talking about the relations among semantics, pragmatics, and truth. As Stalnaker (2003) stresses, in Quinean fashion, it is hard to separate the task of evaluating hypotheses in these domains from the task of getting clear about what the hypotheses *are*. But after some stage-setting, I suggest that we combine Stalnaker's (1970, 1978, 1984, 1999, 2003) externalist account of content with Chomsky's (1965, 1977, 1993, 2000*a*) internalist conception of meaning.

On this view, the meaning of a declarative sentence is *not* a function from contexts to contents. Linguistic meanings are intrinsic properties of expressions that constrain without determining truth/reference/satisfaction conditions for expressions relative to contexts. As we shall see, this independently motivated conception of meaning makes it easier to accept the attractive idea that asserted contents are sets of metaphysically possible worlds. This is to reject certain unified pictures of semantics and pragmatics. But we should be unsurprised if some of the facts that linguists and philosophers describe are interaction effects, reflecting *both* sentence meaning *and* asserted content. While it can be tempting to think of semantics as conventionalized (or "fossilized") pragmatics, with meaning somehow analyzed in terms of assertion, I think that meaning is more independent of—and probably a precondition of—assertion and truth. This may be at odds with some of what Stalnaker says about semantics. But the important points are in the spirit of his work: it isn't obvious *what* the study of meaning (or content, or gold) is the study of; but there are better and worse ways of framing relevant questions.

1. KRIPKE-PROPOSITIONS

Let a K-proposition be a set of *ways the world could be*, leaving it open whether and where this notion is theoretically useful. Let A-propositions be whatever the sentential variables in our best logical theories range over. If such variables range over *structured* abstracta with "logical form," as suggested by the study of valid inference since Aristotle, then K-propositions are not A-propositions. But one shouldn't reason as follows: assertions are governed by logic; sets don't have logical form; so asserted contents are not K-propositions. At best, this would be a misleading way of defining 'content' as a word for talking about things with logical form.

If *ways* the world could be are as Lewis (1986) describes them—universes, like the one that includes us, but each with its own spacetime and distinctive inhabitants—so be it. Though it seems less extreme to suppose, with Kripke (1980) and Stalnaker (1976, 1984), that ways *the* world could be are just that: possible states of the one and only universe, which actually includes us, but which might have been configured differently in many respects; where possible states need not be Ludwigian totalities of things. Like Kripke, I am inclined to identify possible world-states with possible histories of *this* universe. But for present purposes, we need not decide what possible states (or configurations) of the world are. And whatever they are, we can use the now standard phrase 'possible worlds' to talk about them, without identifying the actual world—i.e., the *way* the universe is—with the actual totality of things; see Stalnaker (1984). If there is the totality of (actual) things, it would seem to be distinct from the way it is, just as all of the things seem to be distinct from the ways they are. Though let me stress that the possible worlds, as conceived here, include *all* and only the *logically* possible states of the universe. Put another way, the logically possible worlds are none other than the (metaphysically) possible worlds. We can, if we like, talk about restricted notions of possibility corresponding to sets that include some but not all possible worlds. But the logically possible worlds do not include any ways the universe couldn't really be. So the possible worlds do not exclude any logically possible worlds.[1]

Hence, a K-proposition is a set of logically possible worlds. Though let me enter a caveat, using 'Φ' to abbreviate 'Hesperus (exists and) is distinct from Phosphorus'. I am *not* committed to the following biconditional: it is logically possible that Φ iff there is a possible world at which Φ. On the contrary, there

[1] See Kripke (1980, 15–20), who also says that his third lecture 'suggests that a good deal of what contemporary philosophy regards as mere physical necessity is actually necessary *tout court*' (p. 164). For discussion in the context of supervenience theses, see Pietroski (2000, ch. 6). Stalnaker's (2003) notion may be a little broader, since he speaks of ways *a* world might be (though see also p. 215); and he says the *concept* of possibility is to be understood functionally, "as what one is distinguishing between when one says how things are" (p. 8).

is a possible world at which Φ iff the universe could be such that Φ; and I think Kripke argued persuasively that the universe couldn't be that way. Yet intuitively, it is logically possible that Φ. So I conclude that the complex *sentence* 'It is logically possible that Hesperus is not Phosphorus' does not *mean* that relative to some logically possible state of the universe, the embedded sentence is true (as used). This leaves room for a semantic theory, employing a different technical notion of 'π-World', according to which the complex sentence counts as true iff the embedded sentence is true relative to some π-World.[2] My own preference is for a more Fregean theory that associates each 'that'-clause with a linguistically structured entity whose status as a semantic value of 'logically possible' does not depend on its being true relative to some possible state of the universe; see Pietroski (2000). But whatever one thinks about debates concerning what 'that'-clauses and words like 'possible' *mean* in natural language, one can grant that every logically possible world is a way the universe could be without endorsing the following generalization: it is logically possible that... iff 'that...' indicates a nonempty K-proposition.[3] No definition could ensure such a link between logic, 'that'-clauses, and possible states of the universe. These are matters for investigation, not stipulation.

This highlights the question of what K-propositions *are* good for, apart from providing a semantics for invented languages with sentential operators like '□' and '◊', glossed in terms of metaphysical necessity and possibility. It can seem that K-propositions are ill-suited to the study of human communication and thought. Sets of possible worlds are individuated without regard to how speakers/thinkers represent their environment; and two sentences can differ in meaning yet be true (in each context) relative to the same possible worlds. Given plausible assumptions, every property is such that the worlds at which it is instantiated by Hesperus are the worlds at which it is instantiated by Phosphorus. The possible worlds at which two plus two is four are those at which there are infinitely many primes—and these are the worlds at which it is false that Φ. So it can seem like a mistake to characterize meanings or contents in terms of K-propositions.

In my view, this is half right; semanticists need different tools. But as Stalnaker argues, the apparent deficiencies of K-propositions reflect potential virtues that should not be ignored, at least not if representing the universe involves locating the way we take it to be in a space of possibilities. Appealing to K-propositions is a way of talking about possibilities while abstracting away—in so far as such abstraction is possible for creatures like us—from details concerning *how* thinking/speaking creatures like us represent those possibilities. This will be useful

[2] Theorists employing such a notion would have to say what π-worlds are, in enough detail to support their proffered explanations. But perhaps appeal to Ludwigian totalities, or "ersatz" analogs that are formally similar in certain respects, will be useful here.

[3] Or a little more precisely (using '#' as a corner-quote), one need not endorse the following generalization: #it is logically possible that P# is true iff the semantic correlate of #that P# is a nonempty K-proposition. See Peacocke (1999) for related discussion.

when and to the degree that it is useful to distinguish (i) questions about the world represented and the relevant background possibilities from (ii) questions about how it and they are represented.

A set of possible worlds can serve as a mind/language-independent representative of *what* someone is thinking or saying if she thinks or says that Hesperus rises in the evening (or that Phosphorus rises in the morning). In this technical sense, theorists can treat sets of possible worlds as potential contents: abstract "things" that can be asserted, judged true, doubted, etc. Though to make it explicit that this is a technical sense, we can call K-propositions *kontents*, and then ask whether appealing to them in accounts of human thought and communication is indeed as useful as Stalnaker and others contend. But one cannot object simply by noting, in various ways, that K-propositions play the role they are designed to play as abstractions.

Episodes of asserting that two plus two is four differ from episodes of asserting that there are infinitely many primes, presumably in the differing details of *how* the universe gets represented in such episodes. In my view, we shouldn't abstract away from such details when providing theories of *meaning* for natural language, but such abstraction may be perfectly appropriate when talking about *truth*-conditions. Likewise, asserting that Hesperus is Venus differs from asserting that Phosphorus is Venus. But as we'll see in section three, natural language often marks distinctions (relevant to theories of meaning) that are metaphysically otiose. And it may be a *virtue* of appealing to kontents that it leads us to diagnose such distinctions in terms of asserting something in different ways—say, by using sentences with different meanings. (I defer discussion of "two-dimensional" kontents to an appendix.) If the semantic properties of natural language expressions are irredeemably human and internalistic, appeal to K-propositions will be ill-suited to certain theoretical tasks in linguistics. Though such appeal may help us articulate and answer various questions concerning the *use* of natural language—especially with regard to heavily world-directed uses, like making truth-evaluable assertions and reporting beliefs. It may be that K-propositions are what we need, and all we're likely to get, for purposes of characterizing a substantive mind/language-independent notion of truth-conditions.

As Stalnaker (2003) emphasizes, philosophers should be especially sensitive to the desirability of a framework that lets us distinguish questions about the truth (or plausibility) of what a person said, from interpretive questions concerning what she said, and questions concerning the meanings of her words. He says that while we "cannot separate semantics from substantive questions before we begin to theorize," philosophers can still try "to separate, in context, questions about how to talk, or about how we in fact talk, from questions about what the world is like (2003, 4–5)." I find much to agree with in these passages where Stalnaker expresses his affinity with Carnap's (1950) project of framing metaphysical questions in terms of a distinction between internal and external questions. But I want to challenge the idea—common to Carnap, Quine, and many others—that semantics is somehow not substantive. Of course, one can define 'semantics' as one likes. But there are substantive constraints on how signals of a human language can be naturally associated with meanings. And this is relevant in the present context.

2. CHOMSKY-MEANINGS

In philosophy, as in common parlance, there is a tendency to equate semantic facts with facts that illustrate Sassurean arbitrariness and the presumably conventional aspects of language use: the French word 'chien', synonymous with 'dog' (and not 'cat'), is typically used to talk about dogs (as opposed to cats); and so on. Similarly, if exactly one of two metaphysicians uses 'proposition' to signify sets of possible worlds, we are apt to say "That's just semantics." Theoretically interesting questions are unlikely to turn on such idiosyncratic facts concerning *which* meaning certain speakers happen to associate with a given perceptible sign.[4] But there is still plenty for semanticists to do, since the interesting facts often concern ways in which natural language *cannot* be understood. Moreover, as discussed by Chomsky and many others, the constraints seem to reflect a "human language faculty," as opposed to general principles of reasoning, communication, convention, or learning. If this is correct, perhaps semantics should be characterized in terms of this faculty, whose mature states correspond to natural languages.

Extending a familiar slogan: if theories of meaning are theories of understanding, and these turn out to be theories of a mental faculty that associates linguistic signals with meanings in constrained ways, then we should try to figure out (in light of the constraints) what this faculty associates signals with.[5] Following Chomsky (2000*a*), I don't think theories of meaning for natural language will be theories of truth, in large part because I find it implausible that mature states of the language faculty associate signals with truth/reference/satisfaction conditions; see Pietroski (2003*a*, 2005*b*). But my aim in this section and the next is just to show that there is motivation and conceptual room for a more internalistic conception of the language faculty, according to which it associates certain signals with instructions for building concepts (much as it associates certain concepts with instructions for generating signals). For these purposes, I assume some familiarity with transformational grammar—and in particular, with the idea that sentences can have unpronounced elements, including traces of displacement operations.

Consider the following sequence of lexical items: *hiker*, *lost*, *kept*, *walking*, *circles*. This string of words might well prompt the thought indicated with (1), as opposed to the less expected thought indicated with (2):

[4] In my view, gestures towards (remotely plausible) causal/functional-role/teleological theories are just that. Nobody has a good *theory* of why 'chien' stands for dogs as opposed to cats.

[5] Compare Dummett (1975), McDowell (1976).

(1) The hiker who was lost kept walking in circles
(2) The hiker who lost was kept walking in circles

But (3) has only the meaning indicated with (3*b*), the yes/no question corresponding to (2).

(3) Was the hiker who lost kept walking in circles
(3a) Was it the case that the hiker who was lost kept walking in circles
(3b) Was it the case that the hiker who lost was kept walking in circles

We hear (3), unambiguously, as synonymous with (3b) *and not* (3a). Likewise, (4)

(4) Was the child who fed some waffles at breakfast fed the kittens at noon

can only be the yes/no question corresponding to the bizarre (6).

(5) The child who was fed some waffles at breakfast fed the kittens at noon
(6) The child who fed some waffles at breakfast was fed the kittens at noon

And we know that natural language is not hostile to ambiguity, given examples like (7).

(7) Solicitors who can duck and hide whenever visiting solicitors may scare them saw every doctor who lost her patience with patients.

If a string of words can be understood as a sentence, but *not* as having a sentential meaning easily expressed with a good sentence formed from those words, then this negative fact calls for explanation, especially if the actual meaning is more "cognitively surprising" than the nonmeaning (see Chomsky [1965], Higginbotham [1985]). Standard explanations for the nonambiguity of (3) posit a constraint that (one way or another) precludes extraction of auxiliary verbs from relative clauses; see Ross (1967), Travis (1984). One hypothesizes that the transformation indicated in (3β) is licit, while the transformation indicated in (3α) is not.

(3α) Was {[the [hiker [who__ lost]$_{RC}$]] [kept walking in circles]}
└──────────*──────────┘
(3β) Was {[the [hiker [who lost]$_{RC}$]] [__ kept walking in circles]}
└──────────────────────┘

If a string of words fails to have *any* coherent interpretation, that is a special case of nonambiguity. While (8) and (10) are fine, (9) and (11) are each somehow defective.

(8) The hiker who was lost whistled
(9) Was the hiker who lost whistled
(10) The vet saw each dog that was found
(11) Was the vet saw each dog that found

Though (11) is closer to word-salad. We can start to explain this by noting that the word-string in (9) corresponds to each of the potential transformations indicated in (9α) and (9β);

(9α) Was {[the [hiker [who__ lost]$_{RC}$]][whistled]}
└────────*────┘
(9β) Was {[the [hiker [who lost]$_{RC}$]] ☹ [__ whistled]}
└──────────────────────┘

where '☹' indicates the anomaly of asking whether a *hiker* was whistled—in the way a song can be whistled. This defect is not so severe that it keeps us from hearing (9) as a strange question. (The hiker who lost was whistled? Do you mean that someone whistled to the lost hiker?) But we cannot even hear (11) as a defective way of asking the question indicated with (11α).

(11α) Was {[the vet][saw [each [dog [that__ found]$_{RC}$]]]}
└──────────────*────┘

Likewise, the violation in (9α) is severe enough to keep (9) from being ambiguous.

Constraints on transformations bear on pronunciation as well as interpretation. English allows for contractions like those in (12–14), with blanks indicating wh-traces.

(12) Who do you *want-to/wanna* kiss__
(13) What do you think__ *is-up/'s-up* there
(14) What do you think it *is-doing/'s-doing*__ up there

But in cases like (15–16), contraction is apparently blocked by wh-traces.

(15) Who do you *want__ to/*wanna* kiss Chris
(16) What do you think it *is__ up/*'s-up* there

One can imagine a language that allows contraction in (16), or one that treats (3) as ambiguous. But in natural language, pronunciation is related to interpretation via grammatical structures that constrain both, even if the constraints are arbitrary with regard to both. We can easily *produce* the contracted form of (16), just not as a signal with the same (coherent) sentential meaning as the uncontracted form. And we can certainly comprehend the question, expressible with (3a), that cannot be asked with (3). Our inability to impose this sensible interpretation on (3) suggests that natural languages associate word-strings with complex meanings in ways that satisfy constraints on interpretation *independent of* any limitations imposed by (logic and/or) cognitive systems other than the language faculty. Likewise, such associations satisfy constraints independent of those imposed by the cognitive/biological systems responsible for articulation and perception of linguistic signals. Moreover, the relevant constraints on interpretation and pronunciation apparently overlap, suggesting common language-specific factors.

Indeed, evidence that natural languages are governed by "autonomous" constraints—due to the nature of human language, as opposed to perceptual/articulatory/conceptual capacities—is the original (and perhaps best) evidence for a substantive language faculty; see Chomsky (1965, 1986). I offer additional examples in section three. But at this point, we need to sharpen the terminology. Let's say that a *language* is a system for associating signals with interpretations. A *human* language is one that normal human children can acquire in an environment not atypical for members of our species. In so far as 'system' is ambiguous, between a set of abstract rules and a mechanism that instantiates such rules, 'language' is ambiguous. So following Chomsky (1986), let's use 'I-language' to talk about the relevant aspects of human minds/brains: each speaker of a human language has an I-language, by virtue of which she can associate endlessly many linguistic signals—like Japanese *sounds*, or ASL *signs*—with interpretations. This leaves it open what interpretations are. Idealizing, we can think of children as coming equipped with a language faculty, whose initial state can be changed (through experience and growth) within constraints that we can try to discover; where this faculty typically settles into one or more stable adult states, each of which is an I-language. The picture is familiar.

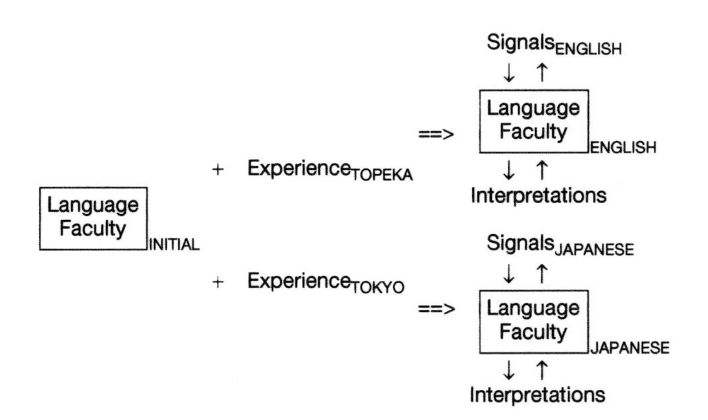

We can go on to ask in which respects, and to what degree, I-languages are: determined by human biology, as opposed to contingent aspects of experience; "encapsulated" from other aspects of human psychology; shared by speakers within a given linguistic community; etc. But the idea is that each "setting" of the language faculty corresponds to the grammar of a possible human language. The range of possible settings is determined by the nature of the human language faculty, which thereby determines which logically possible languages are "options" for a child prior to relevant experience; see, for example, Chomsky (1981, 1986).[6]

Of course, an I-language may be related to publicly perceptible signals only indirectly. It may even be misleading to talk about the language faculty generating *re*presentations of signals, whose linguistically important properties may not be specifiable independently of the structures generated. Similar remarks apply to interpretations, which remain far less well understood than pronunciations. And in any case, it is hardly obvious that I-languages associate signals with (representations of) truth/reference/satisfaction conditions. One might say instead that each complex expression of a spoken I-language links a pair of instructions for creating a complex sound and a complex concept, leaving it open how (and at what "distance") these instructions are related to entities in the public domain. Indeed, recent trends in linguistics suggest that a grammatical expression may just *be* a pairing (generable by the language faculty) of a "phonological instruction" to articulatory-perceptual systems with a "semantic instruction" to conceptual-intentional systems; see Chomsky (1995, 2000*b*).

This invites the thought that linguistic meanings are human *Begriffsplans*: blueprints, produced by the language faculty, for constructing concepts from lexicalized elements.[7] On this view, the meaning of 'brown dog that barked loudly' is an instruction for how to build a complex concept out of resources indicated with the constituent words. A theorist might offer a first-pass proposal about certain aspects of this instruction by using notation like the following.

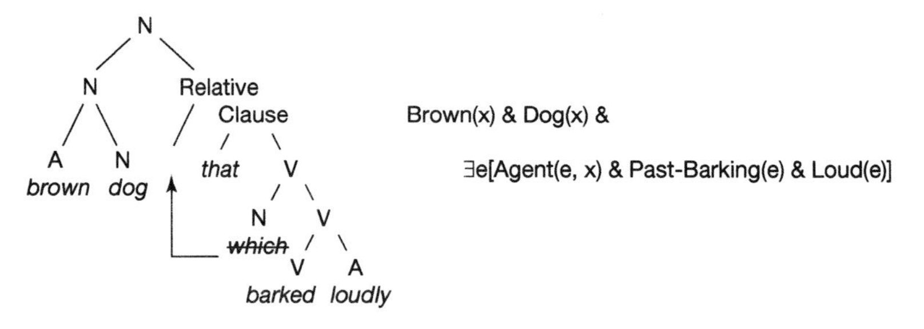

[6] If only for simplicity, let's assume that interpretations are constant across languages, and let's pretend that all speakers of a "named language" like English have the same I-language.

[7] While 'Begriffsplan' is compact, 'Begriffskonstruktionsanleitung' (concept-construction-instruction) displays the point even better. Within analytic philosophy, it was long held that

The logical formalism on the right encodes a hypothesis about how the phrase is understood, as a conjunctive monadic predicate with "thematic" structure. The tree diagram encodes hypothesized grammatical structure, including a displaced covert wh-element, and typology. But the formalism is not self-interpreting. One has to say what symbols like 'x', '∃', 'e', and 'Agent' mean *here*—in the context of theorizing about the human language faculty. And interpreting the variables as ranging over a domain of "real world" entities may not be the best way of describing the instructions. For purposes of characterizing *Begriffsplans*, it may be a mistake to abstract away from *how* humans represent the universe. Correlatively, the notation above may reflect two intrinsic aspects of a natural language expression *and not* a proposal about how a mere piece of syntax generated by the language faculty is related to a domain in terms of which one can provide a plausible theory of truth (that can also serve as a theory of meaning) for English.

This leaves ample room for appeal to K-propositions in descriptions of *how humans use* language and concepts to represent the universe. One can ask how ways-we-take-the-world-to-be are located in a space of ways the world could be. But one shouldn't assume that this is a primary question for the study of linguistic meaning, or that theories of the language faculty should be specified in terms of notions appropriate to investigations based on the kind of abstraction embodied in appeal to K-propositions. This should be obvious, especially in light of Chomsky's (1986) tripartite distinction of research questions: what do adults know about their languages; how do children acquire such knowledge; and how do humans put such knowledge to use? But the point bears repetition. Theories of *understanding* and theories of *use*, if such there be, may require different kinds of abstraction and idealization. Meaning constrains use, and one shouldn't expect meaning to be determined by observable facts about how language is used.

Quinean Translators and Davidsonian Interpreters are misguided. They try to figure out what expressions mean, without relying on a language faculty, given only evidence of the sort available to children. But such evidence does not include the (typically negative) facts that support nativist hypotheses about how children (despite their limited experience) figure out what expressions do and don't mean. The "idealized field linguist" is an empiricist monster—half scientist, half child—operating without prior assumptions about the constrained character of human languages, but also blind to the evidence that reveals semantically relevant constraints. If any such alien searches for sources of linguistic stability across human speakers, it will be led to exaggerate the role of our shared environment

expressions of natural language can't be systematically associated with (A-propositions) or concepts. But given developments in the study of language and logic, it now seems clear that the most famous cases of alleged mismatches between grammatical form and logical form were misdiagnosed, that the resources for rediagnosing other cases are considerable, and that positing such mismatches is problematic (especially in light of how children can understand complex constructions). For review, see Pietroski (2003*b*); see also Neale (1990, 1993), Ludlow (2002).

and underestimate the role of our shared biology. Since the alien ignores the most interesting facts, it will conclude that semantics is relatively superficial—a matter of contingent/conventional associations of grammatically structured signals with features of the environment. But if theories of understanding are best viewed as theories of how the *language faculty* associates signals with meanings, as opposed to theories that an alien interpreter would adopt on the basis of limited evidence, perhaps we should take meanings to be human *Begriffsplans* and try to characterize them in terms appropriate to the language faculty.

3. COMPLEX CONCEPTS AND MENTALESE

The view on offer is, in some respects, like Fodor's (1975, 1978, 1987, 1998) version of the much older idea that linguistic expressions signify mental representations. But there are significant differences. Fodor combines skepticism about truth-conditional semantics for spoken languages with acceptance of the idea that theories of meaning need to be theories of truth, or some such symbol–world relation.[8] Nonetheless, he maintains, there are systematicities that require explanation in terms of a compositional semantics for *some* language that humans use; see Fodor and Lepore (2002). Fodor has independent reasons for positing a language of thought. So he suggests that a spoken language like English has a syntax but no semantics. The idea is that the language faculty associates signals with syntactic structures that must (to be meaningful) be paired, in contexts, with expressions of Mentalese: strictly speaking, a sentence of English does not itself have a linguistic meaning. According to Fodor (1998), sentential utterances are—in complicated, context sensitive ways that preclude theories of truth for spoken languages—associated with tokenings of Mentalese sentences; where these contextually embedded tokenings do have compositionally determined truth conditions (ceteris paribus, *bien sûr*).

From this perspective, expressions of a spoken language are devices for perceptibly indicating concepts whose semantic structure (if any) is independent of the human language faculty. Of course, Fodor doesn't think the relations between syntactic structures and signified concepts are wholly arbitary, as though each spoken sentence were a new word. So he can't really think, despite what he sometimes says, that spoken languages respect no semantic constraints of their own; see Matthews (2003). But in so far as Fodor's view *is* at odds with the idea that spoken sentences have meanings of their own, it makes a mystery of many systematic constraints on possible interpretations of such sentences; see

[8] I endorse the former but not the latter. In my view, Katz and Fodor (1963) rightly eschewed Lewis's (1972) stipulations about what semantics must be. But my claim is not that meanings are concepts. Recall that Strawson (1950) urged us to characterize the meaning of a referential device R in terms of "general directions" for using R, *and not* in terms of some entity allegedly denoted by R.

Higginbotham (1994). One wants to know why there is no context in which the Mentalese correlate of (3) is the correlate of (3*a*).

(3) Was the hiker who lost kept walking in circles
(3*a*) Was it the case that the hiker who was lost kept walking in circles
(3*b*) Was it the case that the hiker who lost was kept walking in circles

Why *can't* the string of words in (3) be used to indicate the Mentalese sentence with which we think the thought actually expressed with (3*a*)? It is not enough to say that (3α) is ungrammatical. For if (3β) is intrinsically *meaningless*, why can't it be used to signify the meaning of (3*a*), perhaps along with other meanings? (Compare 'bank'.)

(3α) Was {[the [hiker [who__ lost]$_{RC}$]] [kept walking in circles]}
└──────────*──────────┘
(3β) Was {[the [hiker [who lost]$_{RC}$]] [__ kept walking in circles]}
└──────────────────────────┘

More generally, given a theory of Mentalese, one might be able to specify an algorithm that pairs sentences of Mentalese with Chomskyan syntactic structures in a descriptively adequate way. But one would want still to know why spoken languages are repeatedly interpreted in accordance with this algorithm as opposed to others; cf. Montague (1974). And not all theoretically interesting limitations on ambiguity are due to constraints on displacement of lexical items.

While (17) has readings indicated with (17*a*) and (17*b*), it does not have the third reading.

(17) The senator called the millionaire from Texas
(17*a*) The senator called the millionaire, and the millionaire was from Texas
(17*b*) The senator called the millionaire, and the call was from Texas
(17*c*) #The senator called the millionaire, and the senator was from Texas

One can hypothesize, plausibly, that (17) is structurally ambiguous as between (17α) and (17β);

(17α) {[The senator] [called [the millionaire [from Texas]]]}
(17β) {[The senator] [[called [the millionaire]] [from Texas]]}

where (17α) is a sentence with 'the millionaire from Texas' as the direct object of 'called', and in (17β), 'from Texas' is an adjunctive phrase modifying 'called the millionaire'. But even given that (17) cannot be structured so that 'The senator' and 'from Texas' form a phrase, there remains the question of why (17β) fails to have the meaning expressed with (17*c*). One can imagine a language where (17β) means that the senator satisfied two conditions: he called the millionaire, and he was from Texas. The phrase [millionaire [from Texas]] can actually be used as a predicate that imposes a conjunctive condition on individuals. So *why* can't the phrase [[called the millionaire][from Texas]] also be used this way? One can say that (17β) has only the meaning/Mentalese-correlate indicated with (17*b*). But this is the fact to be explained.

As a start, one might follow Davidson (1967*b*, 1985) in representing the meaning of (17*b*) as follows: ∃e{The(x):senator(x){the(y):millionaire(y)[Called(e, x, y) & From(e, Texas)]}}. Then one can try to explain why in (17β), 'from Texas' must be understood as a predicate of the 'e'-variable, as opposed to the 'x'-variable. Alternatively, one could formalize (17*b*) as follows: ∃e{The(x):senator(x) [Agent(e, x)] & Past-Calling(e) & the(y):millionaire(y)[Theme(e, y)]}. Then one can hypothesize that *each* phrase in (17β) must be understood as a predicate of the variable associated with the verb, perhaps via thematic relations associated with certain grammatical relations or prepositions (see Parsons [1990], Schein [1993, forthcoming], Pietroski [2002, 2005*a*]). But in any case, it is hard to see how the *2-but-not-3*-way ambiguity of (17) can be accounted for without some such eventish constraint on the interpretation of spoken language.

As another example of interesting nonambiguity, note that the famous (18) is roughly synonymous with (18*a*), not (18*b*).

(18) John is eager to please
(18*a*) *John* is eager that *he* please relevant parties
(18*b*) *John* is eager that relevant parties please *him*

By contrast, (19) is roughly synonymous with (19*b*), not (19*a*).

(19) John is easy to please
(19*a*) It is easy for *John* to please relevant parties
(19*b*) It is easy for relevant parties to please *John*

So there is evidently a constraint on how lexical meanings interact with more general facts about how the unpronounced arguments of 'please' are understood in (18) and (19). One can imagine a language with the words 'eeger' and 'eezy', such that strings homophonous with (18) and (19) are ambiguous—or a language in which the homophone of (18) has only the meaning given with (18*b*), and the homophone of (19) has only the meaning given with (19*a*). But for whatever reason, the human language faculty does not permit these ways of composing expressions that include lexicalization of notions like eagerness and easiness.

Such constraints tell against the idea, also criticized by Stalnaker (1984), that speakable expressions acquire semantic life only by virtue of being paired with expressions of Mentalese. I do not doubt that we have structured mental representations independent of the language faculty, or that these representations are associated with pronounceable expressions in very complex ways. But prima facie, such associations are constrained by what the spoken expressions mean; the associations don't determine what the expressions mean, except perhaps with regard to genuinely arbitrary aspects of meaning (for example, that 'chien' is associated with the concept DOG, not CAT). Moreover, if Fodor is right, the human mind exhibits two different kinds of syntax. For absent structural *mismatches* between Mentalese and English sentences, it is hard to see how there could be a compositional semantics for only the former. But one wonders why nature—faced with the problem of getting Mentalese pronounced—would invent a mismatching syntax, thereby saddling children with the task of figuring out how adults associate one kind of syntax with another (as *part* of the task of figuring out what adults are saying).[9]

By contrast, the idea of meanings as *Begriffsplans* fits nicely with the idea that the language faculty allows for a certain kind of integration and expansion of hominid psychology. Various studies suggest that spoken language is implicated in the adult capacity to form—in ways that prelinguistic children and other very clever animals cannot—certain kinds of complex thoughts whose conceptual constituents lie in disparate cognitive domains; see Hermer and Spelke (1994, 1996), Hermer-Vazquez *et al.* (1999), Spelke (2002), and Carruthers (2002). And whatever else the language faculty lets us do, it lets us *lexicalize* mental representations and *combine* lexical items; where lexicalization involves creating an expression of a certain grammatical type, and the permissible modes of combination depend on grammatical (as opposed to "cognitive") type. This is one of the points Chomsky illustrated with (20),

(20) Colorless green ideas sleep furiously

which is bizarre in many ways, but still a sentence that is not completely incomprehensible. And it would hardly be surprising if hominids without an (activated) language faculty had certain concepts they could not combine, even if such combination would be useful.

Consider the disparate concepts required to understand a phrase like 'every child bigger than the brown dog that did not bark loudly'. One needs a system of "common denominators" to combine these concepts in a coherent fashion—and not just because size concepts are associated with relations between individuals, while the concept of barking is not. We may naturally think about colors as properties *of* surfaces, and think about dogs as things that *have* surfaces, with the result that brute concatenation of the relevant concepts would be unnatural for

[9] See n. 7; cf. Jackendoff (1990). One does not need the mismatch hypothesis to say that in many contexts, a speaker who uses sentence Σ to make an assertion is also "entertaining" one or more sentences $S_1 \ldots S_n$, and that the truth or falsity of the speaker's utterance (of Σ) depends—in complicated ways that frustrate attempts to provide theories of truth for natural languages—on the meanings of Σ *and* $S_1 \ldots S_n$.

us. And this kind of mismatch may be common, especially if human cognition is significantly modular. One can think of a *Begriffsplan* as providing, via resources that the language faculty makes available, a way of building complex concepts from elements that otherwise would not fit together.

Of course, the plausibility of this view depends on the plausibility of specific proposals. And one might be suspicious if semantic theories for natural languages *had* to be characterized in terms of the hypothesis that combining expressions typically signifies *function-application*, as in theories inspired by Frege and Montague. For one might doubt that the prelinguistic hominid mind traffics in such a general and abstract notion. Moreover, if concepts can be lexicalized as words whose (Semantic) Values are functions from Values of other lexical items to potential Values of complex expressions, then the idea of meanings as *Begriffsplans* is trivialized. But as the simple examples above already suggest, one can characterize many linguistic meanings compositionally without resorting to descriptively omnipotent notions like function-application. In fact, one can handle all the usual textbook cases and more in terms of predicate-conjunction, existential closure, and a few thematic relations corresponding to certain grammatical relations and prepositions; see Pietroski (2005*a*), drawing on Boolos (1998), Higginbotham (1985), Parsons (1990), Schein (1993), and Larson and Segal (1995). It remains to be seen how much natural language semantics can be done in these terms. But recoding extant work is often easy. And if study of the language faculty suggests that meanings are constrained *Begriffsplans*, we shouldn't immediately conclude that something has gone wrong, on the grounds that some current semantic theories apparently require less restricted conceptual resources. The correct response may be to offer more constrained semantic theories; see Chomsky (1965) for still relevant discussion.

4. AUTONOMOUS CONSTRAINTS, WORLDLY TRUTHS

Examples like (1–20) do not themselves establish that the language faculty is a largely innate cognitive system specific to human language, as opposed to a general learning device, with I-languages as products of this device in response to linguistic experience. But such examples are "opening acts" in more elaborate and detailed nativist arguments, which often involve convergence of many independent lines of research concerning both adults and young children.[10] For present purposes, I take it as a premise that the best poverty of stimulus arguments remain unrebutted, and that more detailed investigation bolsters the central point: there is a huge gap between (i) experience of the sort that each normal child makes

[10] See e.g., Hornstein and Lightfoot (1981), Jackendoff (1993), Baker (2001), Crain and Pietroski (2001, 2002), Laurence and Margolis (2001), Crain *et al.* (forthcoming).

use of, for purposes of acquiring an I-language, and (ii) linguistic constraints of the sort respected by all normal children. In which case, conceptions of linguistic meaning should be evaluated accordingly: whatever meanings are, children associate them with signals in substantively constrained ways.

If this is correct, as Chomsky has long argued, then it is prima facie implausible that natural language meaning is as tightly related to *truth* as Davidson (1967*a*, 1984) and others have conjectured.[11] If human I-languages are determined largely by innate aspects of human biology, then absent a benficient deity, it seems unlikely that such languages pair sentential signals with truth conditions. Once we drop the idea that meaning is determined by what Radical Interpreters would say, in favor of studying meaning by investigating the language faculty, why think that the theoretical notions (and idealizations) required to characterize linguistic meanings are apt for purposes of providing theories of truth? Or putting it the other way around, why think the notions appropriate for characterizing the conditions in which assertive utterances are true or false will also be suitable for purposes of characterizing the language faculty? This seems too convenient to be true, even if we think of truth conditions as conditional specifications of truth values, as opposed to propositions or "states of affairs."

One can, of course, hypothesize that truth theories provide the best conception of what I-languages associate signals with; see Higginbotham (1986), and Larson and Segal (1995). But we should be clear that this is a bold conjecture about the language faculty. For the empirical virtues of truth-conditional semantics may have little if anything to do with truth *per se*. One can interpret theories that associate sentences with **t** or **f** (relative to assignments of values to variables) as theories that treat sentences as devices for doing something—evaluable in a binary fashion—with a concept built in accordance with a certain *Begriffsplan*. Since assertively uttering a sentence is a paradigm of such use, linguistic meanings constrain the truth conditions of assertive utterances. But it does not follow that meanings of expressions, as opposed to asserted contents, should be specified in terms of truth/reference/satisfaction (TRS) conditions.

Correspondingly, it is very important to distinguish two claims: (a) we can combine our best linguistic theories, which incorporate insights from the Davidsonian program, with truth-theoretic models of I-languages; (b) I-languages associate linguistic signals with functions from contexts to TRS conditions. While (a) may be true, it amounts to little more than the claim that theorists can stipulate certain interpretations for the formalism we use to talk about the syntactic structures of natural language. And even if such stipulation is useful for certain limited purposes, related to the phenomenon of semantic implication, (b) is a

[11] Though, in my view, Davidson was importantly right about the basic *structure* of semantic theories, the need for "event" variables, and the use of an extensional metalanguage; see Higginbotham (1985, 1986), Larson and Segal (1995).

much stronger claim about how the language faculty is (or comes to be) related to the environment. I won't review the arguments that lead me to think (b) is false.[12] But I do want to note one kind of skepticism that is often ignored.

As noted above, it's hard to see how we can even start accounting for the interesting facts concerning (17) without an event analysis of some kind.

(17) The senator called the millionaire from Texas

Historically, event analyses were offered as attempts to specify (partial) truth theories for natural languages; since without them, it seemed that adverbial constructions presented an immediate stumbling block. But as further study revealed, it is unclear what event-variables in *semantic* theories range over; and various considerations suggest that such variables *don't* range over things in terms of which a theory of truth for English would specify a truth-condition for (17). It turns out that for purposes of providing theories of understanding, we need a notion of 'event' sensitive to distinctions that seem to be metaphysically otiose. Higginbotham offers an example.

If a drinker downs a pint continuously over thirty seconds, there was an event of drinking a pint of beer in thirty seconds, and there was an event of drinking beer for thirty seconds. In eventish formalism: ∃e[Drinking(e) & Theme(e, a pint of beer) & In(e, 30 seconds)]; and ∃e[Drinking(e) & Theme(e, beer) & For(e, 30 seconds)]. One is inclined to say that here, we have one event and two descriptions. But if the 'e'-variable ranges over things that can satisfy multiple descriptions in this way—and one expects the ontology for a theory of truth to be language-independent in this sense—then there was an event of drinking beer *in* thirty seconds: ∃e[Drinking(e) & Theme(e, beer) & In(e, 30 seconds)]. Yet in natural language, 'He drank beer in thirty seconds' is anomolous; see Tenny (1994) for extended discussion. This suggests that natural language cares about the difference between (i) drinking a pint of beer, and (ii) drinking beer, even when the beer drunk amounted to a pint. And this seems to be a distinction without difference so far as truth is concerned.

Schein (1993) argues that semanticists must likewise distinguish the facing of Carnegie Hall by Carnegie Deli from the facing of Carnegie Deli by Carnegie Hall, the preceding of 3 by 2 from the succeeding of 2 by 3, and so on. With enough ingenuity and tenacity, one can find ways of introducing the structure needed to account for how speakers understand the relevant sentences, and then ignoring unwanted aspects of that structure when it comes to preserving the idea that theories of meaning are theories of truth; see Schein (2002) for illuminating

[12] See Pietroski (2003*a*, 2005*b*), drawing on Chomsky (1977, 2000*a*), for discussion of examples like the following: France is hexagonal, and it is a republic; the red book is too heavy, though it was favorably reviewed, and the blue one is boring, though everyone is reading it; if you ask the average man's wife whether he likes round squares, she'll say that he doesn't, but I think he does. See also Moravscik (1975), Hornstein (1984), McGilvray (1999).

discussion. But if such measures are required at every turn, perhaps nature is telling us something. And as soon as we turn to thinking about how grammatical relations are related to thematic roles,

(21) The rock broke the window

suggests that rocks can be Agents in the sense of 'Agent' that matters for theories of meaning; see Baker (1997) for discussion and defense; see also Pietroski (1998). While this need not be a problem for internalistic theories of the language faculty—in which 'Agent' can play whatever role it needs to play—one wonders whether an honest theory of *truth* for English can really specify a truth-condition for (21) in terms of the rock being an Agent.

Moreover, and more importantly, it is unclear what appeal to TRS conditions *adds* to explanations of the negative facts that animate study of the language faculty. We can, if we like, say that 'eager' is satisfied by states of eagerness, while 'easy' is satisfied by easinesses. But if we intend this as part of a serious hypothesis about how the language faculty lexicalizes certain notions, we owe an explication of phrases like 'satisfied by' and 'states of eagerness'. Given that predicates of natural language appear to be vague and sensitive to communicative import, one can hardly assume both that 'satisfied by' means what it means in Tarski's theory, and that claims like *'dog' is satisfied by dogs* are truistic; see Pietroski (2005*b*). But set this aside, along with concerns about the covert indices that will be required by a substantive theory of truth that accommodates all the ways truth can depend on context; see Stanley (2000, 2002), cf. Cappelen and Lepore (2005). The crucial point here is that to account for the relevant negative facts, we need to say enough to make it clear that English *differs* from the imagined language, whose speakers can say things like the following: 'eeger' is satisfied by states of eegerness; and 'John is eeger to please' means that John is eeger to please.

Partly for this reason, Higginbotham (1989) urges truth-conditionalists to provide "elucidations" of lexical meanings. This strikes me as the right move to make, in order to defend a truth-conditionalist conception of the language faculty. But until we have a significant number of theoretically interesting and confirmed elucidations specified in truth-conditional terms, locutions like 'satisfied by' seem to be placeholders for appropriate theoretical notions. And recall that one *could* assign a perfectly coherent truth condition to (9α).

(9α) Was {[the [hiker [who__ lost]$_{RC}$]][whistled]}
|______________*______|

So *if* a truth theory helps explain why it is odd to say that hiker (as opposed to a song) was whistled—and here is a place where elucidations might well help—then the truth-conditional interpretation that could be assigned to (9α) will be *less* defective than the one assigned to (9β).

(9β) Was {[the [hiker [who lost]$_{RC}$]] ☹ [__ whistled]}

All of which suggests what we should have expected: explanations of the negative facts illustrated with (9) flow from claims about *how* I-languages associate signals with meanings, not from proposals about how to characterize meanings in terms of TRS conditions.

In this context, let me stress again that constraints on displacement are interesting in part because they are so bizarre from a logical point of view. (Why forbid extraction from a relative clause?) Natural language quantifiers are also governed by such constraints. For example, (22) has only the reading indicated in (22*a*),

(22) It is false that every senator lied
(22*a*) ¬∀x:senator(x)[Lied(x)] (22*b*) ∀x:senator(x)[¬Lied(x)]

suggesting that 'every senator' is displaced but only so far; see Pietroski and Hornstein (2002) for discussion and further references. So it looks like the language faculty is a system operating in accordance with its own principles, without regard for the relation between quantification (as discussed by Frege and Tarski) and truth. Expressions of natural language have the properties they have. Speakers use those expressions, sometimes as devices for making truth-evaluable assertions. And meaning constrains use in subtle ways. But the more we learn about the language faculty, the less plausible "truthy" conceptions of the faculty become. Or so it seems to me. (Chomsky offers such remarks regularly. I make no claim to originality here.)

5. SEMANTICS IS NOT CONVENTIONALIZED PRAGMATICS

In this final section, I want isolate a point of disagreement with Stalnaker, in order to stress a more significant point of agreement. Stalnaker (1999) says that "we should separate, as best we can, questions about what language is used to do from questions about the means it provides for doing it (p. 2)," and that

> The line—or more accurately a number of distinct lines—between semantics and pragmatics shift and blur. But I think there is one line that is worth continuing to draw and redraw: between an abstract account of the functions and purposes of discourse and a theory of the devices that particular languages and linguistic practices use to serve those functions and accomplish those purposes (p. 16).

That seems right and important. But along with many linguists and philosophers, Stalnaker also holds that

a principal goal of semantics is to explain how the expressions used to perform speech acts such as assertion are used to convey information—to distinguish between possibilities—and how the way complex sentences distinguish between possibilities is a function of the semantic values of their parts (2003, 172).

I think this obscures the line worth drawing. For it conflates, under the heading 'semantics', questions about the use of natural language with questions about the nature of linguistic expressions. It presupposes not only that a *sentence* can distinguish between possibilities, but that this sentential property is compositionally determined. And it isn't clear that these are genuine explananda—facts to be explained, as opposed to simplifying presuppositions whose falsity can be ignored for certain purposes—much less explananda for theories of the *devices* that natural language provides. There are certainly facts about what speakers can use sentences to do; and use is severely constrained by aspects of meaning that are compositional. But if semantics is primarily about what *expressions* mean, then semanticists can and should abstract from various details concerning how expressions are used. (And while certain Gricean principles may apply to discourse among rational beings, regardless of what their expressions mean, there are also principles governing what expressions of natural language cannot mean regardless of how they are used.) One should not insist that appeal to kontents explains the fact that propositions are structured in ways that sets are not. But equally, one should not insist that meaning be related to use in certain ways.

In stressing another valuable distinction, between "descriptive" and "foundational" semantic theories, Stalnaker (2003) says that the former specify Semantic Values for expressions of a language "without saying what it is *about the practice of using that language* that explains why" the expressions have those Values (p. 166, my italics). Here, he is presumably thinking about conventional facts: 'chien' signifies dogs as opposed to cats, etc. I would just omit the italicized phrase, which isn't needed to make the distinction. Though I wouldn't quibble about this, if remarks about use were not so regularly combined (in philosophy and elsewhere) with the suggestion that semantics is fundamentally contingent/arbitrary/conventional/learned. Stalnaker says that we should "all agree that it is a matter of contingent fact that the expressions of natural language have the *character and* content that they have" (p. 195, my italics). But is it obviously a contingent fact that 'John is eager to please' has the semantic character it does? This sentence is *not* apt for use as a way of saying that John is eager for us to please him. Prima facie, this negative fact is due to deep properties of the sentence, not superficial features of how the words are related to features of the environment. And perhaps many aspects of the actual character of a sentence are essential to it, much as the actual atomic number of gold is essential to it.

I'm not sure it even makes sense to think about an expression of *natural* language as having a meaning that would violate principles of Universal Grammar. (Is this just to imagine a brain that associates certain *signals* with interpretations in a nonhuman way?) And one can avoid tendentious conceptions of expressions, while endorsing what Stalnaker (2003) takes to be a central aspect of Kripke's thought, viz., that the contents of speech and thought are determined by the things with which speakers and thinkers interact.[13] We can define a useful and externalistic notion of kontent to capture this idea. But we shouldn't assume that linguistic meaning can be determined in like fashion. One is free to define 'meening' and 'kontext' so that the meening of a sentence is a function from kontexts to kontents. But then one can't assume that an adequate theory of meening, whatever that would be, is an adequate theory of understanding. Understanding, in so far as we can have theories of it, may have more to do with the language faculty than with kontents. We can define 'outerstanding' in terms of kontents. But externalist stipulation must come to an end at some point.

Stalnaker (1999) says, "The attempt to do semantics without propositional content is motivated more by pessimism about the possibility of an adequate account of propositions than it is by optimism about the possibility of explaining the phenomena without them" (p. 3). This may be true of many philosophers following Quine and Davidson. But Chomsky is optimistic about the possibility of explaining semantic phenomena as phenomena that reflect the nature and operation of the language faculty—and hence, as phenomena unlikely to be explained in terms of propositions. With this in mind, I have tried to argue here that a Chomskyan internalism about semantics is compatible with the following idea: Stalnaker's account of propositions is adequate *for purposes requiring appeal to propositions, as opposed to meanings.*

Of course, this makes no sense if we assume that "Syntax studies sentences, semantics studies propositions" (Stalnaker [1999, 34]). Fodor, along with many others, combines this common assumption with the idea that propositions are linguistically structured. Stalnaker combines it with an opposing conception of propositions, as sets of possible worlds, and is led to say that the subject of semantics "has no essential connection with languages at all, either natural or artificial" (p.33). Fodor is led to say that expressions of natural language are themselves essentially meaningless.[14] But another possibility is to reject the slogan

[13] One can also agree that meanings, as abstracta, are not in the head; see Stalnaker (2003, 204–10).

[14] Soames (2002) is an interesting case deserving separate treatment. But given his views about propositions, Soames is led to say (in the absence of confirming evidence) that *sentences* carry information, that 'Hesperus is bright' and 'Phosphorus is bright' carry the *same* information, and that these sentences are synonymous. Though Soames may be right that this is the best option, all things considered, for those who take semantics to be the study of propositions and how they are related to sentences.

"semantics studies propositions," leaving it open whether this is a false hypothesis or an unhelpful stipulation.

Prima facie, semantics is the study of linguistic meaning, whatever that turns out to be. If it turns out that sentences of natural language are not systematically associated with propositions, whatever *they* turn out to be, we can conclude that natural language semantics is a branch of human psychology concerned with the language faculty. We are free to invent an enterprise, Psemantics, defined as the study of how *sentences* of spoken languages (as opposed to certain human *actions*) are related to: sentences of Mentalese, kontents, Russellian propositions, or whatever. Inquirers are free to pursue this enterprise instead of, or in addition to, studying the language faculty (independent of prior assumptions about how it is related to propositions). But if fixation on Psemantics keeps getting us into trouble, that may be nature telling us something.

APPENDIX: DIAGONALS OBSCURE NONTRIVIAL ASPECTS

Suppose you do take semantics to be the study of propositions, which you take to be sets of possible worlds, and you grant that 'Hesperus is Phosphorus' and 'Hesperus is Hesperus' differ in meaning. If you also suspect (*pace* Kripke) that any truth knowable *a priori* is in some sense necessary, you may be tempted to identify linguistic meanings with (functions from contexts to) "diagonal" K-propositions described in terms of the apparatus of "two dimensional" modal logic; see Chalmers (1996), Jackson (1998). Stalnaker's (2003) good arguments to the contrary, which illustrate the distinction between descriptive and foundational semantic theories (see also Kaplan [1989]), are independent of Chomsky-style considerations about negative facts. But appeal to diagonal kontents can seem like an attractive way to preserve a unified conception of semantics and pragmatics. So it is worth being clear that this only hinders discussion of natural language constraints on signal-meaning association.

Let w1, w2, and w3 be distinct possible worlds in which the sentence 'You saw us' is used to make an assertion in accordance with the following matrix.

	w1	w2	w3
w1	T	F	T
w2	F	T	F
w3	T	F	F

Suppose that at w1, Chris is talking to Pat, with Hilary as the (only) other relevant party. Then the assertion is true iff Pat saw Chris and Hilary; and it is true at w1 and w3, but not w2. Let w2 be a world at which Pat is talking to Chris, with Hilary as the other relevant party; and let w3 be a world at which Pat is talking to Chris, with the other relevant party being a fourth person, Sam. Then at w2, Pat uses 'You saw us' to make an assertion that is true iff Chris saw Pat and Hilary; and at w3, Pat uses this sentence to make an assertion that is true iff Chris saw Pat and Sam. The matrix represents a function **M**, which Stalnaker (1978) calls a "propositional concept," from worlds to sets of worlds: **M**(w1) = {w1, w3}; **M**(w2) = {w2}; **M**(w3) = {w1}. Let the "diagonal" of **M**, D^M, be {w: w ∈ **M**(w)}. In our simple example, the diagonal is {w1, w2}.

Now consider the matrix below, this time taking w1, w2, and w3 to be possible worlds in which the relevant speaker uses 'I am here' to make an assertion.

	w1	w2	w3
w1	T	F	F
w2	F	T	F
w3	F	F	T

Suppose that at w1: Chris is at location L and speaking to Pat, who is at location L*. Then at w1, the assertion is true iff Chris is at L; and the kontent of this assertion, indicated with the first line, includes w1. Let w2 be a world where Pat is speaking to Chris from a third location L**, while Chris is at L*. Then the assertion is true iff Pat is at L**; and its kontent includes w2. Let w3 be a world where Pat is speaking to Chris from L, while Chris is at L*. Then the assertion is true iff Pat is at L; and its kontent includes w3. Call this matrix **CAP**, to highlight the contingent *a priori* status of the assertions. The diagonal, D^{CAP}, is the "universal" set {w1, w2, w3}.

This provides a theoretical model of a phenomenon that might otherwise seem puzzling: when a speaker says 'I am here' (in typical circumstances), she *can't* be wrong, even though she is making a claim that is only *contingently* true. Thus, D^{CAP} can reflects something that users of 'I am here' have in common; and this is worth noting. But one shouldn't reason as follows: the meaning of a sentence is constant across contexts; so any (semantically relevant) property of a sentence S constant across contexts is a good candidate for being the meaning of S.

Let's recycle the first matrix, using it to evaluate 'You are easy to please'.

	w1	w2	w3	
w1	T	F	T	w1: Chris talking to Pat; other relevant party, Hilary
w2	F	T	F	w2: Pat talking to Chris; other relevant party, Hilary
w3	T	F	F	w3: Pat talking to Chris; other relevant party, Sam

The assertion at w1, true at w1 and w3, is true iff it is easy for Chris and Hilary to please Pat. The assertions at w2 and w3, made by Pat, are true (respectively) iff: it is easy for Pat and Hilary to please Chris; and it is easy for Pat and Sam to please Chris. But focus on two things that Chris *cannot* do by saying 'You are easy to please' at w1. First, Chris cannot make an assertion that is true iff it is easy for Pat and Hilary to please Chris; although Chris could make such an assertion with 'I am eager to please'. Second, Chris cannot make an assertion that is true iff it is easy for Pat to please Chris and Hilary; although Chris could make such an assertion with 'You can easily please (us)'. While the second fact is more theoretically interesting, appeal to diagonals at best helps to characterize the first.

Given a language like English except that the sounds of 'you' and 'I' signify first and second personal pronouns, respectively, Chris could use the sound of 'You are easy to please' to make an assertion that Chris cannot make with the homophonous English sentence. But this just illustrates a conventional (and theoretically uninteresting) feature of English. By contrast, Chris would need to speak a nonhuman language in order to use the English words as a way of saying that the addressee can easily please relevant parties. Put another way, if we represent sentential meanings with propositional matrices, we must associate 'You are easy to please' and 'It is easy for relevant parties to please you' with the *same* matrix, while associating 'It is easy for you to please relevant parties' with a *different* matrix. This last sentence should be associated with the matrix below, given how things are at w1, w2, and w3.

	w1	w2	w3
w1	F	T	F
w2	T	F	T
w3	F	T	T

This raises the question of *why* 'You are easy to please' has the matrix it does, as opposed to the one immediately above. And the force of this question is heightened when we note that the superficially similar sentence 'You are eager to please' has a different semantic character, according to which an assertion made by using it is true iff the addressee is eager to be one who pleases relevant parties. (Indeed, it is easy to describe the worlds so that 'You are eager to please' corresponds to the matrix immediately above; though at this point, few readers will be eager for me to do so. Note also that 'The goose is ready to eat' is ambiguous.)

The idea of identifying linguistic meanings with diagonals seems hopeless once one considers the semantic relations among sentences like 'John persuaded Bill to leave', 'Bill intended to leave', and 'John expected Bill to leave'. With enough effort, one can probably specify an algorithm according to which the second is true at every world where the first is true, while false at some but not all worlds where the second is true; and each sentence can be used to make all and only the assertions it can be used to make. But why think that such an algorithm would be characterizing the meanings of the sentences, as opposed to describing certain features of the assertions one could make with independently meaningful sentences? At best, appeal to diagonals provides a way of capturing certain trivial examples of *a priori* "knowledge" corresponding to contingent/conventional aspects of natural language semantics. The real work lies with describing more interesting aspects of meaning in theoretically perspicuous ways; see Chomsky (1965).[15]

REFERENCES

Baker, M. (1997). 'Thematic Roles and Grammatical Categories', in L. Haegeman (ed.), *Elements of Grammar* (Dordrecht: Kluwer), 73–137.

Barber, A. (ed.) (2003). *Epistemology of Language* (Oxford: Oxford University Press).

Boolos, G. (1998). *Logic, Logic, and Logic* (Cambridge, Mass.: Harvard University Press).

Cappelen, H., and Lepore, E. (2005). *Insensitive Semantics* (Cambridge, Mass.: Blackwell).

Carnap, R. (1950). 'Semantics, Empiricism, and Ontology', *Revue Internationale de Philosophie* 4: 208–28.

Carruthers, P. (2002). 'The Cognitive Functions of Language', *Behavioral and Brain Sciences* 25: 261–316.

Chalmers, D. (1996). *The Conscious Mind* (Oxford: Oxford University Press).

Chomsky, N. (1957). *Syntactic Structures* (The Hague: Mouton).

____ (1965). *Aspects of the Theory of Syntax* (Cambridge, Mass.: MIT Press).

____ (1977). *Essays on Form and Interpretation* (New York: North Holland).

____ (1981). *Lectures on Government and Binding* (Dordrecht: Foris).

____ (1986). *Knowledge of Language* (New York: Praeger).

____ (1993). 'Explaining Language Use', *Philosophical Topics* 20: 205–31.

____ (1995). *The Minimalist Program* (Cambridge, Mass.: MIT Press).

____ (2000*a*). *New Horizons in the Study of Language and Mind* (Cambridge University Press).

____ (2000*b*). Minimalist Inquiries. In R. Martin, D. Michaels, and J. Uriagereka (eds.), *Step by Step: Essays on Minimalist Syntax in Honor of Howard Lasnik* (Cambridge, Mass.: MIT Press).

Crain, S., and Pietroski, P. (2001). 'Nature, Nurture, and Universal Grammar', *Linguistics and Philosophy* 24: 139–86.

____ (2002). 'Why Language Acquisition is a Snap', *The Linguistic Review* 19: 163–83.

Crain, S., Gualmini, A., and Pietroski, P. (forthcoming).

Davidson, D. (1967*a*). 'Truth and Meaning', *Synthese* 17: 304–23.

____ (1967*b*). 'The Logical Form of Action Sentences', in N. Rescher (ed.), *The Logic of Decision and Action* (Pittsburgh: University of Pittsburgh Press).

____ (1984). *Inquiries into Truth and Interpretation* (Oxford: Oxford University Press).

____ (1985). Adverbs of Action', in B. Vermazen and M. Hintikka (eds.), *Essays on Davidson: Actions and Events* (Oxford: Clarendon Press).

Dummett, M. (1975). What is a Theory of Meaning? in S. Guttenplan (ed.), *Mind and Language* (Oxford: Oxford University Press).

[15] This paper has roots in a series of conversations with my thesis supervisor, to whom I owe much. The intervening years have only deepened my appreciation of Bob's work as a philosopher and teacher. My thanks also to Dan Blair, Susan Dwyer, and Norbert Hornstein for helpful comments on earlier versions of this material.

Evans, G., and McDowell, J. (eds.) (1976). *Truth and Meaning* (Oxford: Oxford University Press).
Fodor, J. (1975). *The Language of Thought* (New York: Crowell).
____ (1978). 'Propositional Attitudes', *The Monist* 61:501–23.
____ (1987). *Psychosemantics* (Cambridge, Mass.: MIT Press).
____ (1998). *Concepts: Where Cognitive Science Went Wrong* (New York: Oxford University Press).
Fodor, J., and Lepore E. (1992). *Holism, A Shopper's Guide* (Oxford: Blackwell).
____ (2002). *The Compositionality Papers* (Oxford: Oxford University Press).
Hermer, L., and Spelke, E. (1994). 'A Geometric Process for Spatial Reorientation in young children', *Nature* 370: 57–9.
____ ____ (1996). 'Modularity and Development: The Case of Spatial Reorientation', *Cognition* 61.195–232.
Hermer-Vazquez, L., Spelke, E., and Katsnelson, A. (1999). 'Sources of Flexibility in Human Cognition', *Cognitive Psychology* 39: 3–36.
Higginbotham, J. (1985). 'On Semantics', *Linguistic Inquiry* 16: 547–93.
(1986). 'Linguistic Theory and Davidson's Program', in E. Lepore (ed.), *Truth and Interpretation* (Oxford: Blackwell).
____ (1989). 'Elucidations of Meaning', *Linguistics and Philosophy* 12: 465–517.
____ (1994). 'Priorities in the Philosophy of Thought', *Proc. Aristotelian Society* supp. vol. 20: 85–106.
Hornstein, N. (1984). *Logic as Grammar* (Cambridge, Mass.: MIT Press).
Hornstein, N., and Lightfoot, D. (1981). *Explanation in Linguistics* (London: Longman).
Horty, J. (ms). *Frege on definitions: a case study of semantic content* (University of Maryland).
Horwich, P. (1997). 'The Composition of Meanings', *Philosophical Review* 106: 503–32.
Jackendoff, R. (1990). *Semantic Structures* (Cambridge, Mass.: MIT Press).
____ (1993). *Patterns in the Mind* (New York: Harvester Wheatsheaf).
Jackson, F. (1998). *From Metaphysics to Ethics: A Defense of Conceptual Analysis* (Oxford: Oxford University Press).
Kaplan, D. (1989). 'Demonstratives', in J. Almog, J. Perry, and H. Wettstein (eds.), *Themes from Kaplan* (New York: Oxford University Press).
Katz, J., and Fodor, J. (1963). 'The Structure of a Semantic Theory', *Language* 39: 170–210.
Kripke, S. (1980). *Naming and Necessity* (Cambridge, Mass.: Harvard University Press).
Larson R., and Segal, R. (1995). *Knowledge of Meaning* (Cambridge, Mass.: MIT Press).
Laurence, S., and Margolis, E. (2001). 'The Poverty of Stimulus Argument', *British Journal for the Philosophy of Science* 52: 217–76.
Lewis, D. (1972). 'General Semantics', in D. Davidson and G. Harman (eds.), *Semantics of Natural Language* (Dordrecht: Reidel).
____ (1986). *The Plurality of Worlds* (Oxford: Basil Blackwell).
Ludlow, P. (2002). 'LF and Natural Logic', in Preyer and Peters (2002).
Matthew, R. (2003). 'Does Linguistic Competence Require Knowledge of Language?' in Barber (2003).
McDowell, J. (1976). 'Truth Conditions, Bivalence and Verificationism', in Evans and McDowell (1976).

McGilvray, J. (1999). *Chomsky: Language, Mind and Politics* (Cambridge: Polity Press).
Montague, R. (1974). *Formal Philosophy* (New Haven: Yale University Press).
Moravscik, J. (1975). *Understanding language* (The Hague: Mouton).
Neale, S. (1990). *Descriptions* (Cambridge, Mass.: MIT Press).
—— (1993). 'Grammatical Form, Logical Form, and Incomplete Symbols' in A. Irvine and G. Wedeking (eds.), *Russell and Analytic Philosophy* (Toronto: University of Toronto).
Parsons, T. (1990). *Events in the Semantics of English* (Cambridge, Mass.: MIT Press).
Peacocke, C. (1999). *Being Known* (New York: Oxford University Press).
Pietroski, P. (1998). 'Actions, Adjuncts, and Agency', *Mind* 107: 73–111.
—— (2000). *Causing Actions* (Oxford: Oxford University Press).
—— (2003*a*). 'The Character of Natural Language Semantics' in Barber (2003).
—— (2003*b*). 'Quantification and Second-Order Monadictity', *Philosophical Perspectives* 17: 259–98.
—— (2005*a*). *Events and Semantic Architecture* (Oxford: Oxford University Press).
—— (2005*b*) 'Meaning Before Truth', in Preyer and G. Peters (2005).
Preyer, G., and Peters, G. (eds.) 2002: *Logical Form and Language* (Oxford: Oxford University Press).
—— (2005). *Contextualism in Philosophy* (Oxford: Oxford University Press).
Quine, W. V. O. (1960). *Word and Object* (Cambridge, Mass.: MIT Press).
Ross, J. (1967). *Constraints on Variables in Syntax* (Doctoral dissertation, MIT).
Schein, B. 1993. *Events and Plurals* (Cambridge, Mass.: MIT Press).
—— (2002). 'Events and the Semantic Content of Thematic Relations', in Preyer and Peters (2002).
—— (forthcoming). *Conjunction Reduction Redux* (Cambridge, Mass.: MIT Press).
Soames, S. (2002). *Beyond Rigidity* (Oxford: Oxford University Press).
Spelke, E. (2002). 'Developing Knowledge of Space: Core Systems and New Combinations' in S. Kosslyn and A. Galaburda (eds.), *Languages of the Brain* (Cambridge, Mass.: Harvard University Press).
Stalnaker, R. (1970). 'Pragmatics', *Synthese* 22: 272–89; reprinted in Stalnaker (1999: 31–46).
—— (1976). 'Possible Worlds' *Nous* 10: 65–75.
—— (1978). 'Assertion', *Syntax and Semantics* 9: 315–32; reprinted in Stalnaker (1999: 78–95).
—— (1984). *Inquiry* (Cambridge, Mass.: MIT Press).
—— (1999). *Context and Content* (Oxford: Oxford University Press).
—— (2003). *Ways a World Might Be* (Oxford: Oxford University Press).
Stanley, J. (2000). 'Context and Logical Form', *Linguistics and Philosophy* 23: 391–424.
—— (2002). 'Making it Articulated', *Mind and Language* 17: 149–68.
Strawson, P. (1950). 'On Referring', *Mind* 59: 320–44.
Tarski, A. (1983). *Logic, Semantics, Metamathematics* (Indianapolis: Hackett).
Tenny, C. (1994): *Aspectual Roles and the Syntax-Semantics Interface* (Dordrecht: Kluwer).
Travis. L. (1984): *Parameters and the Effects of Word-Order Variation* (Doctoral Dissertation, MIT).

4

Idiolects

Richard G. Heck, Jr.

1. LANGUAGE

Students of language should presumably study language, but there is room for disagreement about how the ordinary notion of a language should be refined for the not so ordinary purposes of scientific or philosophical inquiry. One central question is to what extent, or in what ways, the appropriate object of study is socially or communally constituted. And one answer, of course, is that it isn't. On this view, the primary objects of study are *idiolects*. An idiolect, in this sense, belongs to a single individual, in the sense that one's idiolect reflects one's own linguistic capabilities and, in that sense, is fully determined by facts about oneself. An alternative is to hold that the primary objects of study should be *common languages*, so called because they are the common property of a community of individuals. It need, of course, be no part of this view that common languages are individuated as we individuate languages in ordinary discourse: Perhaps so, but perhaps not. The important claim is that such languages are in some way socially constituted, in the sense that their properties—most importantly, for present purposes, their semantic properties—are fixed not by the linguistic capabilities of a single individual but only by relations among the linguistic capabilities of an entire community of individuals.

Noam Chomsky has argued, to my mind convincingly, that what he calls 'E-languages' are of no concern to theoretical linguistics (as practiced in the generative tradition).[1] It is rather with 'I-languages' that linguistics is concerned. Now, despite the suggestive prefixes, the distinction between E-language and I-language is not the distinction between 'external' and 'internal' languages—between common languages and idiolects. Rather, the 'E' and 'I' stand for 'extensional' and 'intensional', in the sense in which those terms are sometimes used in the philosophy of mathematics: E-language is to I-language as product is to

[1] Arguments for this conclusion can be found in some of Chomsky's earliest writings, for example, Chomsky (1965). A more recent exposition is in Chomsky (1986). I shall henceforth drop the parenthetical qualification.

process.[2] The notions of E-language and I-language are thus like the two senses of the technical term 'theory': In one of these, a theory is a set of sentences, the theorems of the theory; in the other, a theory is a set of principles, of axioms and rules of inference. An E-language is a set of grammatical sentences, or a function (in extension!) from sounds to meanings; an I-language is a set of syntactic or semantic principles that *generates* an E-language, much as a set of axioms and rules generates a set of theorems.[3] And what Chomsky argues—or just observes, really—is that E-languages in this sense play no significant role in linguistic theory. Syntacticians are not trying to specify the set of grammatical sentences of a given language but rather to specify the *syntactic principles* that determine whether a sentence is grammatical. Nor should semantics, if practiced within the generative tradition, attempt to specify a mapping from sound to meaning. Rather, semantics should attempt to describe the principles—the composition rules and semantic axioms—that generate such a mapping.

Chomsky is, nonetheless, thoroughly dismissive of the notion of a common language. Part of his reason is that the notion has, he thinks, never been adequately characterized. He is clearly right on this score. Even if we do not presume that English, German, and the like should be the primary objects of study, but say only that the important question should be whether the objects of study should *in some way* be the property of a linguistic community, no one, so far as I know, has ever managed to say with any precision how the boundaries of linguistic communities should be drawn. But maybe that problem isn't as serious as it seems, or maybe it can be solved. The more serious problem is that it is hard to know what common languages are if they are not E-languages. English, for example—or the dialect thereof spoken in Cambridge, or the language spoken at First Church in Cambridge, or whatever might turn out to be the relevant notion—is an E-language. For what might one mean by *the* set of principles that generates the grammatical sentences of such a 'language'? As Quine long ago pointed out (Quine, 1970), if there is one such set of principles, there are many. Even if we assume that each speaker must use some set of principles to decide whether a given expression is a grammatical sentence and, if so, what it means, different speakers may use different such sets. If E-languages are what matter to us, then such sets of principles are mere means to ends, the ends of determining whether a sentence is grammatical and what it means, and sets of principles—that is, I-languages—are the property of individual speakers, not of linguistic communities. That is why so many philosophers who take E-languages

[2] The terminology used in Chomsky (1965) seems more appropriate, ultimately: there I-languages are called *grammars* and E-languages are called *languages*. If one so speaks, however, one then has to say that linguistics is not concerned with languages, which sounds paradoxical. It is in part for that reason, it seems, that Chomsky changed terminology.

[3] In fact, or so Chomsky has argued, there is no particular reason to think that grammars do, in any straightforward sense, generate E-languages, since there appears to be no hard-and-fast distinction between grammatical and ungrammatical sentences—none of significant interest to theoretical linguistics, anyway. But I shall waive this point here.

to be the proper objects of philosophical investigation—Michael Dummett, for example—are inclined to dismiss as 'merely empirical' the questions studied in generative linguistics.

Chomsky takes the attitude of such philosophers to be based upon prejudice, and he may to some extent be right.[4] But it is not just based upon prejudice. To understand Dummett's position, for example, one must understand that he is concerned with the *use* of language—something that is, Chomsky readily admits, in a sense alien to linguistic theory. Dummett is, as Frege was, particularly impressed by the fact that speakers are able to communicate with one another, and he takes the central problem of the philosophy of language to be to explain how linguistic communication is possible. For that, he thinks, it will be enough if speakers agree about what utterances of sentences do or would mean: It is not required that they arrive at their beliefs about what sentences mean in the same way. Even if Martian psychology is radically different from human psychology, that need not prevent them from communicating with us. Why should it matter what principles Martians use to arrive at their judgements about what English utterances mean? If they agree with us, say, that "Snow is white" means that snow is white, isn't that enough? Dummett thinks it is. That is, he thinks that whether speakers share an I-language cannot matter so far as the *use* of language is concerned: What matters is that speakers share an E-language.[5]

Now, as said, E-languages are not common languages: One can draw the distinction between I-language and E-language even for the languages of individuals. But, Dummett claims, the fact that a group of speakers *do* share an E-language can only be explained if what they share is a *common* language, in the sense explained above. He writes:

> [O]n Frege's theory, the basic notion is... that of an idiolect, and a language can only be explained as the common overlap of many idiolects. This Putnam is quite right to say is wrong.... An English speaker both holds himself responsible to, and exploits the existence of, means of determining the application of terms which are either generally agreed among the speakers of English, or else are generally acknowledged by them as correct.... [T]his has to be so if words are to be used for communication between individuals. (Dummett, 1978, 425)

Dummett's reasons for this claim are similar to Putnam's, which derive from the now famous (but then recent) elm-beech example on which Dummett is here commenting. After rehearsing that example, Dummett concludes that "there is no describing any individual's employment of his words without account being

[4] This prejudice Chomsky sees as an heir of behaviorism. He characterizes it as "methodological dualism", which he opposes to "methodological naturalism". For discussion, see Chomsky (2000*b* and 2000*c*).

[5] This reasoning, I believe, is extremely common. For Dummett's version of it, see Dummett (1993, 176–8); for Donald Davidson's, see Davidson (1984); for David Lewis's, see Lewis (1985, 177–8).

taken of his willingness to subordinate his use to that generally agreed as correct" (Dummett, 1978, 425).

So, as said, Dummett does have reasons to think that common languages should be the focus of the philosophical study of language. Let me summarize his train of thought. First, the central preoccupation of the philosophical study of language is the phenomenon of linguistic communication. Second, successful communication requires only that individuals share E-languages, not that they share I-languages. And third, the E-languages individuals share cannot be characterized simply as overlapping idiolects, that is, in terms that ignore the existence of other speakers. Rather, "one cannot so much as explain what an idiolect is without invoking the notion of a language considered as a social phenomenon" (Dummett, 1978, 425). In particular, idiolects should be understood as comprising partial, and partially incorrect, collections of beliefs about the common language.

Note that the crucial question here is one of explanatory priority. Dummett's claim is that common languages are the proper objects of philosophical study because common languages are, in an important sense, more fundamental than idiolects, at least from the point of view of philosophers interested in the use of language for communication. Dummett need not deny—and I do not think he would deny—that the empirical study of language, as exemplified by generative linguistics, might most profitably focus its attention on idiolects. Nor need he deny that, as a matter of empirical fact, actual human speakers' beliefs about what their words mean are the result of cognitive though sub-conscious processes of just the sort Chomsky says linguists are attempting to characterize. Dummett's claim, rather, is that the sorts of facts that are centrally relevant to the general theory of communication—most centrally, facts about what lexical primitives mean—can only be understood properly if common languages are taken as fundamental, because these sorts of facts are socially constituted. That they are is supposed to be shown by the fact that there are generally acknowledged linguistic norms—norms of the common language—to which ordinary speakers hold themselves accountable and without which communication as we know it would be impossible.

I am not endorsing Dummett's argument, though I am prepared to accept the first claim. *My* central preoccupation, *qua* philosopher of language, is indeed the use of language in communication. Even if it were not, I think it would remain an important question whether the general theory of communication—the theory of language-use—can operate with the notion of an idiolect or instead needs the notion of a common language. The second claim—that successful communication requires only E-langauges to be shared—is, I think, extremely tempting, though ultimately mistaken, but I shall not argue that point here.[6] It is the third

[6] The argument would have two parts. First, I would argue that this second claim requires that there be no 'residue', so to speak, at the conscious level, of the sub-conscious processes by means of

claim that will be my focus. What I want to argue is that the existence of the sorts of normative phenomena at which Dummett gestures—in particular, speakers' willingness to hold their usage responsible to that of others—does not imply that idiolects cannot be explained without invoking common languages.

Before I even begin to address this issue, however, let me emphasize that there are several senses in which human language might, in some interesting sense, be 'social' that are not at issue here. One, mentioned to me by Justin Broackes, concerns cultural identity, an important aspect of which may be how one speaks: what words one uses, and with which meanings, how one pronounces them, and the like may all be parts of one's cultural identity. That is true not just among teenagers but among various sorts of ethnic groups (say, Irish Americans) and even among groups whose identity is defined more by geography than by heredity (such as Liverpudlians). Cultural norms may enforce conformity along semantic and phonological dimensions, as well as others. I do not deny that. Nor would I—or, I think, Chomsky—wish to deny that there may be some broadly scientific study of the history and evolution of languages in the everyday sense in which English and German are languages, and perhaps such a study should operate not with idiolects but with langauges characterized as socially determined objects that persist through change. As David Wiggins points out (Wiggins, 1997),[7] there is much to be said, and much that has been said, about the development of Italian and the role Dante played in that development. But it does not follow that, within the general theory of communication, common languages must be treated as more basic than idiolects, and that is what is at issue here.

Perhaps most importantly, I do not deny that, given that there are such things as dictionaries, speakers may treat dictionaries as in some sense authoritative. There may even be good reasons of various sorts that one ought to hold one's usage responsible to what some dictionary says—perhaps different reasons, and different dictionaries, for different speakers. The issue, however, is not whether one should use the dictionary. What is at issue here does not depend upon the actual existence of dictionaries. Dummett's argument is meant to apply not just to the languages of the industrialized West but also to dying languages that have no written form and only a few hundred speakers, and it is important to keep this fact firmly in view. What Dummett is claiming is that the very possiblity of communication among speakers of *any* language demands not only that there must be facts of the sort a correct dictionary for the language might record, and not only that speakers of the language must regard these facts as normative standards governing their own usage, but that an individual speaker's own understanding

which speakers arrive at their beliefs about what utterances mean. Second, I would argue that there is such a residue. See Heck (2004) for more on this issue.

[7] Since I am going to be disagreeing with almost everything Wiggins says, let me say now that, in my opinion, (Wiggins, 1997) is an absolutely terrific paper. I recommend it to everyone interested in these issues as containing the clearest exposition I know of the position I am opposing.

of the language cannot be characterized independently of these norms.[8] That is what I deny.

The remainder of the paper is structured as follows. Section 2 introduces a conception of communication—or, better, of communicative exchanges—that guides the discussion. In the discussion that follows, I argue that what we need to explain communicative success is, in the first instance, the notion of what an expression means *to a speaker* rather than the notion of what an expression means *in a language*. There is still room for Dummett to argue that common languages must be invoked in the explanation of successful communication, but I shall argue that, in the sort of explanation Dummett envisages, the reference to common languages is explanatorily idle. Section 3 offers an explanation of the normative phenomena that so impress Dummett, first considering an earlier attempt by Alexander George. Section 4 returns to the question how communicative success should be explained, offering an account that makes no reference to common languages and identifying what I take to be the crucial issue here.

2. COMMUNICATION AND COMMON LANGUAGES

Why does Dummett think common languages are implicated in the explanation of successful communication?

Consider the following simple example of a communicative exchange. Say I'm standing outside my office and a student enters the alcove, heading toward Charles Parsons's door. Knowing that Charles is presently teaching, and not wanting the student to waste her time waiting, I say to her, "Prof. Parsons is teaching". She responds, "Oh, thanks," and goes on her way.

Let me call attention to several features of this example:

(1) The student, whom we may call Janet, acquires a belief as a result of our communicative exchange, namely, the belief that Prof. Parsons is teaching. It is because she acquires this belief, and infers that he is therefore not in his office, that she walks away without knocking on his door. Moreover, under the right circumstances, Janet's newly acquired belief will constitute knowledge.[9]

(2) The belief Janet acquires is one I myself possess, and it is because I wish to convey the information that Prof. Parsons is teaching to Janet that I say that Prof. Parsons is teaching. Moreover, my saying that Prof.

[8] Compare the quotation from Dummett (1978, 425), above. Dummett's argument is thus—to mangle an old joke of Kripke's—a transcendental deduction of the existence of Platonic dictionaries.

[9] What the right circumstances are and why her belief would constitute knowledge are, of course, much debated issues. It is my firm belief that nothing here turns upon how they are resolved, but of course I could be wrong.

Parsons is teaching is an intentional action on my part, one that is under my rational control.

(3) Janet recognizes the intentional character of my speech. In particular, she recognizes that it is with the intention of conveying to her the information that Prof. Parsons is teaching that I said that Prof. Parsons was teaching. It is for this reason that she thanked me.

That speech is a form of rational action is something I regard as extremely important. It will come to the fore only in section 4, but it will constantly be in the background and so is worth emphasizing.[10]

What has been said to this point leaves out an important feature of the exchange, namely, that it was a *linguistic* exchange. We can imagine variants of the example that would not have had this feature. As Janet approached Prof. Parsons's door, I might have looked at her and shook my head. She might then have acquired the belief that Prof. Parsons was not in his office, thanked me, and gone her way. My shaking my head would yet be an intentional act performed because I wanted to convey to Janet that Prof. Parsons was not in his office. A fourth feature of the example, then, is that I use language in my attempt to convey information to Janet:

(4) I say that Prof. Parsons is teaching by uttering the sentence "Prof. Parsons is teaching".

Like my saying that Prof. Parsons is teaching, my uttering the sentence "Prof. Parsons is teaching" is an act that is under my rational control.

This sort of case, then, involves three apparently related events: (i) my uttering the sentence "Prof. Parsons is teaching"; (ii) my saying that Prof. Parsons is teaching; and (iii) Janet's coming to have the belief that Prof. Parsons is teaching. How precisely are these events related? Why does my uttering *this* sentence lead to Janet's forming *this* belief instead of some other belief, or none at all? It will be worth my saying a few words about why I think this question is the right one to ask.

Before I do so, however, let me emphasize that, while I take the example I am discussing to be reasonably representative of ordinary linguistic communication, I am not claiming, either on Dummett's behalf or my own, that everything we would regard as linguistic communication is, in one way or another, a variation on this theme. What is distinctive of this particular sort of case is that, as noted at (2), a speaker expresses a belief in language and, as a result of her doing so, her audience comes to hold what we would ordinarily regard as the very same

[10] For further discussion of the rationality of language-use, see Heck (2006).

belief.[11] Not all cases of successful linguistic communication are like that: there are cases involving implicatures and the like, and there may be other sorts of cases, too. Nonetheless, I think that cases like the one I am treating as exemplary are in some important sense fundamental. I cannot defend that claim in detail at this time, but I will return to the matter from time to time below.

H. P. Grice, at one point in his career, anyway, held that successful communication in this case would consist not in Janet's coming to believe that Prof. Parsons is teaching, but rather in her coming to believe that I, the speaker, believe that Prof. Parsons is teaching (Grice, 1989*c*). If one were to hold that view, then the question I want to raise could be raised by asking why Janet comes to believe that I believe that Prof. Parsons is teaching, rather than that I believe that pigs dance in Peru, or nothing at all. Now, Grice is of course right that communicative success in this case does not require that Janet should in fact form the belief that Prof. Parsons is teaching: It is no failure of the communicative process if Janet refuses to trust me. (Perhaps it is so important that she see Prof. Parsons that she isn't going to trust anyone.) But I do not think the right way to accommodate this observation is to take communicative success to require merely that she come to believe that I believe that Prof. Parsons is teaching.[12] It seems clear that, struck by a bout of paranoia, Janet might refuse even to believe that I am speaking sincerely and yet that there should be no failure of the communicative process. Janet might understand perfectly what I have said, namely, that Prof. Parsons is teaching. It might, indeed, be only because she understands what I have said that she will neither believe me nor even believe that I am speaking sincerely. Had my remark instead concerned, say, a slippery patch on the floor, perhaps she'd have been happy to take evasive action.

The obvious thing to say, then, would be that communicative success, in the sense in which I am interested in it, requires only that Janet should know what I have said—in this case, that she should know that I have said that Prof. Parsons is teaching. But I would prefer not to raise the question in that familiar form, in large part because this direct approach seems to me unlikely to deliver the illumination we seek. Not only are my intuitions about when someone knows what someone else has said weak and unreliable, one would expect them easily to be corrupted by my theoretical commitments regarding the nature of communication, the notion of what is said, and the like. My intuitions about whether someone could acquire knowledge as a result of a particular communicative

[11] Whether we really should regard it as the same belief—that is, as a belief with the very same content—is another matter. In fact, on my view, the belief does not really have to have the same content for communication to have succeeded. It is a difficult question exactly how the beliefs have to be related, an issue I have addressed in relation to proper names and demonstratives in Heck (1995 and 2002). That issue is a subsidiary one here, but it is what led me to the present discussion of idiolects.

[12] For detailed criticism of this sort of approach, see McDowell (1998*c*) and Rumfitt (1995). Michael Rescorla has extended these arguments in unpublished work.

exchange are not only stronger and more reliable but are, at least to some degree, independent of the sorts of theoretical issues just mentioned and so are less vulnerable to corruption. So it seems to me that we need another approach, and my suggestion, in effect, is that we should study the cause by studying its effects. Why, after all, does it so much as matter whether one knows the meaning of an utterance someone else has made? Well, in a certain familiar sort of communicative exchange, it matters because what one takes an utterance to mean will determine the content of the belief one acquires if one takes the speaker at her literal word. If one misidentifies what the speaker has said—if one thinks that her utterance of "Prof. Parsons is teaching" means that pigs dance in Peru, say—then one risks acquiring a belief that has no particular likelihood of being true, even if one's informant has spoken knowledgeably. So it seems a reasonable and potentially fruitful strategy to study knowledge of meaning by considering the role it plays in the communicative exchanges it makes possible.[13]

The question I should like to discuss is therefore this one: What is the relation between my uttering the sentence "Prof. Parsons is teaching" and Janet's coming to believe that Prof. Parsons is teaching? In particular, why does my uttering this sentence lead her to acquire this belief? The question is not why she *acquires* the belief, rather than declining my offer of information, but why, if she accepts my offer of information, she will acquire *this* belief instead of some other; and the question does not merely seek a causal explanation but a sense of why it is rational for Janet to form this belief (a fact that is related to its potentially constituting knowledge). Of course, there are many different beliefs Janet might reasonably form on the basis of my having uttered the sentence "Prof. Parsons is teaching". Some of these depend upon what else she believes. But I take it that, among these beliefs, there is one—namely, the belief that Prof. Parsons is teaching—that is the source of the others, which arise from it by various inferences. Or again, Janet may form quite different beliefs if she takes me, say, to have flouted conversational maxims: She may, in particular, come to believe something I have implicated, and it may be my intention that she should do so. Here again, however, I take it that Janet needs to recognize what I have said—namely, that Prof. Parsons is teaching—if she is to be able to determine what I have implicated. So, to put it as precisely as I can put it, the question I want to raise is why the belief that Prof. Parsons is teaching is the one Janet would acquire if she were to accept my offer of information—the one I made when, intending to be understood literally,

[13] Attempting to justify a similar approach in Heck (1995), I suggested that, although it is certainly true that the success of a particular communicative exchange does not require (say) that Janet come to believe that Prof. Parsons is teaching, nor even that I believe he is, it is nonetheless true that such a transfer of information is the normal goal of many ordinary communicative exchanges. See Strawson (1971) for a related concern, which Strawson puts by saying that it is difficult to know how we might characterize the act of *saying* without invoking something like communicative intentions. I would like to think I no longer need to hold that sort of view, but I admit to still finding it attractive. Much thanks to Brett Sherman for help here.

I uttered "Prof. Parsons is teaching"—entirely because she accepted that offer, literally understood. But I shall speak in terms of the simpler formulation.

An obvious suggestion is that what explains Janet's acquiring the belief that Prof. Parsons is teaching when I utter "Prof. Parsons is teaching" is that, as a sentence of English, this sentence *means* that Prof. Parsons is teaching. If so, then facts about what sentences mean in our common language, English, are implicated in the explanation of successful communication, just as Dummett thinks. This obvious suggestion, however, cannot be right. What the sentence in fact means cannot explain why Janet forms the belief she does. Had Janet believed that "Prof. Parsons is teaching" meant that Benny Parsons was racing, she would have formed the belief that Benny Parsons was racing, rather than the belief that Prof. Parsons was teaching. What explains the belief Janet forms is not what the sentence means but what she *takes* it to mean.

Similar remarks can be made about the speaker—me, in our example. What Janet comes to believe does not, of course, depend upon what I take my words to mean. But if I am successfully to convey my belief that Prof. Parsons is teaching to Janet, I must utter a sentence that expresses that belief. Had I intended to inform Janet that Benny Parsons was racing, uttering the sentence "Prof. Parsons is teaching" because I believed that was what it meant, then, while Janet might still have formed the belief that Prof. Parsons was teaching, she would not have acquired the belief that I had intended to communicate to her. Moreover, her new belief would not have constituted knowledge, even if I myself knew that Prof. Parsons was teaching, for it would not have been formed in the right sort of way. So again, what a sentence actually means, as a sentence of English, cannot on its own explain communicative success. What the sentence means to the speaker and hearer must also be involved. But, of course, it does not follow that what the sentence means in English does no explanatory work, and there is reason to think that what sentences mean to individual speakers cannot wholly explain communicative success, either.

Suppose that, by some weird accident, Janet and I are under the same misimpression: by some coincidence, we both think that "Prof. Parsons is teaching" means that Benny Parsons is racing. Then while Janet will, in this case, acquire the belief I intend to communicate, that belief still would not constitute knowledge, even if I know that Benny Parsons is racing. Something still seems wrong about how she acquires the belief. It does not, therefore, seem enough to explain communicative success simply to note that Janet and I both *believe* that "Prof. Parsons is teaching" means that Prof. Parsons is teaching. Successful communication requires something more than mere agreement about what the uttered sentence means. It requires more than mere coincidence of belief, or, as we might put it, following Dummett, a mere overlap of idiolects.

What, then, distinguishes these two cases? Well, the obvious suggestion this time is that the crucial difference lies in what the sentence actually means, as a sentence of English: in the one case, our shared belief about what the sentence means is *true*, whereas in the other case it is not. But the obvious suggestion is once again mistaken.

Suppose Janet utters the sentence, "I bought a goat today". Suppose, however, that I do not hear the first sound of the word "goat" clearly, so that I am unsure whether she has uttered that sentence or instead "I bought a coat today". Knowing that Janet has been wanting a goat for a long time, and hearing the joy in her voice, I might come to believe that she has uttered the sentence "I bought a goat today", and my belief may be true, even justified. But it still need not be knowledge. If not, then I do not *know* that she said she bought a goat (rather than a coat) and so cannot come to *know*, though I might reasonably come to believe, that she bought a goat. But then, if Janet does not *know* that "Prof. Parsons is teaching" means that Prof. Parsons is teaching, but only truly believes that it does, then she will not *know* that I have said that Prof. Parsons is teaching, but only truly believe that I have done so, and so she cannot come to *know* that Prof. Parsons is teaching on the basis of my telling her so. Successful communication—at least in so far as it involves the potential transfer of knowledge—therefore seems to require not just that the communicants have *true beliefs* about what the uttered sentence means, in that context, but that they *know* what the uttered sentence means.

Tyler Burge, in his discussions of testimony, seems to deny an important presupposition of the preceding discussion (Burge, 1993, 1999). I am supposing here that beliefs acquired by testimony—that is, beliefs that are based upon what someone has said—are, in part, justified by beliefs about what words were uttered and what those words meant. In particular, on my view, Janet's belief that I said that Prof. Parsons is teaching is (in an epistemologically relevant sense) based upon her beliefs that I have uttered the sentence "Prof. Parsons is teaching" and that this sentence means that Prof. Parsons is teaching. According to Burge, however, beliefs about utterances and about the meanings of uttered sentences, though perhaps *causally* necessary if one is to know what has been said, play no justificatory role with respect to beliefs acquired through testimony. Rather, the perception of utterances plays a merely *enabling* role, much as the perception of a written proof, though it may be necessary if one is to think through the proof, is no part of what justifies one's belief in the proof's conclusion.

Now, I accept the traditional distinction upon which Burge is relying, but I do not think that it applies to the case of language comprehension. I have discussed this issue elsewhere (Heck, 2006) and so will be brief here. It is true, of course, that we rarely notice the words other speakers have uttered. But Burge's point is not phenomenological; it is epistemological. Consider again, then, the example just mentioned of Janet's utterance of "I bought a goat". Given the context, it may be clear enough that she must have meant to say that she bought a goat, not a coat. And frequently, that would be the end of the matter, since it need not matter what sentence Janet uttered nor, for that matter, what she actually said. But sometimes it does. In that case, reason to think she meant to *say* that she bought a goat would be good reason to think that she meant to *utter* "I bought a goat". That, in turn, may be good reason to suppose she did utter "I bought a goat" and so did say that she bought a goat. But what Janet *did* say depends upon which sentence she actually uttered, and ordinary speakers know as much. If I am given reason to suppose that Janet actually uttered the sentence "I bought a coat", then I am thereby given reason to suppose she said that she bought a coat, since, if she did utter that sentence, then that is what she said, whether she meant to say it or not.[14] In so far as we want to distinguish the context of discovery from the context of justification, then, beliefs about the words someone has uttered seem more at home in the context of justification than in the context of discovery. As said, we often hardly notice the words our communicative partners have uttered, since our interest is rarely in the words themselves. It does not follow, however, that one's belief about what has been said is not, in an epistemologically relevant sense, based upon one's belief about—or perhaps better, one's perception of—the words uttered. Ordinary speakers are aware that what someone has said is determined by the words she has uttered. When the question is seriously raised what someone has said, we know that, while quite diverse evidence, including evidence about the speaker's intentions, might be brought to bear upon the question, all such evidence must ultimately bear upon what sentence was uttered, for it is the sentence uttered that determines what was said.

The foregoing gives us some reason, then, to suppose that successful communication depends upon both the speaker's and her audience's knowing what the uttered sentence means in the context in which it is uttered. Now, knowing that the meaning of the uttered sentence is (in our example) that Prof. Parsons is teaching is, of course, more than truly believing that it means that Prof. Parsons is teaching. But one can only know if one truly believes. And if such beliefs are to be so much as *capable* of being true, then it would certainly appear as if there must be something the sentence really does mean: Without that, there would be nothing for speakers to be right or wrong about. And so it would seem that what Janet and I both need to know, if we are to communicate successfully, is what the sentence I uttered means in the language we both speak—in English, or in the dialect thereof spoken in Cambridge, or whatever the right 'common language' might turn out to be. Say it's English. Then for me and Janet to know the meaning of the uttered sentence is for us to know what it means in English.[15] If so,

[14] I am assuming here again that Janet has no unusual beliefs about what the relevant words mean. The importance of this assumption will emerge later.

[15] When one puts what is at issue in these terms—whether common languages are the objects of linguistic beliefs—it seems as if the view I am attacking may also be held by Jim Higginbotham. See Higginbotham (1989). But I am not sure Higginbotham does hold any position my arguments here would undermine.

then, once again, reference to common languages and what sentences mean in them is essential to the explanation of successful communication.

Although, so far as I know, Dummett never argues in quite these terms, I think the foregoing constitutes a reasonable interpretation of his thought on these matters. I shall now argue, however, that the reference to English—or, more generally, to common languages—is explanatorily idle.

Consider the following well-worn example. I am one of those unfortunate people who used to believe that the word 'livid' meant *flushed* rather than *bluish-gray*. So I might have uttered the sentence "Jones was livid" meaning to convey the belief that Jones was red-faced. Now, certainly, what I took the word to mean is not what a dictionary will tell you it means. And in that sense, of course, I was mistaken about what it means. On the other hand, however, as the familiarity of the example suggests, I was hardly alone in this misconception. (Perhaps you too shared this misconception until very recently.) Imagine, then, that most of the people with whom I regularly communicated were under the same misimpression as I was. And suppose I wished to convey to my friend Steve my belief that Jones was, as I would then have put it, 'livid', that is, flushed. Steve shares my belief that "Jones was livid" means that Jones was flushed. So if I say to him, "Jones was livid", and he takes me at my word, he will form the belief that Jones was, as he would put it, 'livid', that is, flushed. Is there a failure of communication here? I do not see why we must say there was. But if not, then our joint failure to know what the word 'livid' means in English—and so what the sentence "Jones was livid" means in English—did not frustrate our attempt to communicate. If not, then knowing what the uttered sentence means in the 'common language' cannot be a condition of successful communication and knowledge of meaning is not what explains successful communication.

Now, I myself just argued that mere agreement about meaning is not enough to explain successful communication. In particular, while Steve and I agree about what the word 'livid' means, in the sense that we both think it means *flushed*, mere agreement of this sort is not enough to explain our communicative success. So one might wonder how this example differs from the one discussed earlier, in which both Janet and I were under the misimpression that "Prof. Parsons is teaching" meant that Benny Parsons is racing. The difference is that, in that case, I stipulated that it was an *accident* that the two of us agreed about the meaning of the sentence. And in this case, too, if one has the intuition that it is just an accident that Steve and I agree about what the word 'livid' means, then one should, I think, also have the intuition that we have not really succeeded in communicating with one another. But it need not be an accident that Steve and I agree about what the word 'livid' means. Of course it might be. But there is no reason to suppose it must be. It does not, in particular, follow from the fact that 'livid' does not mean *flushed* in English that it is an accident that Steve and I agree that 'livid' means *flushed*. The belief that 'livid' means *flushed* is sufficiently widespread that the third edition of *Webster's New World Dictionary*, published in 1988, actually notes that it is sometimes used to mean *red*.[16] We can imagine, then, that we stand in the middle of a process of language-change. The word 'livid' might be on its way to meaning *flushed*, rather than *bluish-gray*, at least in some of its uses, as a result of people's using it to mean just that. How else would one imagine such a change occurring? And such changes do, of course, occur. In the midst of such a change, or even at the beginning of such a change, one would expect to find groups, some perhaps quite isolated, in which the word is already consistently used with what will one day become one of its meanings in the 'common language'. Within such groups, communication will proceed quite normally, divergence from 'standard usage' notwithstanding.

One might suggest, however, that Steve and I do know what the word 'livid' means in our common language. It is just that our common language is not English but rather the language spoken by some smaller community of which we are part. But how should this smaller community be specified? The point is not just that it is hard to say. It is that once we start shrinking the linguistic community, there will be no unique community to which both Steve and I belong. We both belong to many linguistic communities, and which one is relevant may depend upon which words are being used. Although Steve and I agree about what the word 'livid' means, we may disagree about what some other words mean, in much the same way we disagree with other speakers about what 'livid' means. Maybe Steve applies 'fish' to all aquatic creatures (so that Steve will say 'Whales are fish'),[17] though I apply it only to aquatic creatures that share biological characteristics I cannot myself enumerate. Maybe Steve is part of a linguistic community all (or most) of whose members use the word 'fish' as he does and so communicate perfectly well using the word in that way. If so, then I see no reason not to say about Steve's use of 'fish' within that community what I just said about our joint use of 'livid'. If so, then, as I claimed, there is no single linguistic community to which both he and I belong, any more than there is a single non-linguistic community to which we both belong. Moreover, we do, of course, talk to others who do not share our belief about what 'livid' means, say, Janet, who agrees with *Webster's* that it means *bluish-gray*. So when I say to Janet, "Jones was livid", communication will fail. In this case, of course, the explanation seems simple enough: We just do not agree about what 'livid' means. But is one of us wrong about what it means in our 'common language'? Which language—whose language—would that be?

[16] Many people with whom I've discussed these issues have confessed that they thought the word 'livid' meant *angry*.

[17] The third edition of *Webster's New World Dictionary* also lists this usage as a 'loose' one.

I do not deny, of course, that there are intuitions here that must be respected: I do not deny, in particular, feeling a strong pull to say that *I* am the one who is wrong about what 'livid' means. The next section offers an account of these intuitions. I shall return, in section 4, to the question how communicative success should be explained.

3. SEMANTIC NORMS

I used to think that 'livid' meant *flushed*. Had I then said to Janet, "Jones is livid", she, not sharing my misconception, would have formed the belief that Jones was livid, that is, bluish-gray (or pale), rather than the belief I was trying to communicate to her, namely, that Jones was red-faced. My communicative intention would thus have been frustrated. Indeed, Janet, upon discovering that Jones was not pale, and that I had known he wasn't, might have accused me of lying. Now, I wasn't lying. I made every effort to speak the truth. Nonetheless, there is a strong intuition that it is my fault that Janet acquired a false belief. I told her that Jones was livid, and he was not.

The intuition that facts about what sentences mean have normative force is deeply ingrained. I am at fault, not Janet, because I *misunderstood* the word 'livid' and so used it *wrongly*. But if there is no 'common language' that we both speak—if there is no such thing as what the word 'livid' means in that language—what content is there to the claim that I was using it wrongly? If, moreover, I am not wrong about what the word 'livid' means, why should I feel compelled to change my usage? But we often do feel so compelled. Upon learning of my disagreement with the dictionary, I did not continue to use 'livid' as I had. Nor would I have regarded it as reasonable or rational for me to do so. Similarly, consider Bert, who insists that the word 'arthritis' applies to all rheumatoid ailments.[18] Bert says "I have arthritis in my thigh", believing that what he says means that he has a rheumatoid ailment in his thigh, which he may well have. But as the literature's response to such examples makes plain, there is a strong intuition that what Bert has said is false, whether or not he has a rheumatoid ailment in his thigh, because the word 'arthritis' in fact applies only to rheumatoid ailments of the joints. If that is not what the word means in our common language—because there is no common language and hence nothing that the word means in it—how can we say that Bert speaks falsely?

Chomsky, for his part, accepts these consequences almost gleefully. He does not, of course, say that there are no linguistic norms. He would insist, for

[18] Let me emphasize that it is important to my treatment of this example that Bert actually thinks the word applies to all rheumatoid ailments, rather than being ignorant about what its extension is. So the case is a Gödel-Schmidt case not a Feynman case (Kripke, 1980, 81ff). Feynman cases raise different issues, which I hope to address elsewhere.

example, that I can indeed be wrong about whether a particular sentence is grammatical. Whether the sentence is grammatical is not determined by whether I think it is but by the grammatical principles I tacitly know. There is no reason to suppose that what those principles imply will always be transparent to me. (Garden path sentences, such as "The horse raced past the barn fell", are a typical sort of example.) For similar reasons, I could be wrong about what some sentence means. If I think that "No eye-injury is too trivial to ignore" is a sensible thing to put on a sign in a hospital—as did the people who apparently put it on just such a sign—then I am again just wrong.[19] These possibilities of error derive, however, from the familiar gap between competence and performance: That these beliefs are erroneous can be explained without invoking anything beyond my own linguistic competence. What Chomsky means to deny is that there are any sources of linguistic norms—any sources, that is, of what can rightly be called 'error'—that lie outside a speaker's own linguistic competence. Perhaps better: Any such norms as there may be are merely hypothetical, depending for their force upon, say, the speaker's desire to conform her usage to that of others. That is not, of course, to say that such a desire cannot be, or even typically is not, rational, but only that it is not required by anything that is specifically linguistic.

But does rejection of the notion of a common language really lead to such radical conclusions? I am going to argue that it does not.

The following passage is reasonably representative of Chomsky's many discussions of this matter:

> If Bert complains of arthritis in his ankle and thigh, and is told by a doctor that he is wrong about both, but in different ways, he may (or may not) choose to modify his usage to that of the doctor's.... If my neighbor Bert tells me about his arthritis, my initial posit is that he is identical to me in this usage. I will introduce modifications to interpret him as circumstances require; reference to a presumed 'public language' with an 'actual content' for *arthritis* sheds no further light on what is happening between us, even if some sense can be given to the tacitly assumed notions. (Chomsky, 2000*a*, 32)

Chomsky here denies that Bert means by "I have arthritis in my thigh" what the doctor means. If not, then there is, presumably, no peculiarly linguistic reason Bert need modify his usage to conform with the doctor's. Of course, there may be reasons he should do so, even quite compelling ones, but, if there are, these reasons derive not from the nature of Bert's language but from something external to it. It is, moreover, important to see that Chomsky's main point is not that no sense can be made of the notion of a public language: It is that, even if some sense can be made of it—which he, of course, doubts—appeal to such a notion would be unexplanatory.

[19] As a little thought will show, the sentence actually means that no eye-injury is so trivial that one would not be justified in ignoring it: That is, it's OK to ignore all eye-injuries, even the trivial ones. (I learned of the example from George Boolos, who told me he had heard it from Jonathan Bennett, but I have not located an original source.)

In his intriguing defense of the primacy of idiolects, Alexander George argues that we should distinguish what a word means in a person's idiolect from what she *thinks* the word means in her idiolect. If there is such a distinction, then we get much of the benefit of the appeal to common languages without having to invoke them. In particular, we can make sense of the phenomenology of linguistic error, which George describes thus:[20]

> When I came to believe that 'livid' does not mean *red* but rather *bluish-gray*, I then took myself to have been in error, to have made a mistake about the meaning of the word 'livid' in my language. When I thought back to times I had used the word, I felt I was discovering what I had actually said, as opposed to what I thought I had. On [Chomsky's] view, however, such reactions are out of place. (George, 1990, 289)

The intuition here is that there is a difference between what my language is and what I take it to be. But we can have that difference, the thought is, even if there no such thing as a common language. The gap George wants us to acknowledge is not, it should be noted, the one generated by the distinction between competence and performance. On the contrary, what he wants is a distinction that would allow me to be wrong about what the word 'livid' means in my idiolect not because of a performance error but rather because of a competence error: To do so, he thinks, we need to find a way to say that what I think the word 'livid' to mean need not be what it means, *even in my own idiolect.*

The question immediately arises, of course, what determines what 'livid' means in my idiolect, if it is determined neither by what I (tacitly) take it to mean nor by what it means in a communal language I allegedly speak. George quickly admits that he does not know the answer to this question. He suggests, however, that the meaning of a word "is sensitive to considered changes to one's linguistic beliefs that one would make as a result of communication with others or observation of them" (George, 1990, 292). As George notes, there is considerable unclarity in a notion so characterized, largely due to unclarity surrounding the relevant counterfactuals. But let that pass. The more significant worry, it seems to me, is that a notion so characterized is of no explanatory use, even for the limited purposes for which George wants it. George wants to be able to say that I am mistaken about the meaning of the word 'livid' if, and only if, what I believe it means is not what it in fact means, even in my own idiolect. But when unpacked, that turns out, according to George, to mean (very roughly) that I am mistaken just in case I would change my view about what 'livid' means if I had the right sorts of interactions with other speakers. If so, however, then we cannot say, as George would apparently also like to be able to say, that I should change my view *because it is wrong*. Rather, it is wrong only if I am (and in suitably similar circumstances would be) prepared to change it.

[20] George calls the view in question the 'no-error view'. I've interpolated the reference to Chomsky, which is implicit in the paper.

George comes close to acknowledging this point, and it may just not bother him. The reason is that, like Chomsky, George does not think that idiolects, in his sense, are of any interest to linguistics (George, 1990, 295). His purpose is not to rehabilitate a notion of common language—or some ersatz—that will be suitable to figure in scientific investigations of language. He is simply trying to gain some purchase on normativity from within the perspective on language that informs contemporary linguistic theory. Nonetheless, George does seem to want his notion of an idiolect to do some explanatory work. Thus, he concludes his paper as follows:[21]

> [C]ertain features of grammars over which speakers might have some control confront a seemingly objective reality that determines the correctness of the relevant linguistic beliefs. This reality partly consists in actual and potential considered transformations of individual grammars consequent upon communicative interactions among their bearers. That we take ourselves to be constrained in the interpretations we can place upon expressions if we are to proceed correctly is [an] important aspect of the perceived objectivity and independence of language. (George, 1990, 297)

But why should what I *would* do if I had certain sorts of interactions with others so much as seem to constitute an "objective reality"? How can I regard what I would do in such circumstances as correct if what is correct is determined by what I would do? Here—to borrow from Wittgenstein—it really does look as if what seems right is right.

In discussing the passage from Chomsky above, I remarked that he is not claiming, of course, that one cannot have reason to change one's usage. His claim, rather, is that the reasons one can and often does have are not *linguistic*: They have nothing particular to do with the nature of language but are reactions to external forces, for example, peer pressure. This observation gives us another reason to be dissatisfied with George's response to the problem of linguistic error. One can have a variety of reasons to change one's linguistic beliefs as a result of interaction with others. Changes one would make because of peer pressure, however, do not seem appropriate constraints on the identity of one's idiolect. Now, George seems aware of this point, too, writing that such changes must be "considered", by which he means that they should "result from our reflection on the correctness of our [linguistic] beliefs" (George, 1990, 289). But the problem, once again, is that it is difficult to see how to understand such reflection if correctness is what George says it is. If my beliefs are only correct if I would not make considered changes to them as a result of interaction with others, and if what I am trying to decide, when reflecting on the correctness of my beliefs, is whether to make just such changes, I seem to be going in a circle.

[21] The quotation is consistent with the suggestion that George thinks that there really is no correct way to proceed and that such constraints are ultimately illusory, though we do "take ourselves" to be so constrained and do suppose that there is a right and wrong about how we use our words. The overall tone of the paper suggests to me, however, that this is not George's view.

The problem to which we keep returning derives from George's characterizing the semantic properties of idiolects in purely dispositional terms. That these properties are explanatorily impotent should thus be no surprise. An object's possessing a property that is characterized in wholly dispositional terms cannot explain its exercise of the very dispositions in terms of which that property is characterized: That a pill is dormative cannot explain why it makes you sleepy if its being dormative *just is* its tending to make people who take it sleepy.[22]

Consider again the example of Bert and his doctor. Suppose this time not that Bert says to the doctor, "I have arthritis in my thigh", but that the doctor says to Bert, "You do not have arthritis in your thigh". Bert thinks to himself, "Man, what a relief!" In fact, Bert has absolutely nothing about which to be relieved, for the doctor continues, "You do, however, have a very serious rheumatoid ailment in your thigh".

Chomsky would have us say that Bert is, if he wishes, free to continue using the word 'arthritis' as he does. And, in some sense, presumably, he is. But my intuitions are somewhat different in this case than they were in the earlier one. When it is Bert who is speaking, there is at least some temptation to say, with Humpty Dumpty, that the question is who should be master: If Bert hardheadedly insists that he's always used the sentence "I have arthritis in my thigh" to mean that he has a rheumatoid ailment in his thigh and he isn't going to stop now—that may be silly, but it doesn't seem incoherent. But if, on the other hand, Bert insists that he's always understood people who've told him "You do not have arthritis in your thigh" to mean that he does not have a rheumatoid ailment in his thigh, and he's gonna keep right on understanding them that way—then I think he's not just being silly and is being incoherent. As the example makes plain, Bert's failure to understand the doctor makes him liable to form false beliefs not just about what the doctor has said but about whatever may happen to be the subject of conversation: health, sports, politics, or the weather. The intuition that Bert misunderstands the word 'arthritis' is thus, pretty much as George suggested, the intuition that Bert would have good reason to change how he understands the word if he had certain sorts of knowledge about other speakers:[23] Understanding 'arthritis' as he does makes the beliefs Bert forms on the basis of what others say to him using that word liable to be false—or, at least, deprives those beliefs of any likelihood of truth, even when conditions are otherwise ideal (for example, his informant is speaking knowledgably).

If the goal were to rehabilitate a notion of common language that will do serious explanatory work, then we would, I am happy to admit, remain far short of it: The question how the boundaries of linguistic communities are to be drawn

[22] Of course, the pill's being dormative would explain its making you sleepy if dormativity had a categorical basis, as fragility familiarly does. But it is definitive of the position George is defending to deny that the relevant counterfactuals have a categorical basis.

[23] Of course, Bert might be so hard-headed that he won't change how he uses any word for any reason, but the intuition is that there are certain sorts of semantic *norms*.

remains, as do the various sorts of relativity about which I complained earlier—relativity to conversational partners and even to the words one is using. But the goal was simply to account for a certain intuition—the intuition that Bert misunderstands the word 'arthritis'—without appealing to common languages, and I say I have done that. The intuition has its source in the fact that, when language is used for the purpose of exchanging information, how one understands an utterance determines the content of the belief one acquires if one accepts it as true and so whether the belief so formed is likely to be true, even under otherwise ideal conditions.

It should be no surprise that this account of semantic norms makes reference to communication: There is obviously no question of conforming one's usage to that of other speakers unless there are other speakers to whose usage one might conform, and there will be no need for conformity unless one is interacting with those speakers linguistically. And assuming that Bert is communicating with his doctor does not trivialize the account I have offered: Even if we do so assume, it is not immediately obvious why Bert should then be subject to any semantic norm. If my view were that Bert is subject to semantic norms only insofar as he wants to understand his doctor and be understood by him, then that, I would be happy to agree, would be entirely empty. But that is not my view. My view is that Bert is subject to semantic norms insofar as (i) Bert is exchanging information with his doctor and (ii) Bert wants to have true beliefs. If someone wanted to insist that (i) and (ii) are conditions whose satisfaction is not required by linguistic competence and so that semantic norms are not purely *linguistic* norms, I would not disagree. I do not see why it should matter, one way or the other.[24]

One might now object, however, that no reason has been given to suppose that Bert uses the word 'arthritis' *wrongly*. When Bert discusses his health with his doctor, they misunderstand *each other*. Why think the doctor is right and Bert is wrong? It seems to me, however, that the strength of the intuition that Bert is wrong varies with the degree of his idiosyncracy. It is because Bert disagrees not just with the doctor but with just about everyone else, too, that we have such a strong sense that he is mistaken. But if one thinks about a case like that of the word 'livid', intuitions aren't nearly so strong. At least, mine aren't. In fact, I'm not at all sure, as I look back on my earlier usage of the word, that I feel as if I know now what I was actually saying then: If my conversational partners also

[24] Thanks to Tad Brennan for some pressure here that took a long time to take effect.

For what it's worth, I am inclined to think that the desire that one's beliefs be true—if one should call it that—is a precondition of rationality. And, in any event, semantic norms are no worse off than epistemic norms, or even logical norms, if they are conditioned by that desire. The more interesting question concerns the status of the assumption that Bert is exchanging information with his doctor. As it happens, I am inclined to think that communication most fundamentally is a means of exchanging information. But even if that is not so, it doesn't matter for present purposes: Absent the assumption that Bert is exchanging information with his doctor, I simply do not have the intuition I have set out to explain, namely, that Bert ought to use his words in a certain way.

took the word 'livid' to mean *flushed*, then they interpreted me as saying just what I then intended to say.

4. COMMUNICATIVE SUCCESS

In section 2, I argued that communicative success is not to be explained by communicative partners' knowing what uttered sentences mean in some common language they all speak. I argued that this proposal will not do, because there is no independent standard to which to hold beliefs about meaning. At least, if there is one, it is not the meaning of the words in some common language. Mere agreement about meaning, on the other hand, is insufficient to support successful communication, at least if communicative success is supposed to make it possible for one to come to know what one is being told. But where is the middle ground? It is not enough for speakers to agree about what a given sentence means, because mere agreement can be accidental: And if it is a mere accident that Steve and I agree about what "Jones was livid" means, then the truth of Steve's newly acquired belief will itself be accidental, even if I know that Jones was livid. Requiring that both Steve and I must *know* what "Jones is livid" means will serve to prevent such accidents, but that proposal requires there to be something the sentence really does mean in our common language, and there is no common language that we share. But there is actually no need to appeal to common languages to implement the main idea behind this proposal: It is enough to require that Steve and I *knowingly agree* about what the sentence means. If so, it will be no accident that Steve's newly acquired belief is true if mine is.[25]

Note that I did not say that Steve and I must *know that we agree* about what "Jones is livid" means to communicate successfully.[26] That would suggest, surely wrongly, that one must have the concept of *agreement* to be able to communicate. Of course, one might well want to know what this notion of *knowing agreement* is supposed to be, and I confess that I am not entirely sure. Part of what is wanted is that I should know what Steve means by "Jones is livid", and he should know what I mean. But, for reasons I shall not rehearse, it is likely also necessary that I should know that Steve knows what I mean, that he should know that I know what he means, and so forth (Grice, 1989a): What I should probably say, then,

[25] There are various reasons to think knowledge may not be quite the right notion here. For one thing, as I'll note below, one's apprehension of meaning is most immediately perceptual, and perceptual states are not the sort of thing that can be classed as knowledgeable or otherwise. But if it turns out that we need some analogous notion that applies to perceptual states, one should be available.

[26] Brett Sherman has urged me to say what it might have seemed obvious I was about to say, namely, that Steve and I must *non-accidentally* agree about what the sentence means, rather than to say we must know we agree and then derive the non-accidental character of our agreement from that fact. The problem, however, is that it's not obvious that non-accidental agreement about meaning is enough to secure knowledge in ordinary communicative exchanges.

is that it is *common knowledge* between me and Steve that each of us means that Jones is flushed by "Jones is livid".[27] But I do not want to pursue that issue here. I therefore leave my view a bit underspecified: Communicative success depends upon and should be explained in terms of communicative partners' knowingly agreeing about what uttered sentences mean.

One might object that this view makes no distinction between ordinary cases of communication and the following sort of case. Suppose Bob and Patrick have agreed that today is opposite day: They are always to say the opposite of what they really mean. In particular, when Bob says "Gary is not home", he knows that Patrick will take him to mean that Gary is home. If so, then it would seem that Bob and Patrick knowingly agree that Bob's utterance of "Gary is not home" means that Gary is home. Now, in a sense, that's fine with me: Communication may indeed succeed in such cases; one wants to be able to explain why it succeeds; and that seems a pretty good explanation. But in another sense, something does seem to have been left out, namely, what distinguishes this case from the normal case. One might suspect that what distinguishes them is that "Gary is not home" just doesn't *mean* that Gary is home, however Bob and Patrick might have agreed to interpret it, and when one talks about what the sentence means, one is talking about what it means in English, or whatever Bob and Patrick's common language might be. As David Wiggins puts it:

> Normally, it may be said, [a speaker] aims to be understood. But that isn't quite right. A speaker aims to be understood, but not as saying just anything. He aims to get across a certain thing, but he aims, if possible, to say this thing *and* to be understood as saying it.... If there were no such thing as a [common] language, what would be the difference between someone's simply being understood... by an audience as wanting to communicate that such and such and someone's successfully... *saying* this or that? (Wiggins, 1997, 504)

Without the notion of a common language, Wiggins suspects, we cannot distinguish merely managing to get it across that Gary is home from getting it across by *saying* that Gary is home. This distinction certainly needs respecting, but I do not think we need to invoke common languages to respect it.

As John McDowell has emphasized (McDowell, 1998*a*; 1998*b*), when we hear another speak, we do not hear just her words, adding an interpretation in thought. We literally *hear* what she says in her words. To hear what someone has said—in the semantic as well as the phonetic sense—is to perceive, and like other forms of perception, this one is in some ways similar to belief. When I see a scene before me, it looks as if objects really are arranged a certain way in the space around me. Visual perception, that is to say, is representational not only in the sense that it has truth-evaluable content but also in the sense that it

[27] See Lewis (1986) for the classic discussion of common knowledge. The notion also figures importantly in Grice's work.

represents the world as actually *being* a certain way.[28] Perception of speech in a language one understands is, in this respect, similar. If my wife tells me, "There are moths in the closet", my perceptual system represents my wife not just as having uttered these words but as having *said* that there are moths in the closet. Similarly, when the doctor says to Bert, "You do not have arthritis in your thigh", Bert's perceptual system represents the doctor as having *said* that he does not have a rheumatoid ailment in his thigh. Bert's perceptual system thus misleads him in this case, for the doctor has said no such thing.

I can't change how a sentence sounds to me simply by deciding that it should sound some other way, any more than I can make myself see a scene otherwise just by deciding that things ought to look differently—and that is as true when 'sounds' has semantic import as it is when it just has phonological import. In that sense, judgements of meaning have their source outside our conscious minds. Their source, however, is not outside our minds altogether, in facts about what our expressions mean in some common language. Their source is, rather, our *unconscious* minds: what we tacitly know about meaning. I hear the sentence "Gary is not home" as meaning that Gary is not home, and use it to say that Gary is not home, because that is what my I-language determines that this sentence means to me. What distinguishes Bob and Patrick's conversations on opposite day from their conversations on other days is thus the source of the judgements about meaning on which those conversations depend. In normal cases of linguistic communication, the source of these judgements is one's linguistic competence. In abnormal cases, as on opposite day, these judgements have some other, partially non-linguistic, source.[29]

This last point is actually overstated. Our judgements about what utterances mean almost never have a purely linguistic source. The reason is that almost every sentence we utter contains one or another expression whose interpretation depends upon the context in which it is uttered. So to be able to interpret almost any utterance, one needs to know relevant facts about the context in which the utterance is made. What we ought to say, then, is that what distinguishes normal communication from communication on opposite day is that, in the normal case, judgements of meaning depend as little as possible upon non-linguistic sources. A couple of assumptions are being made here: First, that opposite day does not itself present us with an example of context-dependence; and Second,

[28] For more on this matter, see Heck (2000).

[29] Similar remarks serve to distinguish literal communication from that involving implicatures. If Prof. Smith writes in his letter of recommendation that Jones is punctual and has good penmanship, and conspicuously declines to say anything else, then both Smith and his reader may know that Smith's utterance in some sense meant that Jones is not well-qualified for graduate study. But there is, as Grice taught us, an important sense in which Smith's utterance does not mean that, and this difference should be registered in our account of communicative success. Here again, I would say that, while Smith and his reader may well know that Smith's utterance in some sense means that Jones is a poor student, the source of this knowledge lies not in their linguistic competence but at least partially elsewhere.

that context-dependence can be brought under sufficient theoretical control that the qualifier 'as little as possible' can be given a clear meaning. The first assumption, though worth noting, should be uncontroversial. The second is not at all uncontroversial. A first stab at explaining what 'as little as possible' is supposed to mean might be the following: only in so far as is necessary if the utterance is to be interpeted at all. But that constitutes only a very little bit of progress. To defend—or even to explain—this second assumption in any detail, one would have to take a stand on the the very large and very difficult question how context affects what is said. I have no such stand to take. But if one were to adopt Jason Stanley's view that context affects what is said only in so far as it fixes the assignment of values to free variables present in the logical form of the uttered sentence (Stanley, 2000), then 'as little as possible' could be taken to mean: only in so far as is necessary to assign values to variables present at LF. I do not, I should emphasize, think this second assumption requires any view as strong as Stanley's to be correct. But it does require that it should be possible to isolate the contribution context makes to determining what is said—lest too many ordinary cases turn out to be too much like opposite day—and Stanley's view illustrates what sort of requirement that is.

So, to a slightly better approximation, my suggestion is as follows: Communicative success depends upon and should be explained in terms of communicative partners' knowingly agreeing about what uttered sentences mean; what distinguishes the ordinary case, in which one is not only understood as saying something but actually says it, is that speakers' judgements about meaning derive entirely from their linguistic competence, except in so far as these judgements depend upon such knowledge of contextual features as is necessary if the utterance is to be assigned any meaning at all.

The following might seem like a counterexample.[30] Suppose my four-year-old friend Sophie says to me, "I have a brother". If I believe her, and if she does have a brother and knows that she does, then one would naturally suppose I might thereby come to know that Sophie has a brother. It is far from clear, however, that Sophie and I now have the same belief, because it is far from clear that we mean the same thing by 'brother'. For me, *brother* is a biological notion: Someone is your brother only if the two of you have the same biological parents. Sophie knows nothing of biological relationships. Her concept of *brother* is more a social one. So our beliefs have different contents, and that makes it natural to suppose that our words do, as well. If so, then we do not even agree about what Sophie's utterance of "I have a brother" means, let alone knowingly agree, but we nonetheless seem to have communicated, at least in the sense that

[30] Thanks to Stephen Schiffer for this example, which he mentioned as I struggled to understand a question someone else had asked at NYU. I do not know the original questioner's name but thank her anyway for asking the question.

I have acquired knowledge as a result of what Sophie said to me. If so, then successful communication does not even require agreement about meaning, let alone known agreement.

This example undeniably has some appeal. But it seems to me that it derives its appeal from the fact that the word 'brother' *isn't* always used to express a purely biological notion, even by adults. When I say that my niece Julie has a brother, or my brother calls Julie his daughter and Alex his son, or I say that Don and Shannon are their parents, that's all true, even though both Alex and Julie were adopted. There are perfectly good adult uses of such words as 'brother' and 'parent' on which they denote certain sorts of social relationships: Hence the term 'birth-mother'.

In the last couple of decades, developmental psychologists have constructed an increasingly compelling case that the concepts young children associate with certain words are importantly different from the concepts adults associate with those same words. And, as it happens, the concepts such children associate with words with which well-educated adults associate biological notions are very often social or functional.[31] In some of those cases, the meaning with which the child uses the word does not survive into adult language, and so examples constructed using such words would not suffer from the flaw I just claimed to find. So let's set aside the ambiguity of the word 'brother' and assume that it expresses a social notion for Sophie and a biological one for me. If so, then I don't think I *can* come to know that Sophie has a brother from her telling me that she has (as she would put it) a 'brother'. By hypothesis, Sophie would truly have believed that she had a 'brother' even if Sebastian were adopted, and I presume she would still have said so, too. But then it is only because Sebastian happens not to be adopted that my belief is true if hers is, and I'm just lucky not to be wrong. Granted, it's not sheer luck—as it would be had I taken Sophie's utterance to mean that she had a Barbie—but the truth of my belief nonetheless seems too accidental to count as knowledge. If so, then communication wasn't successful after all, and there's nothing that needs explaining.

The matter could undoubtedly use further discussion, but it won't get it here.[32]

[31] See Carey (1985) for an extensive discussion of young children's understanding of biological notions. A similar account of young children's understanding of psychological notions can be found in Wellman (1990). One could presumably construct examples similar to Schiffer's with many of the concepts Carey and Wellman discuss. Carey and Wellman's work is not, of course, uncontroversial, and some philosophers have raised doubts about how they individuate concepts: See, e.g., Fodor (1998). The really hard questions about how concepts are to be individuated don't affect the discussion here, however, since the concepts in question differ extensionally.

[32] One quick remark, however. First, suppose I knew that Sophie's parents had strange religious beliefs that barred them from raising any children but their own biological offspring. Then perhaps I could come to know that Sophie had a biological brother, since that rules out his being adopted. I mention the point because there may be similar cases in which the intuition that I know is strong enough to survive the challenges in the last paragraph. My knowing might then be explained along these sorts of lines, by appeal to relevant background knowledge.

5. TALKING TO STRANGERS

I communicate on a regular basis with all sorts of people I've never before encountered, and communication with such people frequently succeeds: When the police officer says to me, "Your car is parked in a fire lane", I can come to know that my car is parked in a fire lane and take appropriate action. If so, then, on my view, the officer and I must knowingly agree about what her utterance means and so, minimally, I must know what her utterance means. But on what basis might I claim such knowledge?[33]

There are several options here. One would be to insist that I have inductive evidence that what sound like sentences of my language are used with the same meanings with which I would use them. Another option is to say that I have a default entitlement to suppose that what sound like sentences of my language mean what I would mean by them: The question would then arise why it is rational for me to rely upon this entitlement (Burge, 1993, 1999). Yet another option is externalist: One might deny that I have any warrant—even an entitlement—for such beliefs and claim that they constitute knowledge because they are reliable, or what have you. Fortunately, we need not choose among these options here.[34] All of them, in one way or another, require that, typically, what sound like sentences of my language are used with the meanings with which I would use them. To put the point differently: Any reasonable answer to this question must presume that different speakers tend to use the same sentences the same way. And one might well suppose that it is here that the friend of idiolects has a problem: If there is no such thing as a common language that both the officer and I strive to speak, why should there be any convergence at all? That, I suggest, is what is worrying Dummett when he writes:

> An English speaker both holds himself responsible to, and exploits the existence of, means of determining the application of terms which are either generally agreed among the speakers of English, or else are generally acknowledged by them as correct. . . . [T]his has to be so if words are to be used for communication between individuals. (Dummett, 1978, 425)

Now, Dummett is well aware, of course, that people do not always understand or use their words the same way. But he is suggesting that it is only because speakers of English recognize shared norms governing their usage that we avoid the chaos of Babel.

I will now argue for two claims: First, that the appeal to common languages cannot do the work Dummett seems to think it can do; and second, that the

[33] Let me register two very large debts here: One is to Chris Peacocke, who forced me to address this question directly; the other is to Kyle Stanford, who helped me understand what my view was.

[34] As it happens, I don't much care for the first option, but I am no epistemologist.

convergence of usage that is necessary if we are to be able to communicate successfully with strangers can be explained without any appeal to common languages.

To take the first point first, people who speak what anyone would have to regard as different languages frequently communicate successfully. Speakers of different dialects often communicate successfully, and between these dialects there may be significant syntactic and phonological differences, as well as semantic differences at the lexical level: If I may borrow an example from Gabriel Segal, in British English, something is rightly called a 'pie' only if it has a pastry top, whereas in American English, a 'pie' need not have a top at all. That does not imply that Bostonians can't communicate successfully with Londoners. Communication will succeed so long as the parties to a given exchange knowingly agree about what the various sentences uttered during that exchange mean, and such agreement will be common so long as the speakers' dialects do not differ too much in the areas that are typically used for communication between them. But then speakers' knowingly agreeing about what a particular sentence means cannot be explained in terms of their speaking a common language: There need be no common language they speak, for their dialects can differ arbitrarily outside the parts they typically use to communicate.

The situation is no different, in principle, when I speak to the proverbial man on the street. In order for us to communicate successfully, we do not need to speak a common language. We need only agree about what the sentences that occur in our particular communicative exchange mean. And that's a good thing, since there is, so far as I can see, no reason to suppose that I and all of my everyday conversational partners actually do speak the same language, in any reasonable sense: For all I know, the officer thinks 'livid' means *flushed* or, as may even be more common, *angry*. One could, of course, insist that one of us must be wrong, but I've already examined and dismissed that claim. One could try isolating some 'core language' and insist that successful everyday communication depends upon our all speaking *it*. But it should be obvious by now that the part of the language on which there must be agreement if communication is to be successful in a particular case will vary from episode to episode: No stable 'core' need exist, and there is no obvious reason to suppose one does.

The most that can be demanded, then, is that the officer and I speak a 'common language' as regards the sentences used in communication between us: We must—to modify Dummett's remark—'hold ourselves responsible to means of determining the application of terms we actually use, means that are either generally agreed among us or else are generally acknowledged by us as correct'. But what does that add to the claim that we must know that we mean the same thing by those sentences?

The appeal to common languages will not, therefore, help us understand why speakers' usage converges in the way it must if successful communication is to be possible. I shall now sketch an alternative explanation.

In a passage quoted earlier, Chomsky writes that, once Bert is informed that he and the doctor use the word 'arthritis' differently, Bert "may (or may not) choose to modify his usage to that of the doctor's" (Chomsky, 2000*a*, 32). That strongly suggests that what Bert means by the word is under Bert's control, and I doubt that is true.[35] When I was very young, I had a small piece of purple fuzz on my arm. Pointing at it, I asked my mother what it was, and she said, "That's a muscle". So for some time I went around using the word 'muscle' to apply to small pieces of fuzz. Later, I found out what had happened and learned that most people use the word 'muscle' to apply (as I'd now put it) to muscles, not to fuzz. That information did not, however, lead to an immediate change in what the word meant to me. Over time, of course, it did, but for quite a while, as I recall, I continued to *hear* the word 'muscle' as meaning *piece of fuzz*: I might have had reason to say to myself, "I know that Dad didn't just say that working out has given him bigger pieces of fuzz, but that's nonetheless how it sounded". I was thus the victim of a kind of semantic illusion, one driven by features of my semantic competence that were not entirely within my control. And as with visual illusions, these sorts are at least somewhat persistent: just as knowing that the lines in the Müller-Lyer illusion are the same length doesn't make them look the same length, believing that 'muscle' means *muscle* doesn't make it sound like it does.

Changing what the word 'muscle' meant to me was not something I could accomplish simply by force of will. Fortunately, however, I was not doomed to suffer from illusion forever and so be forced to make do by consciously reinterpreting other speakers. Repeated exposure to other speakers who used the word 'muscle' to mean *muscle* eventually led me to hear the word 'muscle' as meaning *muscle*, too. And, importantly for our present purposes, it is not obvious that I could have *prevented* this change simply by force of will, either. It seems to me that repeated exposure to speakers who meant *muscle* by 'muscle' would have led to such a change by itself: Whether I wanted to do so or not, I eventually would have begun to hear the word 'muscle' as meaning *muscle*.

Much the same thing can be said about my own usage of the word 'muscle'—that is, how I used it in my own speech, as opposed to how I heard it when it was used by others. Even after I had discovered that I was different from other boys, if I had wanted to inform someone about pieces of fuzz, it was only with some effort that I could stop myself from using the word 'muscle'. But, eventually, that changed. The reason is very simple: There is not even the possibility of a gap between what one means by a word when one utters it oneself and how one understands it when it is uttered by somone else;[36] the semantic and

[35] As it happens, George speaks of "features of grammars over which speakers might have *some* control" (George, 1990, 297, my emphasis), strongly suggesting that we do not have total control over what our words mean to us, but he makes nothing of the point.

[36] *Modulo* context-dependence, of course. The possibilities of the sort of reinterpretation illustrated by opposite day, or by Chomsky's talk of 'modifications' to our 'initial posits' (Chomsky,

other linguistic information upon which one relies when one speaks is the *same* information upon which one relies when one interprets the speech of others. The change in how I heard other speakers' utterances of the word 'muscle' and the change in my own usage of it—what I meant by it when I uttered it myself—were, in fact, a single change.

The mentioned features of human linguistic competence thus together encourage, and may even enforce, semantic conformity. If Bert speaks regularly enough with other people about his 'arthritis', and if it becomes sufficiently apparent to him that he does not mean by it what they do, then how he understands and uses the word will change. Of course, whether, and how quickly, Bert's usage of the term changes will depend upon many factors: How idiosyncractic his usage is, how often the word is used, by and with whom, and so forth. But surely that is as it should be: One does not want an 'explanation' of why the community of 'English' speakers is semantically homogeneous. What one wants is an account of why there is sufficient semantic homogeneity that it is possible (if not always advisable) to talk to strangers. Dummett's answer to this question is that speakers have a common goal: they strive to speak a common language, recognizing shared norms as governing their usage. Dummett's view is thus teleological in a way that mine is not. On my view, as a conseqence of how human linguistic capacities function in communication, one's own understanding of a term tends to track that of those with whom one regularly uses it in conversation. In virtue of connections between one's own regular conversational partners and other speakers, one's usage tends also to track theirs, the strength of the tendency being a function of both the length and the nature of the connection.[37] More commonly used terms can be expected to be used more uniformly within a wider population, with less commonly used terms more likely to be understood differently within conversationally isolated groups. And so forth. It's all very messy, but that, again, is surely as it should be.

6. CLOSING

The question whether idiolects or common languages are more fundamental is sometimes framed as the question whether language is ultimately *mine* or *ours*. I have argued that idiolects are necessarily involved in the explanation of successful communication and that the phenomena that make the appeal to common languages seem necessary have their source in general facts about human linguistic capacities and how they are deployed in communication. These same facts allow

2000*a*, 32), or by Davidson's discussion of malapropism (Davidson, 1986), are not counterexamples, since they do not concern how one *hears* what is said.

[37] The presence of mass-media, of course, greatly increases the number of one's regular conversational partners.

us to preserve a sense in which my language is always *our* language, despite the fact that there is no explanatory use to be made of common languages.

As noted earlier, speech is a form of rational action. This form of action is rationalized in part by one's linguistic—and, in particular, semantic—beliefs.[38] So my uttering the sentence "Prof. Parsons is teaching", for example, may be an intentional act, and my reasons for performing it may be something like the following:

(5) I want Janet to believe that Prof. Parsons is teaching.

(6) Janet believes that, in the present context, an utterance of "Prof. Parsons is teaching" means that Prof. Parsons is teaching, and she will deploy that belief in interpreting my speech.

(7) If I utter "Prof. Parsons is teaching" in the present context, Janet will take me to have spoken the literal truth.

(8) If I utter "Prof. Parsons is teaching" in the present context, Janet will (be in a position to) come to believe that Prof. Parsons is teaching. [From (6) and (7)]

(9) I shall utter "Prof. Parsons is teaching". [From (5) and (8)]

The important point about this rationalization is that my reasons for uttering "Prof. Parsons is teaching" make explicit reference not to what I think this sentence means but, rather, to what I think Janet takes it to mean. The same, of course, is true of my interpretation of her speech: If I want to learn from Janet, I need to consider what she takes her utterance to mean.

This rationalization suffers from a problem we encountered earlier: nothing distinguishes it from rationalizations we might give of utterances made on opposite day. On opposite day, I would instead utter "Prof. Parsons is not teaching", and part of my reason for doing so would be that Janet will believe that an utterance of it means that Prof. Parsons is teaching. What distinguishes normal and abnormal cases is again the source of one's judgements about what one's conversational partners will take utterances to mean. In the normal case, the source of the judgement is one's linguistic competence;[39] in abnormal cases, one's judgement relies upon other sources as well. On opposite day, for example, my judgement about what Janet will take my utterances to mean is a product of my linguistic competence together with my knowledge of our agreement to say the opposite of what we mean.

[38] Let me say again that it is possible, indeed likely, that the relevant cognitive states are not all or always beliefs. They may be perceptual or, perhaps, of some other kind. What matters, for present purposes, is that they are conscious, in the sense that they are available for the explanation of intentional action, as here.

[39] As we saw earlier, the source is not *solely* one's linguistic competence, but similar remarks can be made here as were made there.

What expressions mean to me is thus determined by what I tacitly know about the meanings of those expressions. But when this information is deployed, it is used, in the most familiar cases, to arrive at decisions about how one's conversational partners understand utterances, both one's own and theirs. And, as emphasized earlier, how I understand an utterance and how I unreflectively take my conversational partners to understand it are one and the same. From my point of view, then, my language is indeed your language: My language, as I understand it, just is your language, as I take you to understand it.[40]

REFERENCES

Burge, T. (1993). Content preservation. *Philosophical Review*, 102: 457–88.

—— (1999). Comprehension and interpretation, in L. Hahn (ed.), *The Philosophy of Donald Davidson*, 229–250. Open Court, Chicago.

Carey, S. (1985). *Conceptual Change in Childhood.* MIT Press, Cambridge MA.

Chomsky, N. (1965). *Aspects of the Theory of Syntax.* MIT Press, Cambridge MA.

—— (1986). *Knowledge of Language: Its Origin, Nature, and Use.* Praeger, Westport CT.

—— (2000*a*). Explaining language use, in Chomsky (2000*d*), 19–45.

—— (2000*b*). Language from an internalist perspective, in Chomsky (2000d), 134–63.

—— (2000*c*). Naturalism and dualism in the study of language and mind, in Chomsky (2000*d*), 75–105.

—— (2000*d*). *New Horizons in the Study of Language and Mind.* Cambridge University Press, New York.

Davidson, D. (1984). Reality without reference, in *Inquiries Into Truth and Interpretation*, 215–25. Clarendon Press, Oxford.

—— (1986). A nice derangement of epitaphs, in E. Lepore (ed.), *Truth and Interpretation: Perspectives on the Philosophy of Donald Davidson*, 433–46. Basil Blackwell, Oxford.

Dummett, M. (1978). The social character of meaning, in *Truth and other Enigmas*, 420–30. Duckworth, London.

—— (1993). Language and communication, in *The Seas of Language*, 166–87. Clarendon Press, Oxford.

Fodor, J. (1998). *Concepts: Where Cognitive Science Went Wrong.* Oxford University Press, New York.

[40] Extracts from this paper were read at New York University, in January 2003; at the University of Illinois at Chicago, in February 2003; at Brown University, in April 2003; at Yale University, in September 2003; at the University of St Andrews, in February 2004; and at the University of California at Irvine, in June 2004. The opportunity to present this material and receive the audiences' comments greatly improved the paper. Questions from Justin Broackes, Bill Hart, Robert May, Chris Peacocke, Jim Pryor, Agustin Rayo, Stephen Schiffer, Kyle Stanford, Dan Sutherland, Brian Weatherson, Crispin Wright, and a graduate student at NYU whose name I do not know were especially helpful. Thanks, too, to everyone who has attended the various courses and seminars at Harvard University during which I have discussed my evolving views on these issues, especially Bernard Nickel. Finally, conversations with Michael Rescorla and Jason Stanley did much to shape the paper, and I thank them for their ongoing support of my work.

George, A. (1990). Whose language is it anyway? Some notes on idiolects. *Philosophical Quarterly*, 40: 275–98.

Grice, H. (1989*a*). Meaning, in Grice (1989*b*), 213–23.

____ (1989*b*). *Studies in the Ways of Words*. Harvard University Press, Cambridge MA.

____ (1989*c*). Utterer's meaning and intentions, in Grice (1989*b*), 86–116.

Heck, R. (1995). The sense of communication. *Mind*, 104: 79–106.

____ (2000). Non-conceptual content and the 'space of reasons'. *Philosophical Review*, 109: 483–523.

____ (2002). Do demonstratives have senses? *Philosophers' Imprint*, 2:http://www.philosophersimprint.org/002002/.

____ (2004). What is compositionality? *Draft*.

____ (2006). Reason and language, in C. MacDonald and G. MacDonald (eds.), *McDowell and His Critics*. Blackwell, Oxford.

Higginbotham, J. (1989). Knowledge of reference, in A. George (ed.), *Reflections on Chomsky*, 153–74. Blackwell, Oxford.

Kripke, S. (1980). *Naming and Necessity*. Harvard University Press, Cambridge MA.

Lewis, D. (1985). Languages and language, in *Philosophical Papers*, vol. 1. Oxford University Press, Oxford.

____ (1986). *Convention*. Harvard University Press, Cambridge MA.

McDowell, J. (1998*a*). Anti-realism and the epistemology of understanding, in McDowell (1998*d*), 314–43.

____ (1998*b*). In defense of modesty, in McDowell (1998*d*), 87–107.

____ (1998*c*). Meaning, communication, and knowledge, in McDowell (1998*d*), 29–50.

____ (1998*d*). *Meaning, Knowledge, and Reality*. Harvard University Press, Cambridge MA.

Quine, W. (1970). Methodological reflections on current linguistic theory. *Synthese*, 21: 386–98.

Rumfitt, I. (1995). Truth-conditions and communication. *Mind*, 104: 827–62.

Stanley, J. (2000). Context and logical form. *Linguistics and Philosophy*, 23: 391–434.

Strawson, P. (1971). Meaning and truth, in *Logico-Linguistic Papers*, 170–89. Methuen, London.

Wellman, H. M. (1990). *The Child's Theory of Mind*. MIT Press, Cambridge MA.

Wiggins, D. (1997). Languages as social objects. *Philosophy*, 72: 499–524.

5

There Are Many Things

Vann McGee

A. THE MATHEMATICAL PART

Ontology is the most general science. Whereas the specialized sciences restrict their concern to one or another kind of being, ontology is fully comprehensive, taking all things of whatever kind into its domain. One thing ontologists sometimes try to do is to provide inventories that list, although not in specific detail, all the kinds of things there are. Here I want to attempt something a little less ambitious. I don't want to describe all the things there are, but I want to propose that, whatever kinds of things there are, there are a great many of them, so that there is a general principle to the effect that, if it is possible that there are at least so-and-so many things, then there are, in fact, at least so-and-so many things.

The proposed principle is motivated primarily by considerations from metamathematics. It is not the job of mathematics to try to tell us what the world is like; that task belongs to the various empirical sciences. The job of mathematics to provide models of all the possibilities for what the world is like, leaving it to the special sciences to determine which of the models best fit the actual world. The ontological thesis I want to advance is that mathematics has the resources to do the job, so that there is a mathematical model that depicts each possibility. If this principle is correct, it has significant consequences for modal metaphysics. It tells us that we don't need to postulate other possible worlds, since we can describe each of the ways the world might have been by supplying an appropriate mathematical model formed from things that exist in the actual world. This form of possible-world ersatzism[1] has advantages over modal realism, and not just in terms of metaphysical efficiency. We can provide mathematical models not only of all the ways the world might have been, but of all the ways we can conceive the world to be, even if these conceptions turn out not to be genuine possibilities. This liberalization, I shall claim, enables mathematical ersatzism to do even better what modal realism does best, which is to articulate the truth conditions for conditionals.

[1] "Ersatzism" is David Lewis's [1986] name for doctrines that attempt to enlist things that aren't possible worlds to play the roles of possible worlds.

Ontology is an audacious enterprise. It takes considerable temerity for us—weak, shortsighted creatures living in an obscure corner of an unimpressive galaxy—to even attempt to formulate general hypotheses that comprehend everything there is. Once we've formulated ontological theses, we have no real likelihood of being able to prove them. Ontology is first science. There is no more basic science whose principles we can rely on as a basis for proving ontological theses. I myself have certainly not been able to pull ahead of the pack here. I shall commend a certain ontological thesis by pointing to likable features of the thesis that I hope you will find attractive, but if you don't find the thesis attractive, I won't say anything that will make you feel compelled to accept it.

The motivation for the thesis comes from the philosophy of mathematics, in particular, from consideration of the status of Euclidean geometry. For Euclidean geometry as the ancient Greeks practiced it, Aristotle's way of thinking, according to which the geometer studies ordinary bodies regarded from an abstract point of view that ignores features like color, weight, and hardness and only pays attention to size, shape, and position, had a great deal of appeal. In particular, since whenever we kick a soccer ball we have causal contact with a sphere, the view deftly evades the epistemic difficulties that led Plato to the desperate doctrine that mathematical learning is a matter of recovering prenatal memories.

Nineteenth-century geometry grew substantially beyond ancient Greek geometry, and the Aristotelian epistemology no longer fit it. The plan of making Euclidean geometry even more secure than it had been for the ancients by deriving its one dubious axiom, the axiom of parallels, from the other axioms was implemented in a program of assuming the denial of the troublesome axiom and trying to derive a contradiction. No contradiction was forthcoming. Instead, the investigations led to a new geometry, which, while strange, was demonstrably consistent if Euclid's geometry is consistent. Indeed, several alternative geometries were devised, and the question, "Which of these geometries is the correct one?" came to be seen as an empirical question, to be answered by careful measurement. The culmination of this line of inquiry came with Einstein's general theory of relativity, which pointed the way to an exquisitely delicate astral measurement that showed that the axioms of Euclid's *Elements* do not, in fact, correctly describe the world around us.

This development left the geometer with an urgent need to update her job description. Previously, the geometer's task was to describe the structure of space, but now that's the cosmologist's job. The mathematician has neither the inclination nor the ability to describe the spatial structure of the world around us. What she can do is to provide the cosmologist with the full array of possibilities, but it is the cosmologist's job to determine which of the many possible structures of space is physically actual. For the geometer to do her job properly, she must make sure the menu presented to the cosmologist doesn't leave out any possibilities. At the end of the day, the cosmologist needs to be able to assure us that his cosmic geometry mirrors nature, rather than merely reflecting some artificial constraints imposed by the mathematician.

Even after getting the news from the astronomer that physical space is non-Euclidean, the geometer still does Euclidean geometry, although the more versatile methods of analytic geometry have supplanted derivations from Euclid's axioms. The geometer still does Euclidean geometry, and she does it with high seriousness. There's no indication that she's engaged in an elaborate fantasy, like trying to figure out what American history would have been like if the South had won the Civil War or what human life would be like if the oxygen-bearing element in our blood had been copper rather than iron. She still does Euclidean geometry, only now she sees the object of her study is something mathematical, rather than something physical. But what is her study about, exactly? We hear talk about "Euclidean 3-space," regarded as the collection of ordered triples of real numbers, but this isn't much help, since our naive understanding of "real numbers" treats a real number as a ratio of lengths of line segments, with "length" understood according to the dictates of Euclidean geometry. So we're trying to build geometry on top of the theory of real numbers, having first built the theory of real numbers on top of geometry. Not good architecture.

The escape from the brier patch, whose discovery was principally due to Richard Dedekind [1872], proceeds in two stages. The first is to characterize the real numbers in purely algebraic terms. By "characterize," I mean to give a system of axioms that is true of the real numbers and isn't true of any other system that isn't structurally indistinguishable from the real numbers, a "categorical" system of axioms, as they say. The hard part, thinking of the problem in geometric terms, is to make sure that the axioms require every possible ratio of lengths to be represented by a number. For example, the rational numbers, that is, the quotients of integers, have most of the algebraic properties we'd expect from the real numbers, but there isn't any rational number corresponding to the ratio of the diagonal of a square to its sides. Moreover, the standard algebraist's tricks for filling in gaps still leave holes. What gets categoricity is to add to the familiar algebraic axioms the *least upper bound principle*, which says that every nonempty collection of real numbers that is bounded above has a least upper bound. This gives us an axiom system that has, up to isomorphism, at most one model. But how do we know that there's even one model? Back when we had full confidence in Euclid's account of shape and distance, we could get our model by fixing a standard unit of length and identifying a positive real number as the length of a line segment. Without the geometric basis, one has to be more cunning. Dedekind's ingenious solution was to start with the integers, then to get the structure of the rational numbers by treating a positive rational as an ordered pair of relatively prime integers, and finally to identify the real numbers with properly selected sets of rationals.

This phrase, "identify the real numbers with selected sets of rationals," should give us pause. Dedekind's construction with sets of rationals gives us a model of the axioms, a *complete ordered field.* We take this model of the axioms and treat real-number talk as referring to it. But once we have the Dedekind construction, we can see that there are plenty of other complete ordered fields, and taking "real number system" to refer to any of the others would have worked as well. For ordinary mathematical purposes, there is neither a basis nor a need for choosing a preferred complete ordered field, since the categoricity theorem assures us that the same sentences are true in all of them. Our usage is able to provide real-number statements with determinate truth values—a sentence is true if and only if it is true in at least one complete ordered field, which happens if and only if it is true is all of them—without providing real-number terms with determinate referents. If, for some special purpose, say a textbook writer wanting to achieve a uniform notation, a uniquely specified reference function is desired, the author is free to choose arbitrarily.

The contrast with classical geometry is striking. If one person refers to as "points" what another person refers to as "pentagons," one wouldn't say that the difference between them was merely that they had made different arbitrary choices among equally good options. The shape of an object is discovered, not stipulated. What gives the pentagonal tabletop its shape is the way the log was sawn, not the relation of its shape to other geometric entities.

On the modern conception, there isn't anything that is a Euclidean point or pentagon absolutely. A thing is a point or a pentagon relative to a structure, in virtue of its position in the structure. The way model theorists think of it, a structure has two parts, a domain of individuals and an ensemble of functions and relations on that domain. Any collection is suitable to serve as the domain, as long as it has the right number of elements, and once we select a domain, there will be many different structures we can impose upon the domain so as to make the axioms true.

The conception of geometry that sees geometric objects as characterized by structural, rather than spatial, characteristics has methodological, as well as ontological, implications. One disturbing discovery made during careful nineteenth-century investigations of the foundations of geometry was that, for many of Euclid's arguments, the conclusion doesn't follow from the premisses. Imported from Euclid's diagrams are various tacit assumptions about the meeting places of lines that aren't justified either by the hypotheses of the theorem or by the axioms. This observation is disturbing, because for two thousand years Euclid's proofs had been universally upheld as the exemplar of rigorous deductive reasoning, yet his proofs contained holes that generations of meticulous scholars had managed to overlook. One possible response is that, whereas the fact that some of the assumptions were left unspoken was undoubtedly regrettable, little harm was done, since once we make the tacit assumptions explicit, we can see that they're obviously true, and so the theorems are true. With the new, more abstract geometry, this response makes no sense. A geometric sentence counts as true if it is true in all the intended models, where the axioms determine which are the intended models. So it doesn't make sense to talk about a sentence being true even though it doesn't follow from the axioms. For a proof to have cognitive value, the conclusion has to strictly follow from the premisses, without any assistance from spatial intuitions. As Hilbert puts it,[2] deductions should still be legitimate if we replace "point," "line," and "plane" by "table," "chair," and "beer mug."

What I would now like to do is to take what we've seen by looking at non-Euclidean geometry and extrapolate in two directions, one tame and the other wild. The first is that the highly abstract approach to geometry, according to which what make a structure appropriate as a model of Euclidean geometry are not the intrinsic spatial attributes of the things that make up the model, but rather structural characteristics that are manifested equally well by any model of the same isomorphism type, works equally well for the other branches of mathematics. The reason this counts as a modest extrapolation is that when we look at the way the other mathematical specialties are practiced, we see that geometry isn't any less concrete than number theory or set theory or algebraic topology. The slogan, "The referents of mathematical terms are determined at most up to isomorphism," seems to hold across the board.

The bold extrapolation starts with the observation that it is the task of modern geometry to provide models of every spatial possibility and extend it to reach the conclusion that it is the duty of modern mathematics to provide models of every possibility of every sort. Set theorists need to depict all possible ways of forming collections, without regard for what things are actually available to be collected. Group theorists need to investigate every possible symmetry, not just the symmetries that are realized in nature. Each of the subdisciplines has its duties, and in addition, it is the overarching job of mathematics to provide models of every possibility, across the board.

This extrapolation is a stretch to be sure, but it's going with the flow of history. One of the most striking characteristics of modern science, ever since Galileo, has been the continually widening application of mathematical methods. Used to be, mathematics was only useful for land surveys, and now it's the case that whatever one wants to study, one studies it by producing an appropriate array of mathematical models. The point of view I would like to advocate has two components. First, it is the job of mathematics, at least if we take at face value what mathematicians appear to be doing, to provide models of every possibility, not just every geometric possibility but every possibility of every sort. Second, this job is something the mathematician can actually accomplish. Mathematics is not like astrology. It is the job of the astrologer to explain our fortunes and character by natal astral influences. The astrologer cannot successfully carry out his assigned task, because natal planetary positions have no effect on our fortunes and character. The mathematician, I want to claim, is better off than the astrologer, in that

[2] See Reid [1970], 57.

she can really do what her job requires her to do. Embracing this view is, to a large extent, just an expression of an anti-skeptical bias on my part. Unless there are strong forces pushing in the opposite direction, we should incline, it seems to me, to accept at face value the theoretical commitments of stable, successful theories. Some policy along these lines is surely needed to legitimate the acceptance of theoretical entities. By definition, theoretical entities are things with which we cannot have any direct connection, so what could justify our belief in them other than the success of the theories in which they appear? Such an anti-skeptical attitude will incline us to write classical applied mathematics an ontological blank check, for classical applied mathematics is the most stable and successful theory of them all.

That mathematics succeeds in providing a structure that realizes every possibility is a bold hypothesis, and I mean it in the most audacious way. The various alternative theories of the structure of physical space are, as far as I am aware, all metaphysically possible, but this fact made no difference to the way non-Euclidean geometry developed. What mattered was the fact that the alternatives were all epistemically possible. Determining which of the alternatives was actual was something that could only be achieved by detailed empirical investigation. The situation is different with chemistry. Standard metaphysics[3] tells us that it is essential to gold that it have the atomic number 79, which implies that the atomic theory of matter is, at least in its fundamentals, metaphysically necessary (so long as there is such stuff as gold), and anyone who attempted to develop a non-atomic theory of matter would be spelling out a scenario that was literally impossible. That's the way things look today, but we can easily imagine circumstances arising, someday in the future, that force us to reexamine the foundations of twenty-first century chemistry. Should that day arise, the chemist will ask the mathematician to provide mathematical models of the various alternative chemistries, and it won't be the mathematician's task to filter out the theories that are metaphysically impossible. Even if the eventual outcome is to reconfirm the atomic theory, that will be a judgment the chemist makes on the basis of chemical evidence, not something the mathematician decides. The mathematician needs to provide models for all coherent chemical theories, even ones that turn out, in the end, to be metaphysically impossible.

The emboldened hypothesis is that mathematics provides a model of every *logical* possibility, so that, if ϕ is consistent, then there is a mathematical model in which ϕ is true.

How plausible we regard the hypothesis depends on our notion of logical possibility. If we take logical possibility to be consistency by the first-order predicate calculus, we get the hypothesis easily, since Gödel's [1930] completeness theorem assures us that every logical possibility is realized in some model whose domain consists of natural numbers. In talking about categorical axiomatizations

[3] This doctrine is due principally to Saul Kripke [1972].

of geometry, I was presuming a logic richer than the first-order predicate calculus, since no first-order axiom system that has an infinite model is categorical. The categorical axiomatization of geometry (which was first given by Oswald Veblen [1904]) employs the least-upper-bound principle, whose expression requires second-order logic. There are numerous other logics that one might investigate, but here I'd like to focus our attention on the second-order predicate calculus. So we'll look at languages whose non-logical symbols include names and predicates, and whose logical symbols include, in addition to the sentential connectives and the identity sign, both first-order variables, which occupy the same grammatical positions as names, and second-order variables, which occupy the role erstwhile played by predicates. Both kinds of variables can be bound by the quantifiers "∀" ("for all") and "∃" ("for some"), so that the least-upper-bound principle takes the form:

$$(\forall X)(((\exists y)Xy \wedge (\exists z)(\forall y)(Xy \supset y \leq z)) \supset$$
$$(\exists z)(\forall w)((\forall y)(Xy \supset y \leq w) \equiv z \leq w)).$$

Actually, there are two very different ways of thinking about the second-order predicate calculus. The traditional conception has the lower-case variables ranging over individuals and the upper-case variables ranging over collections. This is perfectly satisfactory as long as the things we want to talk about—the domain of the first-order variables—form a well-defined totality that doesn't include collections. When we are doing ontology, however, we intend to talk about everything. If collections aren't within the domain of the first-order variables, then our ontological theory has failed to take account of everything there is, because it has left out collections. If collections are in the domain of the first-order variables, however, we won't be able to utilize second-order logic without falling foul of Russell's paradox.

A more versatile interpretation of second-order logic was developed by George Boolos ([1984] and [1985]), who got the idea by careful examination of the role of plural nouns and pronouns in English. Second-order variables play more or less the role of English plural personal pronouns ("they" and "them"), the same way that first-order variables take after singular personal pronouns ("he," "she," "him," "her," and "it"). The key idea is to extend the notion of reference so that it encompasses common nouns and adjectives as well as proper names. The usual story has it that, in the sentence "Theaetetus is brave," "Theaetetus" refers to the man Theaetetus and "brave," if it can be said to refer at all, refers to a collection or property or universal. Boolos proposes instead that "brave" refers to the creatures who are brave, so that both names and adjectives refer to individuals, but a name is subject to the constraint that it has to denote one and only one individual. Both the "is" of identity and the "is" of predication serve to connect denoting terms with denoting terms; they differ on whether the complement denotes one thing or many. The roles of first-order variables and of monadic second-order variables mirror, respectively, the roles of names and of adjectives. In the Boolos treatment, a variable assignment associates individuals with variables of both types. The assignment of individuals to first-order variables is required to obey the condition that a variable assignment must pair one and only one individual with each first-order variable, whereas the association of individuals with monadic second-order variables is not similarly constrained. The semantic theory (which was fully worked out by Agustín Rayo and Gabriel Uzquiano [1999]) gives a compositional characterization of the conditions under which an open sentence is satisfied by a variable assignment, then goes on to declare that a (closed) sentence is true if and only if it's satisfied by at least one variable assignment, which happens just in case it is satisfied by every variable assignment. In pluralist terms, the least-upper-bound principle reads like this: There aren't any numbers such that, even though there is a number greater than or equal to all of them, there isn't a least number greater than or equal to all of them.

Boolos's theory of so-called *plural quantification* gives us the version of second-order logic in which quantified variables are allowed to take the places of one-place predicates. Provided we have ordered pairs, we can bootstrap our way up to full second-order logic.

The bold ontological hypothesis that every logical possibility is realized in some model was put forward by Alfred Tarski in a 1936 paper that, in spite of its brevity, has proven enormously valuable in illuminating the connected notions of consequence and model. The reason the paper is so short is that it is built on top of a very long earlier paper that shows how to give a compositional theory of truth and satisfaction. In the 1936 paper, Tarski defines a *model* of a theory to be a variable assignment that satisfies the formulas you get from the theory by uniformly replacing all the nonlogical terms that occur within the theory by free variables of appropriate type. A theory is *consistent*, says Tarski, if it has a model; a sentence is a *logical consequence* of a theory if every model of the theory is a model of the sentence, and a sentence is *logically true* or *valid* if it is modeled by every variable assignment.

Focusing our attention on the single-sentence case (which is a lot easier than dealing with theories with infinitely many axioms), for a given sentence ϕ, let $\phi^{\exists}$ be the sentence obtained from ϕ by replacing all the nonlogical terms by variables, then prefixing existential quantifiers to bind the newly introduced variables. If we think of all the nonlogical terms as "theoretical," $\phi^{\exists}$ is the Ramsey sentence for ϕ.[4] Let $\phi^{\forall}$ be the sentence we obtain by prefixing universal quantifiers instead. ϕ is valid, on Tarski's proposal, if and only if $\phi^{\forall}$ is true, and ϕ is consistent if and only if $\phi^{\exists}$ is true. Thus the standard formalization of "Theaetetus is brave or Theaetetus is not brave" counts as valid, on Tarski's criterion, because the sentence "$(\forall Y)(\forall x)(Yx \vee \sim Yx)$" is true. (The two-stage process, first replacing extralogical terms with variables and then assigning values to the variables, served

[4] See Ramsey [1929].

Tarski's purpose of unifying the theory of truth and the theory of consequence, but for other purposes it is more complicated than need be. We shall sometime use the word "model" to mean a function directly assigning values of appropriate type to the extralogical terms, eliminating the intermediate variables.)

Tarski emphasized that it was not his intention to introduce novel notions or to employ the words "valid" and "consistent" in an eccentric way. He intended to explicate the notions of validity and consistency that were already current. But his criterion, as it is written, doesn't succeed in this aim, for we would ordinarily regard both the axioms of a complete ordered field (which characterize the real number system) and the second-order Peano axioms (which describe the natural numbers) as consistent. Yet they cannot both be consistent by Tarski's criterion, for there is no way we can get from a model of the second-order Peano axioms to a complete ordered field by changing the values of the extralogical terms, because there are more real numbers than there are natural numbers, and because reassigning values to the extralogical terms won't change what individuals are in the domain of the model; it will only change how they are described.

A simple emendation avoids this sort of unwelcome example. Allow the universe of discourse to vary from one model to another. Whether this change seriously disrupts what Tarski was saying is a controversial question. One can argue (although not to universal assent)[5] that the amended criterion was what Tarski really had in mind, and that he misstated it either through inadvertence or (as Wilfred Hodges ([1986], 136) has suggested) as an expository simplification.

Given a theory and a predicate "D" we *relativize* the theory to "D" by restricting the first-order quantifiers, replacing "$(\exists v)$" by "$(\exists v)(Dv \wedge$," and replacing "$(\forall v)$" by "$(\forall v)(Dv \supset$," getting from a sentence ϕ a relativized sentence we may write "ϕ^D," and by adding an axiom saying that D is nonempty and an axiom for each name saying that the object named is a member of D. The corrected criterion tells us that a sentence ψ is a consequence of a theory just in case ψ^D counted as a consequence of the relativized theory according to the uncorrected criterion. ϕ is valid, according to the corrected criterion, if and only if $(\forall D)((\exists x)Dx \supset \phi^{\forall D})$ is true, and ϕ is consistent if and only $(\exists D)((\exists x)Dx \wedge \phi^{\exists D})$ is true. The new criterion counts the formalization of "Theaetetus is brave or Theaetetus is not brave" as valid because "$(\forall D)((\exists x)Dx \supset (\forall Y)(\forall x)(Dx \supset (Yx \vee \sim Yx)))$" is true. We include the extra antecedent "$(\exists x)Dx$" in the characterization of logical validity and the extra conjunct in the characterization of consistency to reflect the fact that standard model theory excludes models with the empty domain. The consequences of this more or less arbitrary decision don't run very deep; see Quine [1954].

The corrected characterization of logical validity isn't vulnerable to any off-the-top-of-your-head counterexamples, but is it correct? The left-to-right direction—if ϕ is valid, it is true in every model—is evident enough. If you can falsify

[5] See, e.g., Timothy Bays [2001].

ϕ by picking an appropriate domain and assigning appropriate values from that domain to the extralogical terms, ϕ isn't valid. It's the other direction that's dubious. How do we know that there are enough models to ensure that, if ϕ isn't valid, then there's a model in which ϕ is false?

Because the set of logically valid sentences is closed under the rules Universal Generalization[6] and Universal Specification,[7] the sentence ϕ will be valid if and only if $\phi^{\forall}$ is valid. What's hard is to find some assurance that $\phi^{\forall}$ valid if $(\forall D)((\exists x)Dx \supset \phi^{\forall D})$ is true. Thus it will concentrate our efforts where they are really needed if, in asking about the correctness of the revised Tarski criterion, we restrict our attention to sentences with no nonlogical terms. How do we know that there are enough models to ensure that, if ϕ is a sentence containing no extralogical terms that isn't logically valid, then there is a model in which ϕ is false?

The question, "How do we know there are enough models?" readily reduces to the question, "How do we know there are enough things?" If there are enough individuals, there are sure to be enough models, because the only feature of a model that a sentence with no extralogical terms can depict is its cardinality. If a sentence contains no extralogical terms, it will have the same truth value in any two models of the same size, and, conversely, if a sentence has the same truth value in any two models of the same size, then it is true in exactly the same models as some sentence with no extralogical terms. Specifically, given a sentence ϕ whose truth value in a model only depends on the size of the model, we find a sentence with no extralogical terms true in the same models as ϕ by replacing all the nonlogical terms by variables, then prefixing existential quantifiers, universal quantifiers, or a mixture. This observation is fairly resilient, in that it continues to hold even when we add new logical operators, like the quantifier "there are infinitely many" or "there are at most $\aleph_7$," to the language. This follows from Tarski's [1986] and Mautner's [1946] characterization of the logical operators as those invariant under arbitrary one–one correspondences.

A sentence is true in all models that are sufficiently large (so that, if it's true in a model $\mathfrak{A}$, it's true in all models the same cardinality as $\mathfrak{A}$ or larger) if and only if it's true in the same models as some sentence of the form $(\exists D)((\exists x)Dx \wedge \phi^{D})$, with no extralogical terms. "Same cardinality" can be cashed out in second-order logic on the basis of Hume's ([1739], Book I, part iii, section 1) principle ("When two numbers are so combin'd, as that the one has always an unite answering to each unite of the other, we pronounce them equal"), without requiring that the domain have a cardinal number; this is important since, at least on Cantor's treatment, things have a cardinal number only if there are few enough of them that they form a set. Similarly, a sentence is true in all models

[6] If you have derived $\phi(c)$ from a set of premisses, and if the individual constant c doesn't appear in $\phi(x)$ or in any of the premisses, you may derive $(\forall x)\phi(x)$ from the same set of premisses. There is an analogous second-order rule.

[7] If you have derived $(\forall x)\phi(x)$ from a set of premisses, you may derive $\phi(c)$ from the same set of premisses. There is an analogous second-order rule.

that are sufficiently small if and only if it is true in the same models as some sentence of the form $(\forall D)((\exists x)Dx \supset \phi^D)$, with no extralogical terms. The crux of Tarski's definition of logical validity is that the universe as a whole isn't so small that it makes some sentence of the form $(\forall D)((\exists x)Dx \supset \phi^D)$, with no extralogical terms, true even though it might have been false. *Tarski's thesis* is the statement that, for each sentence ϕ with no extralogical terms, we have

$(\forall D)((\exists x)Dx \supset \phi^D)$is true if and only if_is logically valid.

In order for Tarski's thesis to say anything substantive, there has to be something on the other side for it to push against. The familiar alternative characterization identifies logical validity as a species of analyticity. A sentence is analytic if its truth in assured by the meanings of its words, whereas a sentence is logically valid if its truth is ensured by the meanings of the *logical* words. This slogan has a lot of intuitive appeal, but in practice, in the absence of a general theory of meaning fixation, the problem of partitioning truths into matters of meaning and matters of fact has proven utterly intractable. Of course, we have the platitude that the meaning of a word is determined by its use, but, in general, the uses of a word are so variegated that they resist useful description.

For logical words the situation is perhaps better, for there is at least a plausible conjecture: The meanings of the logical words are given by their inferential role, as described by the rules of inference. I am inclined to endorse this standard story but not its standard ending, which has it that the truth of a valid sentence is ensured by the rules in such a way that a sentence is logically valid if and only if it is derivable by the rules from the empty set of premisses. Every truth about the real numbers is a logical consequence of the axioms of a complete ordered field, but not every such truth is derivable from those axioms. If ϕ is a true, underivable sentence, then the conditional whose antecedent is the conjunction of the axioms and whose consequent is ϕ will be a valid sentence not derivable from the empty set.

Where ϕ is one of our underivable truths of analysis, we can make the axioms and everything derivable from the axioms true and ϕ false by constructing a so-called *Henkin* [1950] *model* in which the second-order quantifiers are interpreted abnormally, so that only certain combinations of individuals are allowed into the range of the second-order variables. Within the Henkin model, the rules of inference are truth-preserving, but what is required for the rules of inference to take on the inferential role of *laws of logic* is not merely that they be truth-preserving, but that the fact that the rules are truth-preserving be ensured by the meanings of the logical words. Within the Henkin model, all sentences of the form $((\forall X)\phi(X) \supset \phi(R))$ are true, but their truth isn't ensured by the meanings of the logical words, since we could make some of the sentences false by changing the meanings of the extralogical terms.

Employment of the rules of inference determines the meanings of the logical words, which is what makes the logically valid sentences true. "Employment of the rules" here doesn't just mean that you are disposed to accept the conclusion when you accept the premisses. It means that your willingness to embrace the conclusion once you've accepted the premisses doesn't depend on anything beyond the meanings of the logical words. To put the matter robotically, your logic module is able to make the inference on its own authority, without any inputs concerning whatever it is the nonlogical terms refer to, and without relying on the presence in your belief box of any theses other than the premisses of the argument.[8]

If we accept the dictum that logically valid sentences are true in virtue of the meanings of the logical constants, and we explicate the dictum by agreeing that the meanings of the logical constants are given by the rules of inference, we can put Tarski's thesis this way: For ϕ a sentence with no extralogical terms, unless ϕ is rendered impossible by the status of the rules of inference as laws of logic, $(\exists D)((\exists x)Dx \wedge \phi^D)$ is true.

Sadly, in cashing out the cognitive content of Tarski's thesis, reformulating it as the doctrine that a valid sentence is one whose truth is guaranteed by the autonomous applicability of the rules of inference is less useful than one might have hoped. In general, the truth of a sentence is a product of two factors, the state of the world and the meanings of the words. For a true sentence with no extralogical terms, we can be more specific. The truth of the sentence is the effect of two causes, the cardinality of the universe and the meanings of the logical words as given by the rules of inference. What one can't do, in our present state of knowledge, is to separate the two components, factoring the explanation of the sentence's truth into a cardinality part and a meaning part, and this is what we would need to do in order to identify the sentences that are true solely in virtue of the meaning component, those for which the rules ensure that they would have been true no matter what the cardinality of the universe had been.

An alternative characterization of logical validity that holds out some promise of being helpful has it that logical validity is the most general notion of necessity. As a doctrine, this way of putting things is hopeless, since it presupposes a theoretical understanding of which notions count as notions of necessity. But the characterization has potential utility as a program guide: whenever you have a concept you can recognize as a notion of necessity, expect to be able to produce a mathematical model of each statement your concept acknowledges as possible.

B. THE METAPHYSICAL PART

I think of Tarski's thesis as a dictum in the philosophy of mathematics, to the effect that mathematics is able honestly to fulfill its traditional role of providing models for every possibility. I think of it as a metamathematical thesis, and I have

[8] See McGee [2000] and McGee [2006].

made it nearly inevitable that you would think of it the same way, simply by calling it "Tarski's Thesis," since it would be silly to use that name to refer to, say, a doctrine in modal metaphysics. Nonetheless, we can think of the thesis as such a doctrine. It's a version of possible-worlds realism, using models in place of possible worlds and talking about logical possibility in place of the expected metaphysical possibility.

We exploit the analogy between logical possibility and metaphysical possibility, but formal similarities should not blind us to deep differences. Logical possibility, on Tarski's conception, is a linguistic concern. To determine whether a sentence is logically possible, we don't need to speculate about how history might have unfolded differently. We take the world as it is, and we ask whether we can make the sentence true by temporarily changing the meanings of the nonlogical words. For metaphysical possibility, we leave the meanings of the words fixed and we ask whether, given what the words mean, the world might have evolved in such a way as to make the sentence true. But in spite of these drastic differences, there is a systematic connection between the two notions. If Tarski's thesis is correct, every alternative possibility for how the world might have been can be simulated by an appropriate actual-world change in the meanings of our words, so that any sentence that could be true can be made true in the actual world by changing the meanings of the nonlogical words.

Metaphysical possibility is a worldly affair, not a linguistic one. Whether a thesis is metaphysically possible doesn't depend on human language and thought (unless the thesis happens to be about language or thought). This feature is fundamental to standard treatments of the topic, which regard metaphysical possibility and necessity as attributes of *propositions,* which are independent of language and thought. The standard theory of propositions, which originates in Frege's "Über Sinn und Bedeutung," has it that the same proposition can typically be expressed in many different languages, and within a language, it can typically be expressed by many different sentences, which can be quite dissimilar grammatically. Personally, I have misgivings about whether one thing can do all the things a proposition is alleged to do, but let me set those aside and attend instead to the contrast between metaphysical and logical possibility. Logical possibility is not an attribute of propositions, at least not if we think of propositions the way Frege thought of them, where the same proposition can be expressed in many different ways, or we think of propositions the way Stalnaker ([1984] and [2003]) thinks of them, as sets of possible worlds. Logical possibility is an attribute of sentences; or if you want to say that it's a property of propositions, you'll require a conception of propositions on which they have a fixed structure that closely reflects the grammar of the sentences that express them.

The difference displays itself dramatically when we contrast Hilbert's and Frege's attitudes toward relative consistency proofs in geometry.[9] Eugenio

[9] Their correspondence appears in Frege [1980], 31–52.

Beltrami [1868] showed that, if you systematically replace the primitive terms of axiomatic geometry by other geometric terms chosen in just the right way, the reinterpreted axioms of non-Euclidean geometry become theorems of Euclidean geometry. As Hilbert understood it, Beltrami's construction showed that the program of proving the axiom of parallels by deriving an absurdity from the other axioms together with its denial had reached an impassable dead end, by demonstrating that the negation of the axiom of parallels is, in fact, consistent with the other axioms (assuming the other axioms are consistent). Frege disagreed. When Beltrami reinterpreted the axioms, he changed their meaning. Beltrami's construction shows that the new "axioms" are true (although they don't have the conceptual status of axioms), but their truth doesn't show us anything about the status of the original axioms. Euclid and Beltrami used the same sentences, but they used them with different meanings. To draw conclusions about Euclid's geometry on the basis of Beltrami's proofs is to commit a fallacy of equivocation.

The crucial difference between them, if I understand them correctly, is that Hilbert understood logical implication to be a relation between sentences. One sentence implies another if the truth of the one ensures the truth of the other solely on the basis of their logical form. The other axioms can be seen not to imply the axiom of parallels because Beltrami's reinterpretation gives us a collection of sentences of the same logical form within which the counterparts of the other axioms are all true and the counterpart of the parallels postulate is false.[10] Frege, by contrast, thought of implication as a relation between propositions, and it makes no sense to talk about the logical form of a Fregean proposition, since a single proposition can be expressed by many different sentences, and the sentences that express it can assume a wide variety of logical forms. Beltrami's so-called axioms were entirely different propositions from Euclid's axioms and their homonymy isn't a logical connection, rather a superficial accident of form, scarcely better than a pun.

Contemporary mathematicians are of one voice in agreeing with Hilbert that Beltrami's construction demonstrates the consistency of non-Euclidean geometry. But saying that Hilbert was right doesn't necessarily mean that Frege was wrong, for they were using different notions of consistency. In one crucial respect, Frege's notion of consistency was closer to the contemporary notion of metaphysical possibility than it was to our current notion of logical consistency. He regarded consistency as a property of propositions, which he conceived as independent of the accidents of linguistic convention and of human psychology.

[10] This way of putting things is overly simple, because it makes no allowance for arguments like the following:

> This sentence contains fewer than ten words.
>
> Therefore, either this sentence contains fewer than ten words or grass is white.

Such arguments don't figure prominently in geometry textbooks, but a general theory of valid inference would need to take account of them.

Metaphysical and logical possibility are in different lines of work. One is an attribute of propositions, the other of sentences. There is, however, a derivative notion of metaphysical possibility that we can apply to sentences. A sentence is metaphysically necessary, contingently true, contingently false, or impossible in a language in a context according as the proposition it expresses in that language in that context is necessary, contingently true, or whatever. If we use the notion of metaphysical possibility in this derivative sense, we can compare the two notions, and we see that metaphysical possibility is more restrictive than logical possibility. Every sentence that is metaphysically possible is logically possible, but not conversely.

We only know how to provide a Tarskian theory of logical possibility for languages for which we have a well-developed semantics, which means, at present, a suitable range of formal languages and those very limited fragments of English whose semantic properties can be satisfactorily represented within the formal languages. Even with this severe restriction, the connection between logical and metaphysical possibility is worth examining, I think, because of the hope that understanding how intensional and extensional notions interact within these little scale-model languages will give us useful guidance when we turn to the projects we really care about, developing semantics of natural languages and developing semantics of the language of thought.

For the modal sentential calculus, extending possible-world semantics to encompass logical as well as metaphysical necessity is really easy. We take a logically possible world to be a model, and we take the metaphysically possible worlds to be those models in which all the metaphysically necessary statements are true. To implement this plan in a substantive theory, we'd have to determine which statements are metaphysically necessary, and that will require hard work. But we surely wouldn't expect a semantic theory to tell us which features of the world are essential and which are accidents. Metaphysical necessity is a worldly matter, and semantics is a linguistic theory. Moreover, even after we have a substantial modal metaphysics, it won't be entirely obvious which models to count as metaphysically possible worlds. The models assign values to the nonlogical terms, but before we can sensibly ask what sentences are true in a model, we have also to know what the logical terms are, and what logical operators to allow into the language—in particular, whether to permit infinitary logical operations—is up for grabs. This is not a question that need to detain us here. What is important here is to recognize that the models we are referring to as "metaphysically possible worlds" are not ways the world might have been but only mathematical representations of ways the world might have been. It is not the function of logically possible but metaphysically impossible statements to describe ways the world might have been that are too wild for the metaphysician's delicate sensibilities. Instead, they tell impossible tall tales whose impossibility can't be shown without looking beyond the behavior of the logical operators.

We use modal locutions in different ways for different purposes, and it's worth the effort to try to unify them, so as better to understand connections among the different modal notions, and so as to exploit formal similarities. If, in attempting to develop possible-world semantics for these alternative modal notions, the "possible worlds" we have to work with are ways the world might be, then our semantics will be obtained by arranging and rearranging metaphysically possible worlds. Inasmuch as, as Kripke puts it ([1972], 304), metaphysical necessity is necessity "in the strictest possible sense," there aren't any possible worlds that aren't metaphysically possible worlds. For characterizing some modal notions, such as nomic necessity, metaphysically possible worlds are all we need. Because nomic necessity entails metaphysical necessity, we can simply take a nomically possible world to be a metaphysically possible world in which the laws of nature are all upheld. On the other hand, metaphysically possible worlds are useless for provability logic,[11] since the same arithmetical statements are true in all possible worlds. To study the logic of provability in Peano Arithmetic, we need to take our "possible worlds" to be models of Peano Arithmetic.[12]

A more interesting, and more important, case is epistemic logic. If we try to obtain an epistemic interpretation of modal logic by taking "$\Box\phi$" to mean "It is known that ϕ" or "So-and-so knows that ϕ," then the endeavor to exploit formal similarities between epistemic necessity and metaphysical necessity falters at its first step. Since even the most basic laws of modal logic fail for epistemic logic, there are no formal similarities to exploit. At minimum, the laws of modal logic require that set of "necessary" truths be closed under first-order consequence, and we don't always know the first-order consequences of the things we know. At least, that's the situation if we use the verb "know" in the ordinary way. In ordinary usage, we would surely want to allow that I don't know whether Goldbach's conjecture is true, and that I know that I don't know whether Goldbach's conjecture is true. But if Goldbach's conjecture is false, its negation can be derived from known arithmetical truth by exhibiting an explicit counterexample. So if we suppose that the set of truths I know is closed under consequence, we can conclude that, if I don't know whether Goldbach's conjecture is true, then Goldbach's conjecture is true. Having proved this, I know it, presumably, hence if I know that I don't know whether Goldbach's conjecture is true, then I know that Goldbach's conjecture is true. Thus the assumption that the set of things I know is known to be closed under consequence leads to the absurd conclusion that, if I know that I don't know whether Goldbach's conjecture is true, then I do know whether Goldbach's conjecture is true.[13]

[11] See Boolos [1993]. [12] See McGee [1991], Ch. 2.

[13] There is an extraordinary usage that gets the outcome that the set of things one knows is closed under consequence by individuating the objects of knowledge very coarsely; see Stalnaker [1991]. It is, I think, an unfortunate usage, since it makes out mathematics to be the least rational of human endeavors, in which the mathematician toils, hour after dreary hour, conducting "investigations"

To get the right formal properties, we need a different reading of "□." We can take □ϕto mean, not that ϕ is actually known, but that ϕ could be known on the basis of available evidence, so that an agent who has the same evidence we have and who has, in addition, unlimited time and a boundless appetite for patient and careful deduction would eventually come to know that ϕ.

With such a reading of "□," there is no obvious formal impediment to the application of possible-world semantics, but the treatment that worked for nomic necessity—regarding the epistemically possible worlds as a subclass of the metaphysically possible worlds—yields unsatisfactory results. Not only didn't Empedocles know that gold is an element and water is not, he wouldn't have known these things even if he had had unlimited powers of logical discernment. The totality of physical evidence available to Empedocles does not imply that gold is an element or that water is not an element, so that, if we read "□ ϕ" as "ϕ could have been known on the basis of evidence available to Empedocles," "∼ □ Gold is an element" and "∼ □ Water is not an element" ought to count as true. "Gold is an element" and "Water is not an element" are, however, (at least on standard accounts) metaphysically necessary. This is because "gold" and "water" are rigid designators, so that each of them refers to the same stuff, be it an element or a compound, in every metaphysically possible world.[14] Consequently, if we take epistemically possible worlds to be a subspecies of metaphysically possible worlds, we shall be unable to find an epistemically possible world in which "Gold is an element" and "Water is not an element" are false.

Simply regarding epistemically possible worlds as a subcategory of metaphysically possible worlds yields outcomes that aren't satisfactory. David Chalmers [2006] gets better results by employing the same worlds for both kinds of necessity, but evaluating truth-in-a-world in two different ways. For metaphysical necessity, we use possible worlds regarded as counterfactual, whereas epistemic necessity is spelled out in terms of possible worlds regarded as actual.[15] In determining whether "Gold in an element" is true in the world as Empedocles believed it to be, we ask not whether in that world gold is an element, but whether in that world the stuff Empedocles would have called "gold" in the world (had he spoken English) is an element. Chalmers' stratagem cleverly eliminates those misattributions of knowledge brought about by rigid designation, but there are other necessary truths we are not in a position to know that the maneuver can't handle so well. Every mathematical statement is, according to standard metaphysics, either necessary or impossible, but even the most dedicated logician wouldn't be able to settle every mathematical question on the basis of what we

that can't possibly teach her anything she doesn't already know. Quite the contrary, it seems to me that mathematics is the paragon of human rationality.

[14] The principal source of this doctrine is Kripke [1972].

[15] Actually, for both purposes Chalmers uses *centered worlds*, possible worlds with a distinguished point that locates indexicals.

now know. Chalmers' response to this difficulty is to introduce a new sort of entity he calls "scenarios." In one sense of "might," possible worlds are the ways the world might be, but in another sense of "might," the ways the world might be are scenarios. The two modal systems don't appear to be very closely connected.

The model-theoretic approach provides a uniform framework that treats logically possible worlds, metaphysically possible worlds, and epistemically possible worlds as things of a single kind. So long as the relevant logic is the first-order predicate calculus, the model-theoretic treatment of epistemic logic gives us precisely the results we want. A sentence is deducible from the things we know if and only if it is true in every model of the things we know. This is Gödel's [1930] completeness theorem.

Once we extend our logic to include second-order predicate calculus, the situation becomes more complicated, since a sentence can be a logical consequence of a second-order theory, and hence true in every model of the theory, even though it's not derivable from the theory. Thus, either the continuum hypothesis or its negation is a logical consequence (under a suitable formalization) of the axioms of a complete ordered field, but no one expects to be able to derive either the hypothesis or its negation from the axioms. If we take epistemically possible worlds to be the models in which all the things we know are true, we find that there are statements that are true in every model which even the most indefatigable logician wouldn't be able to deduce.

There is a natural conflict between the epistemology and the model theory, but we can force them into agreement by changing the class of relevant models, using Henkin models in place of standard second-order models. What's interesting is that we are willing and able to make this move, adopting *ad hoc* and artificial modifications in how our models treat the quantifiers. The model theory doesn't endeavor to describe the epistemological situation. The model theory provides mathematical tools that the epistemologist or the psychologist can use to describe the epistemological situation. All we are doing is providing mathematical models, and if the models we have on hand don't have the features we want, we are free to make up new ones that serve our purposes better.

The Henkin model illustrates both the versatility of the model-theoretic approach to modality and the modesty of its ambitions. Whatever our methods of second-order inference are, we can expect to be able to represent them by an appropriate Henkin model. This means that knowing that we have a Henkin model tells us next to nothing about what our rules of inference are. This is, it seems to me, as it should be. To figure out what methods of inference human beings employ, and thereby to determine what those methods could achieve if they were unhampered by mortality and fatigue, is a task for the psychologist, or perhaps for the epistemologist, and not for the model theorist. The model theory is useful. We can often recognize that we don't actually know something by simple introspection. But how do we recognize that a given proposition is not, in principle, derivable from things we know, without examining infinitely many, arbitrarily long derivations? We are able to demonstrate that a proposition is not derivable by producing an appropriate Henkin model. The model theory is useful, but it is useful as a handmaiden to cognitive science, not as cognitive science.

The crown jewel of possible-world semantics is the theory of conditionals.[16] Before its utilization in the analysis of conditionals, possible-world semantics looked like little more than a vividly poetical characterization of a technical device for providing consistency proofs in modal logic, but possible worlds have now become an indispensable tool of modern metaphysics. In spite of its undoubted success—remember the awkwardness of pre-Stalnaker discussions of counterfactual situations—the possible-world account suffers from a grievous limitation: it has nothing useful to say about conditionals with impossible antecedents. By default, such conditionals are indiscriminately declared "true," whatever their consequent. That would be fine if such conditionals never mattered, but they do. My favorite examples are the conditionals Turing [1939] and Post [1944] investigated in their studies of relative recursiveness, which reached such conclusions as the following: if there were an algorithm for testing which arithmetical sentences are true, there would be an algorithm for solving the halting problem, whereas it's not the case that, if there were an algorithm for the halting problem, there would be an algorithm for arithmetical truth. Current understanding of such counterfactuals is exquisitely well developed, one of the glories of modern mathematical logic; yet our best theory of conditionals can't handle them, because their antecedents are impossible. That there isn't a decision procedure for the halting problem isn't an empirical judgment about the limitations of current programming techniques; however ingenious programmers of the future may become, they'll never devise such an algorithm because such an algorithm is impossible. Even so, to ask what further problems we could solve if we had such an algorithm is a perfectly intelligible question with a mathematically significant answer.

We don't have to look to higher mathematics to find such examples. Everyday applications of *modus tollens* make the same point. Things like, "If that (pointing at the liquid in a flask) had been acidic, it would have turned the litmus paper red," and "If Bruce Wayne weren't Batman, we would have seen them together sometimes." The evidence that Bruce Wayne is Batman is pretty convincing, but it is not so inviolable that we are unable to entertain the contrary hypothesis. We formulate the hypothesis that Wayne and Batman are distinct and we work out the consequences of that hypothesis, but those consequences turn out to be so strongly opposed by the available evidence that we reject the hypothesis. This reasoning isn't impaired by the fact that the hypothesis turns out to be not merely false but metaphysically impossible. The fact that the antecedent is impossible in no way vitiates the conditional's usefulness.

[16] Stalnaker [1968]; see also Stalnaker and Thomason [1970] and Lewis [1973].

The Batman example indicates a problem, but I don't think that the source of the problem is a deficiency in Stalnaker's analysis of conditionals. The Stalnaker analysis works fine. In accepting the Batman conditional, we make a judgment that the nearest possible world in which Batman is different from Wayne is one in which the two of them are seen together sometimes. Since we know that, in the actual world Batman and Wayne are never seen together, we conclude that the closest world in which Batman is different from Wayne is not the actual world, that is, we conclude that Batman is identical to Wayne. This reasoning only makes sense, however, if we allow that there are possible worlds in which Batman is different from Wayne. If we think that there are no such worlds, then we'll still accept, "If Bruce Wayne were different from Batman, we would have seen them together sometimes," but we'll also accept (according to the standard story that treats conditionals with impossible antecedents as always true), "If Bruce Wayne were different from Batman, they would be careful never to be seen at the same place at the same time." Someone who accepts the latter conditional won't regard the fact that Wayne and Batman are never seen together as evidence for their distinctness.

In the imagined situation of people trying to puzzle out Batman's secret identity, ordinary speakers will accept the conditional, "If Bruce Wayne were different from Batman, we would see them together sometimes," and they will reject, "If Bruce Wayne were different from Batman, we still wouldn't ever see them together," and it is on the basis of this attitude that they conclude from the evidence that Wayne is identical to Batman. This is precisely the inferential behavior the Stalnaker analysis predicts, provided it is understood in such a way as to allow possible worlds in which Batman is distinct from Wayne. Such worlds are not allowed if "worlds" are taken to be metaphysically possible worlds, for all true identity statements are metaphysically necessary. If "worlds" are taken to be models, however, worlds in which Batman is distinct from Wayne are perfectly permissible, since the only logical necessary identity statements have the form $\tau = \tau$. Our apparent counterexamples to the Stalnaker semantics aren't really difficulties with the semantics for conditionals, but problems about the background ontological theory within which the semantics is interpreted.

A so-called logically possible world isn't at all like us and our surroundings. It also isn't at all like the way things actually are. (I admit to being a little squeamish about the notion that, in addition to things, there are such further entities as the way those things are.) Logically possible worlds model the ways things might have been mathematically, but a mathematical model of a flying cannonball isn't a flying cannonball. In terms of how they function in semantic theory, models play the role that has traditionally been ascribed to possible worlds; that's why I persist in referring to them as "logically possible worlds." But when we attempt to fit semantic theory into a larger metaphysical conception, we see that models aren't at all the kind of thing that modal realists have traditionally had in mind.

A warning: talk about models cannot be taken at face value. A model is not a thing, nor is it some sort of spooky entity-that's-not-a-thing. Models are made up of ordered pairs, and we talk about "models" as a way of describing collective properties of the pairs. Models play a similar role to that played by proper classes, for those who believe that the only extensional collections there are sets. Such people talk about the class of ordinals, but that's merely loose talk. They really believe that there are such things as ordinals, but there's no such thing as the class of ordinals, and statements that appear to talk about the class of ordinals are merely a convenient way of talking about collective attributes of the ordinals.

Similarly here, ordered pairs that collectively meet certain conditions are said to assign values to the nonlogical terms of the language, in which case they can be said to satisfy or fail to satisfy a sentence. Among the conditions are the requirements that the first component of each of the pairs be a nonlogical term, that each individual constant be paired with exactly one thing, and that the things paired with an n-place predicate all be n-tuples. When pairs assign values to the nonlogical terms of the language, they are said to form a model of the language. "Form a model" is a *plural predicate*,[17] the sort of thing we find in such sentences as "The children carried the piano" and "The Greeks outnumbered the Trojans." The satisfaction relation and the selection function for conditionals are also expressed by plural predicates.

Talk about pairs forming a model is systematically misleading. A model isn't another thing over and above the pairs, nor is it a thing constituted by the pairs. A model isn't a queer kind of entity, because in fact there are no such entities as models. Talk about models is a queer way of talking, because the things we say cannot be taken at face value. When it sounds like we're ascribing a property to a model, what we're really doing is ascribing a collective property to some ordered pairs that together assign values to nonlogical terms. The reason for talking this way is that complex plural constructions in English are so painfully awkward. What is said is apt to mislead, but what is meant is so cumbersome that it's worth the risk.

If we have some pairs that assign values to nonlogical terms and those pairs are few enough that we can make a set out of them, then can take the word "model" to refer to the set of ordered pairs, and once we do that there's no further need for doublespeak. This is the standard set-up in model theory, but limiting our attention to models whose domains are sets will be too restrictive if we intend to use the notion of model to explicate the notions of necessity and possibility. We can't take it automatically for granted that every second-order sentence that's true is reflected in some set-sized model, so if we take "$\Box\phi$" to mean that ϕ is true in every set-sized model, we can't be entirely confident of the schema "$\Box\phi \supset \phi$." The schema is a substantive mathematical thesis that cannot be derived from the standard axioms of second-order set theory.[18]

[17] See Rayo [2002].

[18] See Georg Kreisel [1967] and Stewart Shapiro [1991], §6.3.

The situation isn't improved by adopting a more expansive theory of collections, one that allows, for example, proper classes as well as sets to serve as the domain of a model. The resulting theory will be either false (which is what will happen if there are no proper classes) or inadequate (because it can't take account of models that include every class). The thesis that whenever we have pairs that assign values to nonlogical terms, we can associate an individual with them in the way the elements of a set are associated with the set, so that different models are associated with different individuals, implies Frege's inconsistent Basic Law V.[19] Even the weaker requirement that non-isomorphic models need to be associated with different individuals cannot be sustained, lest we succumb to the Burali–Forti paradox.[20] If we want our so-called model theory to take account of all the ways values could be assigned to extralogical terms, we'll have to cash out talk of models in terms of plural predicates.

We've been talking about formal languages that have individual constants and individual predicates—that is, predicates that take singular noun phrases in their argument places—as their nonlogical terms. We can also consider languages that contain plural predicates, and ask how the truth values of sentences depend on the semantic values of such predicates. This comes up, among other places, in developing the metatheory of an object language that doesn't have plural predicates but does have modal operators, when we ask, for example, how the truth conditions would change if we drew the line separating the nomologically possible from the metaphysically possible but nomologically impossible at a different place. It comes up again when we ask how the truth conditions for conditionals would change if we used a different selection function. I don't know of any way to talk about the semantics of languages that contain plural predicates without going to third-order logic, and I am not aware of any philosophical story that makes it plausible that third-order logic is available in contexts in which the individual quantifiers are all-inclusive. Boolos ([1984] and [1985]) has convinced me that plural quantification is ontologically innocuous, and Rayo [2002] has persuaded me that it's legitimate to go on to embrace plural predication, but no one has done the same for third-order logical devices. The ease with which we talk about "collective properties" makes this problem seem less thorny than it really is. If you are a realist about collective properties, you will think that, if the people live harmoniously together, it's because they collectively participate in the universal *living together in harmony*. But that just means that the realist will accept and the nominalist deny that *living together in harmony* is one of the items in the universe of discourse, and adding items to the universe doesn't help with the problems that propelled us to higher-order logic.

[19] Frege [1893]. The inconsistency and Frege's response to it are given in Russell [1902] and Frege [1902] and [1903], appendix. See also Ch. 3 of the introduction to the 2nd edn. of Whitehead and Russell [1927].

[20] Burali-Forti [1897]; see also Ch. 3 of the introduction to the 2nd edn. of Whitehead and Russell [1927].

The extravagance of the logical resources required for the metatheory of the modal language (as least when we go beyond just giving truth conditions to ask things like how the truth conditions would change if we altered the selection function) is an instance of one of the most perplexing problems in philosophy. In general, giving the metatheory of a language requires conceptual resources beyond those available within the language itself, which would appear to leave us helpless in circumstances, such as trying to give truth conditions for sentences of English, in which we don't have a richer metalanguage available. Alas, even if I had something useful to say about this problem, it wouldn't be appropriate to say it here.

Returning our attention to the project of using models as replacement worlds, we have seen that, at the level of modal sentential calculus, the fact the logically possible worlds aren't ways things might have been, but only mathematical representations of ways things might have been, scarcely makes any difference. When we turn to modal predicate calculus, however, things get quite a bit more complicated. We want to model the fact that, although I might have had a sister, there isn't anyone who might have been my sister. We can find a model in which "Vann has a sister" is true by forming a model in which the pencil sharpener, say, is in the extension of "Vann's sister," but if we go on to give the natural account of satisfaction, according to which an object satisfies "*x* might have been Vann's sister" if and only if there is a possible world in which it satisfies "*x* is Vann's sister," we get the unwelcome outcome that the pencil sharpener might have been my sister. If we were willing either to restrict the domain of the object language or to expand the domain of the metalanguage to, as it were, quantify over unactualized possibles, we could appoint something from outside the domain to play the role of my sister, but such a maneuver leaves us with a modal logic that is inapplicable at the place where it is most needed, in a metaphysical inquiry that attempts to tell the truth about everything.

Instead, we resort to a counterpart relation. The notion of counterpart was introduced by David Lewis [1968] in the context of a grand theory in which the metatheoretic quantifiers range over everything and the object-language quantifiers range only over the local things, but the device is quite versatile. What makes it the case that I might have had a sister is that there is a model of all the metaphysically necessary statements in which there is an individual who, though not a counterpart of anything, stands in the "sister of" relation of the model to my counterpart.

If we want to maintain the logical laws governing the identity relation, we shall have to impose certain constraints on the counterpart relation. The counterpart relation must be reflexive, symmetric, and transitive, and a thing can have at most one counterpart in a world. These are the same conditions Lewis would need to impose if, pursuing the world/time analogy, he were to allow that the same individual could exist in different worlds as well as existing at different times. Had he chosen that route, Lewis could identify a person with the mereological sum of the person's counterparts and so allow that the brother in the other world is, strictly and literally, me. We can't do the same thing with the model-theoretic treatment. The "brother" in the other model isn't me, but the individual that plays my role. In this context, the doctrine that "Aristotle" is a rigid designator tells us that "Aristotle"'s referent in a model that plays the role of a metaphysically possible world has to be Aristotle's counterpart in the model, if he has one.

Lewis develops his account of counterparts as part of a highly unorthodox cosmological theory that postulates a vast infinitude of regions of spacetime that are causally and spatially isolated from us. Most of the time, our attention is focused on our own cosmological neighborhood, and our quantifiers are restricted accordingly, but if we expand our horizons to include the things that can't be reached from here, we find that every possibility is realized somewhere. In describing ways the world might be, Lewis has no need to resort to mathematical representations. We can find a model of every possibility while leaving the meanings of the extralogical terms unchanged, just by varying the domain of quantification.

I think of the story being told here as an ontologically temperate account that reproaches Lewis's [1986] ontologically profligacy. To be sure, the account requires mathematical entities in great profusion, but modern science is filled throughout with mathematical entities; what Tarski's thesis postulates is just more of the same. Its generous acceptance of mathematical *abstracta* is the account's only ontological indulgence. It is inclined to accept at face value what the sciences tell us about what kinds of physical objects there are, whereas Lewis commits himself to the existence of such extraordinary physical things as literate spiders, fire-eating mushrooms, bodies that repel one another gravitationally, and liberal senators newly elected from Georgia.

I think of the model-theoretic approach to modality as ontologically restrained and Lewis's as spendthrift, but one isn't obligated to think of it that way. Tarski's thesis requires that there be a great many things, but it doesn't say anything, one way or another, about what the things are like; in particular, it doesn't deny that the things it requires are concrete. Tarski's thesis is perfectly compatible with Lewis's view that, although there aren't any nonphysical things, there are enough physical things to realize every possibility. Indeed, if we neglect the differences between what Lewis calls "consistency" and what Tarski would regard as *logical* consistency, we can say that Lewis's theory entails Tarski's thesis.

The objects required by Tarski's theory are abstract in the same way as, on Aristotle's conception, the spheres required by geometry are abstract. Geometers don't care whether the spheres they talk about are hard or heavy, and Tarski's thesis doesn't care whether the things that make up the models have mass or take up space.

Where the story I'm telling—not Tarski's thesis itself but the metaphysical framework into which I am attempting to fit it—really does differ from Lewis's account is that I want to accept at face value the zoologists' contention that there aren't any sea-dwelling zebras. I imagine that zoologists, if they were inclined to speculation, would be happy to agree that there might exist on other planets aquatic creatures that closely resemble zebras, but those wouldn't be zebras. Zebras have a common ancestry, and anything that doesn't share that ancestry isn't a zebra, however much it may resemble a zebra in habits and appearance. Even though we've never travelled to other planets, we can be confident that there aren't any sea-dwelling zebras there. We have, on the other hand, looked around the earth pretty extensively, and we can feel confident that, if there were marine zebras here, we would have come across them. Consequently, zoology tells us, and I am inclined to agree, there are no aquatic zebras anywhere. Lewis insists, on the contrary, that the statement that there are ocean-dwelling zebras is strictly and literally true, and the zoologist's insistence, "There are no aquatic zebras anywhere, either on earth or anywhere else," is a manifestation of her provincialism. There are aquatic zebras, but not in any of the places the zoologist refers to as "anywhere." If there had evolved in Australia a species of animal that looked and acted like zebras but were unrelated to African zebras by ancestry, they wouldn't be zebras. The same would be true of zebra-like creatures that emerged on the other side of the galaxy. However, if we look still farther away, to planets immeasurably distant from earth whose inhabitants have no causal connections with African zebras or their ancestors, we can find zebras, and some of them live underwater. Lewis, who makes few concessions to common sense or common science, thinks that the methods of science are pretty nearly incapable of recognizing general truths. Scientists are highly skilled at discerning patterns within the local landscape, but the discovery of general laws requires the metaphysician's farsighted vision. I, by contrast, am inclined to defer to the scientist, allowing that the zoologist means what she says when she tell us, "There are no aquatic zebras anywhere," and that what she says is true.

There is a mismatch between the tools we have available and the task for which we intend to use them. Tarski's semantics were designed for precise formal languages that rigorously formalize pure mathematics. We intend to apply the semantics to modal language that is anything but precise. I say this, but the claim that the formalized languages Tarski studied were fully precise needs to be taken with a grain of salt. Naively, one would have taken a fully precise language to be one that had a unique intended model, but the geometric theories presented by Veblen [1903] and Hilbert [1909] didn't have a unique intended model, but rather a unique-up-to-isomorphism intended model. This is enough to ensure that every geometric sentence has a uniquely determined truth value. The way geographers use the word "hexagonal," there may be no fact of the matter whether "France is hexagonal" is true, because France is a borderline case, but "hexagonal" in the geometers' usage suffers from no such imprecision.

The key idea that we need in order to retrofit Tarski's precision tools for use with modal notions that are completely lacking in precision was introduced by

Bas van Fraassen in his [1966] study of the logic of nondenoting names.[21] A vague language doesn't have a single intended interpretation, or a single isomorphism class of intended interpretations, but an extended family of acceptable interpretations got by adjudicating the disputed cases in various ways. A sentence counts as determinately true, determinately false, or indeterminate according as it is true under all, none, or some but not all of the acceptable interpretations.

The multiplicity of acceptable interpretations makes it possible to eliminate some awkward features that otherwise afflict counterpart theory. We can imagine a situation in which there are two different ships with strong and equally legitimate claims to be the counterpart of the ship actually sailed by Theseus. To say that each of the candidates is the ship of Theseus would be absurd. If the candidates aren't identical to one another, they aren't both identical to the ship of Theseus. Yet we aren't comfortable either preferring one candidate arbitrarily or insisting that, in the envisaged situation, Theseus's ship wouldn't have existed. The van Fraassen semantics provides a welcome alternative. We can say that, in each acceptable interpretation of the language, the phrase "ship of Theseus" picks out one or the other candidate, but the chosen candidate varies. In this way, "Either the ship in Slip A is the ship of Theseus or the ship in Slip B is" will be determinately true, but neither disjunct will be determinately true.[22]

In a similar way, recognition of multiple acceptable interpretations allows us to acknowledge the fact that, in attempting to determine which of the worlds in which a given sentence ϕ is true is most similar to the actual world, we will sometimes encounter ties, without having to relinquish the simple elegance of the Stalnaker [1968] semantics for one of the more complicated systems developed by Lewis [1973], and without frustrating ordinary speakers' inclination to declare that either Bizet and Verdi would have both been French had they been compatriots or they would both have been Italian. There are some acceptable interpretations whose selection functions pick out as the closest world in which "Bizet and Verdi are compatriots" is true worlds in which both composers' counterparts are in the extension of "French," and there are others that select worlds in which the counterparts satisfy "Italian," but all acceptable interpretations fall into one or the other of these two camps. We are going to need multiple interpretations anyway, to cope with the vagueness of "heap" and "bald," and throwing in a few more interpretations to accommodate "if," as recommended by Stalnaker [1981], is no extra burden.

The van Fraassen semantics mitigates some of the clumsiness that arises from the fact that models aren't ways the world might have been but only mathematical representations of ways the world might have been. The fact remains that

[21] See also Kit Fine [1975].

[22] Notice that this example is not an instance of vague identity. The identity sign has the same extension in every model, in keeping with the maxim, "Everything is what it is and not another thing." It is the vague name "ship of Theseus" whose referent varies from one acceptable model to another.

the model-theoretic semantics is a form of ersatzism, and ersatzism is, by its nature, awkward and artificial. Against these defects, we must weigh a rather large advantage. It gives us a treatment of modality that respects what we were taught by Parmenides: Things that are not in no way are.

If I am holding a stick behind my back and am wickedly inclined, I can demonstrate the existence of the stick by thwacking you with it. For intangible things or tangible things that are not in our immediate vicinity, our methods of demonstration have to be less direct. For existence claims that go beyond the concrete and local, the best we can hope to do, in general, is to provide the claim with a substantial role in an attractive theory. That's what I've tried to do here, displaying the usefulness of Tarski's thesis both for metamathematics and for modal metaphysics. Even if, as I hope, you find these consequences attractive, Tarski's thesis is so audacious, not in terms of the kinds of things it requires (in that respect, it is entirely innocent), but in terms of their sheer number, that one would wish for something more. One would like a direct argument. Sadly, I don't have one to offer.[23]

REFERENCES

Bays, Timothy. [2001] "On Tarski on Models," *Journal of Symbolic Logic* 66: 1701–26.

Beltrami, Eugenio. [1868] "Saggio di Interpretazione della Geometria Noneuclidea," *Giornale di Matematiche* 6: 284–312.

Boolos, George S. [1984] "To Be is to Be a Value of a Variable (or to Be Some Values of Some Variables," *Journal of Philosophy* 81: 430–9; reprinted in Boolos [1998], 54–72.

____ [1985] "Nominalist Platonism," *Philosophical Review* 94: 327–44; reprinted in Boolos [1998], 73–87.

____ [1993] *The Logic of Provability*. Cambridge: Cambridge University Press.

____ [1998] *Logic, Logic, and Logic*. Cambridge, Mass., and London: Harvard University Press.

Burali-Forti, Cesare [1897] "Una Questione sui Numeri transfiniti," *Rendiconte del Circolo mathematico di Palermo* 11: 154–64. English trans. by Jean van Heijenoort in van Heijenoort [1967], 104–12.

Chalmers, David J. [2006] "The Foundations of Two-Dimensional Semantics," in Manuel García-Carpintero and Josep Macià (eds.), *Two-Dimensional Semantics*, Oxford and New York: Oxford University Press, 55–140. On-line at http://jamaica.u.arizona.edu/~chalmers/papers/foundations.html.

Davis, Martin, (ed.) [1965] *The Undecidable*. New York: Raven Press.

Dedekind, Richard [1872] *Stetigkeit und irrationale Zahlen*. Brunswick: Vieweg. English trans. by Wooster Woodruff Benam in Dedekind, *Essays on the Theory of Numbers*. New York: Dover, 1963, 1–27.

[23] Versions of this paper were read to philosophy colloquia at the University of St Andrews, at New York University, and at Reed College. I am grateful for the many helpful comments I received.

Fine, Kit. [1975] "Vagueness, Truth, and Logic." *Synthese* 30: 265–300; reprinted in Rosanna Keefe and Peter Smith (eds.), *Vagueness: A Reader*. Cambridge, Mass., and London: MIT Press, 1996, 119–50.

Frege, Gottlob [1892] "Über Sinn und Bedeutung," *Zeitschrift für Philosophie und philosophische Kritik* 100: 25–50. English trans. by Max Black in Frege, *Translations from the Philosophical Writings*, edited by Peter Geach and Max Black. Oxford: Basil Blackwell, 1966, 56–78.

—— [1893] *Grundgesetze der Arithmetik*, vol. 1 (Jena: Pohle). A partial trans. by Montgomery Furth was published in Los Angeles and Berkeley by the University of California Press, 1964.

—— [1902] Letter to Russell, trans. by Beverly Woodward, in van Heijenoort [1967], 126–28.

—— [1903] *Grundgesetze der Arithmetik*, vol. 2 (Jena: Pohle). The appendix that discusses Russell's paradox is reprinted in Montgomery Furth's translation. Los Angeles and Berkeley: University of California Press, 1964, 127–41.

—— [1980] *Philosophical and Mathematical Correspondence*. English trans. by Hans Kaal. Chicago: University of Chicago Press.

Gödel, Kurt [1930] "Die Vollständigkeit der Axiome des logischen Funktionenkalküls," *Monatshefte für Mathematik und Physik* 37: 349–60. English trans. by Stefan Bauer-Menelberg in van Heijenoort [1967], 582–91.

Harper, William L.; Robert C. Stalnaker; and Glenn Pearce; eds. [1981] *Ifs*. Dordrecht, Holland; Boston and London: D. Reidel.

Henkin, Leon [1950] "Completeness in the Theory of Types," *Journal of Symbolic Logic* 15: 81–91.

Hilbert, David. [1909] *Grundlagen der Geometrie*, 2nd edn. Leipzig and Berlin: B. G. Teubner; English trans. by Leo Unger, LaSalle, Illinois: Open Court, 1971.

Hodges, Wilfred [1986] "Truth in a Structure," *Proceedings of the Aristotelian Society* 86: 135–51.

Hume, David. [1739] *A Treatise of Human Nature*. London: J. Noon.

Kripke, Saul A. [1972] "Naming and Necessity" in Donald Davidson and Gilbert Harman (eds.), *Semantics of Natural Language* (Dordrecht, Holland and Boston: D. Reidel), 253–355. Published separately at Cambridge, Mass., by Harvard University Press, 1980; page references are to the original.

Lewis, David K. [1968] "Counterpart Theory and Quantified Modal Logic," *Journal of Philosophy* 65: 113–26; reprinted in Lewis, *Philosophical Papers*, vol. 1. New York and Oxford: Oxford University Press, 1983, 26–39.

—— [1973] *Counterfactuals*. Cambridge, Mass.: Harvard University Press.

—— [1986] *On the Plurality of Worlds*. Oxford and New York: Blackwell.

Mautner, F. I. [1946] "An Extension of Klein's Erlanger Program: Logic as Invariant-Theory," *American Journal of Mathematics* 68: 345–84.

McGee, Vann [1991]. *Truth, Vagueness, and Paradox*. Indianapolis: Hackett.

—— [2000] '"Everything,"' in Gila Sher and Richard Tieszen (eds.), *Between Logic and Intuition*. New York and Cambridge: Cambridge University Press, 2000, 54–78.

——[2006] "There's a Rule for Everything," in Agustín Rayo and Gabriel Uzquiano (eds.), *Unrestricted Quantification: New Essays*. Oxford and New York: Oxford University Press, 179–202.

Post, Emil [1944] "Recursively Enumerable Sets of Positive Integers and Their Decision Problems." *Bulletin of the American Mathematical Society* 50: 284–316; reprinted in Davis [1965], 304–37.

Quine, Willard van Orman [1954] "Quantification and the Empty Domain," *Journal of Symbolic Logic* 19: 177–9; reprinted in Quine, *Selected Logic Papers*, enlarged edn. Cambridge, Mass.: Harvard University Press, 1995, 220–3.

Ramsey, Frank Plumpton. [1929] 'Theories.' Published posthumously in Ramsey, *Foundations of Mathematics and Other Logical Essays* (London: Routledge and Kegan Paul, 1931), 212–36, and in Ramsey, *Philosophical Papers*. Cambridge, New York, Port Chester, Melbourne, and Sydney: Cambridge University Press, 1990, 112–36.

Rayo, Agustín [2002] "Word and Objects," *Noûs* 36: 436–64.

Rayo, Agustín, and Gabriel Uzquiano [1999] "Toward a Theory of Second-Order Consequence," *Notre Dame Journal of Formal Logic* 40: 315–25.

Reid, Constance [1970] *Hilbert*. Berlin and New York: Springer-Verlag.

Russell, Bertrand [1902]. Letter to Frege in van Heijenoort [1967], 124–5.

Shapiro, Stewart [1991] *Foundations without Foundationalism*. Oxford: Clarendon Press.

Stalnaker, Robert C. [1968] "A Theory of Conditionals," in Nicolas Rescher (ed.), *Studies in Logical Theory, American Philosophical Quarterly* monograph series, no. 2. Oxford: Blackwell, 98–112; reprinted in Harper, Stalnaker, and Pearce [1981], 41–55.

——[1981] "A Defense of Conditional Excluded Middle" in Harper, Stalnaker, and Pearce [1981], 87–104.

——[1984] *Inquiry*. Cambridge, Mass., and London, MIT Press.

——[1991] "The Problem of Logical Omniscience, I," *Synthese* 89: 425–40; reprinted in Stalnaker [1999], 241–54.

——[1999]. *Context and Content*. Oxford: Oxford University Press.

——[2003] *Ways the World Might Be*. Oxford and New York: Oxford University Press.

——, and Richmond H. Thomason [1970] "A Semantical Analysis of Conditional Logic," *Theoria* 36: 23–42.

Tarski, Alfred [1935] "Der Wahrheitsbegriff in den formalisierten Sprachen," *Studia Philosophica* 1: 261–405. English trans. by J. H. Woodger in Tarski [1983], 152–278.

——[1936] "Über den Begriff der logischen Folgerung," *Actes du Congrès International de Philosophie Scientifique* 7: 1–11; English trans. by J. H. Woodger in Tarski [1983], 409–20. Page references are to the translation.

——[1983] *Logic, Semantics, Metamathematics*, 2nd edn. Indianapolis: Hackett.

——[1986] "What are Logical Notions?" *History and Philosophy of Logic* 7: 143–54. This is the posthumous text of a 1966 lecture, edited by John Corcoran.

——Andrzej Mostowski; and Raphael M. Robinson [1953]. *Undecidable Theories*. Amsterdam: North-Holland.

Turing, Alan M. [1939] "Systems of Logic Based on Ordinals," *Proceedings of the London Mathematical Society*, 2nd ser., 45: 161–228; reprinted in Davis [1965], 154–222.

Van Fraassen, Bas C. [1966] "Singular Terms, Truth-value Gaps, and Free Logic," *Journal of Philosophy* 63: 481–95.
Van Heijenoort, Jean [1967] *From Frege to Gödel.* Cambridge, Mass., and London: Harvard University Press.
Veblen, Oswald [1904] "A System of Axioms for Geometry," *Transactions of the American Mathematical Society* 5: 342–84.
Whitehead, Alfred North, and Bertrand Russell [1927] *Principia Mathematica*, 2nd edn., 3 vols. Cambridge: Cambridge University Press.

6

Stalnaker on the Interaction of Modality with Quantification and Identity*

Timothy Williamson

0. Logic is sometimes conceived as metaphysically neutral, so that nothing controversial in metaphysics is logically valid. That conception devastates logic. Just about every putative principle of logic has been contested on metaphysical grounds. According to some, future contingencies violate the law of excluded middle; according to others, the set of all sets that are not members of themselves makes a contradiction true. Even the structural principle that chaining together valid arguments yields a valid argument has been rejected in response to sorites paradoxes. In each case, a deviant metaphysics corresponds to the deviant logic. Of course, if one is trying to persuade deviant metaphysicians of the error of their ways, one is unlikely to get far by relying on logical principles that they reject. But that obvious dialectical exigency stably marks out no realm of logic. Each logical principle has persuasive force in some dialectical contexts and not in others. We do better to admit that logic has metaphysically contentious implications, and embrace them—if we know what they are.

Logic and metaphysics are not mutually exclusive. They overlap in the logic and metaphysics of existence, identity and possibility, for instance. The exploration (but not total conquest) of that area was one of the great achievements of twentieth century philosophy. Here, as in so many other areas, Bob Stalnaker has played an exemplary role, as a voice for metaphysical sobriety and the careful archaeology of logical structure. His intervention clarifies and deepens every debate in which he participates. In this essay I will examine the innovative argument of his 1994 paper 'The Interaction of Modality with Quantification and Identity'.

1. Stalnaker begins by considering two languages. One is a first-order language with quantification, predication and an identity sign but no modal operators.

* An earlier version of this paper was presented to a meeting of the Belgian Society for Logic and Philosophy of Science and in classes at Oxford; thanks to the audiences for useful discussion. I thank Agustín Rayo and Yannis Stephanou for detailed and thoughtful written comments. Above all, I thank Bob Stalnaker himself, not just for his answers to some technical questions but for all that I have learnt from him about the geometry of conceptual space.

The other is a propositional language with modal operators but no quantification, predication or identity sign. For each language, he provides a sound and complete axiomatization with respect to what we may call its *standard semantics*; this semantics differs slightly from the usual Kripke semantics in ways noted below, but for many purposes we can ignore the differences. Combining the two languages gives us a language for quantified modal logic with identity. We might hope that combining the axiomatizations for the two languages would give us a correspondingly sound and complete axiomatization with respect to the standard semantics for the combined language. However, Stalnaker proves that the hope is vain, by providing non-standard semantic theories that validate the combined axiomatization but falsify some formulas that are valid on the standard semantics: consequently, those formulas are underivable in that axiomatization. Thus the combined axiomatization, although sound, is not complete with respect to the standard semantics for the combined language.

What is less clear is the significance that Stalnaker attaches to his result. In one case, he says that the non-standard semantics allows for a 'nonstandard conception of individuals and their modal properties' (a form of counterpart theory); whether or not that conception is defensible, 'the issue is a philosophical one that cannot be settled by logical theory' (p. 154).[1] That might lead one to interpret Stalnaker as denying that the formulas invalidated by the non-standard semantics are genuine logical truths. Presumably, the standard semantics would be at fault, for employing too narrow a range of models and thereby validating formulas that are in some sense too substantive to deserve the status of logical truth. But Stalnaker later describes the formulas at issue as 'logical principles that are valid' (p. 157). If he is employing a notion of validity only relative to a semantic theory, then of course every formula of the language is valid relative to some semantic theories and invalid relative to others. In a postscript added in 2002, he describes a formula valid on the standard but not the non-standard semantic theory as 'less central' to the logic in question than is a formula valid on both semantic theories (p. 161). That informal idea of comparative centrality may come closest to what he has in mind.

Let us postpone these questions of philosophical significance, and examine Stalnaker's argument in detail. We start with a description of the syntax and semantics for the combined language, from which those for the two original languages can easily be derived by deletion of inapplicable features. The syntax is slightly unusual, because Stalnaker parses quantification in terms of a predicate abstraction device $^$.

The simple expressions of the language are as follows. What Stalnaker calls the 'descriptive' (non-logical) vocabulary consists of denumerably many atomic sentences (sentence letters), denumerably many *n*-place atomic predicates for

[1] All pages references are to Stalnaker 1994 as reprinted in Stalnaker 2003 unless otherwise specified.

each $n > 0$, and denumerably many individual constants. There are also denumerably many individual variables and parentheses. The logical vocabulary consists of the two-place predicate '=' and $\wedge$, $\sim$, $\Box$, $\forall$ and $\hat{}$. In the usual way, sentences involving other logical symbols, such as $\vee$, $\rightarrow$, $\leftrightarrow$, $\Diamond$ and $\exists$, are read as meta-linguistic abbreviations for sentences made up of the primitive vocabulary. The singular terms are the variables and the individual constants.

The complex expressions are constructed thus. If F is an n-place predicate and $t_1, \ldots, t_n$ are singular terms then $Ft_1 \ldots t_n$ is a sentence. If φ and ψ are sentences then $\sim \varphi$, $\Box\varphi$ and $(\varphi \wedge \psi)$ are sentences. If F is a one-place predicate then $\forall F$ is a sentence. If φ is a sentence and x is a variable then $\hat{x}\,\varphi$ is a one-place predicate. Informally, if we can read φ as '... x ...' then we can read $\hat{x}\,\varphi$ as 'is such that... it...' and therefore $\hat{x}\,\varphi t$ as 't is such that... it...' and $\forall\,\hat{x}\,\varphi$ as 'Everything is such that... it...'.

Here is Stalnaker's standard model-theoretic semantics for the language. A model is a quadruple $\langle W, R, D, v\rangle$, where: W is a nonempty set; R is a binary relation on W; D is a function from members w of W to sets D_w (which may be empty); v is a function that takes each individual constant c to a partial function from members w of W to members $v_w(c)$ of D_w; v takes each sentence letter A to a total function from members w of W to members $v_w(A)$ of $\{0, 1\}$; v takes each n-place atomic predicate F to a total function from members w of W to sets $v_w(F)$ of n-tuples of members of D_w. Informally, we can think of W as a set of possible worlds, R as a relation of accessibility between worlds, D_w as the domain of the world w (the set of individuals that exist at that world), $v_w(c)$ as the denotation of c at w (if any), $v_w(A)$ as the truth-value of A at w (1 if A is true, 0 otherwise), and $v_w(F)$ as the extension of F at w; in particular, $v_w(=)$ is the set of pairs $\langle d, d\rangle$ for all d in D_w. Thus v maps descriptive expressions to intensions of the appropriate type; the identity sign is mapped to its intended intension. The semantic rules assign values to simple and complex expressions relative to a model and an assignment, which is a partial function from individual variables to members of the union of the sets D_w for all w in W. If s is an assignment and x is a variable, $s[d/x]$ is the assignment that maps x to d but otherwise is like s. Given a model $\langle W, R, D, v\rangle$, semantic values are assigned thus (where s is an assignment and w is in W):

If x is an individual variable, $v_w^s(x) = s(x)$.
If φ is an atomic descriptive expression, $v_w^s(\varphi) = v_w(\varphi)$.
If F is an n-place predicate and $t_1, \ldots, t_n$ are singular terms,
  $v_w^s(Ft_1 \ldots, t_n) = 1$ if $\langle v_w^s(t_1), \ldots, v_w^s(t_n)\rangle \varepsilon v_w^s(F)$; otherwise
  $v_w^s(Ft_1 \ldots, t_n) = 0$.
If φ is a sentence, $v_w^s(\sim \varphi) = 1 - v_w^s(\varphi)$.
If φ and ψ are sentences, $v_w^s((\varphi \wedge \psi)) = \min\{v_w^s(\varphi), v_w^s(\psi)\}$.
If φ is a sentence, $v_w^s(\Box\varphi) = 1$ if $v_u^s(\varphi) = 1$ whenever $w\mathrm{R}u$; otherwise
  $v_w^s(\Box\varphi) = 0$.

If F is a one-place predicate, $v_w^s(\forall F) = 1$ if $v_w^s(F) = D_w$; otherwise $v_w^s(\forall F) = 0$.
If φ is a sentence and x is a variable, $v_w^s(\hat{x}\varphi) = \{d \in D_w : v_w^{s[d/x]}(\varphi) = 1\}$.

To obtain the semantics for the non-modal first-order language, delete W and R in the definition of a model, the clause for $\Box\varphi$, and the world subscript (w) throughout, and conceive the constituents D and v of a model accordingly. To obtain the semantics for the propositional modal language, delete instead D from the original definition of a model, the clauses that involve singular terms or predicates, and the assignment superscript (s) throughout.

Stalnaker's axioms and rules for the combined language are as follows, where $\vdash$ expresses provability, which is restricted to closed sentences.

Propositional Logic	If φ is a truth-functional tautology, $\vdash \varphi$
Modus Ponens	If $\vdash \varphi \to \psi$ and $\vdash \varphi$ then $\vdash \psi$
K Schema	$\vdash \Box(\varphi \to \psi) \to (\Box\varphi \to \Box\psi)$
Necessitation	If $\vdash \varphi$ then $\vdash \Box\varphi$
Abstraction	$\vdash \forall\hat{x}\,(\hat{y}\,\varphi x \leftrightarrow \varphi^x/_y)$, where $\varphi^x/_y$ is the result of substituting x for all free occurrences of y in φ (relettering bound variables in φ where necessary to prevent clashes)
Quantification	$\vdash \forall\hat{x}\,(\varphi \to \psi) \to (\forall\hat{x}\,\varphi \to \forall\hat{x}\,\psi)$
Redundancy[2]	$\vdash \forall\hat{x}\,Fx \leftrightarrow \forall F$, where x is not free in F
Existence	$\vdash Ft_1 \ldots t_n \to \exists\hat{x}\,x = t_i$, where F is an n-place predicate
Identity[3]	$\vdash s = t_i \to (Ft_1 \ldots t_i \ldots t_n \to Ft_1 \ldots s \ldots t_n)$ where F is an n-place predicate
Universal Generalization	If $\vdash \varphi \to \psi$ and t does not occur in φ then $\vdash \varphi \to \forall\hat{x}\,\psi^x/_t$

To obtain the axioms and rules for the non-modal first-order language, delete the K schema and Necessitation. To obtain the axioms and rules for the propositional

[2] Only the left-to-right direction of Redundancy figures in the original 1994 axiomatization, supplemented with two extra axiom schemas: the universal instantiation principle $\vdash \forall\hat{x}(\forall F \to Fx)$ and the permutation principle $\vdash \forall\hat{x}\forall\hat{y}\,\varphi \to \forall\hat{y}\forall\hat{x}\,\varphi$. Stalnaker informs me (p.c.) that the change in the 2002 version was just to improve the economy of the axiomatization: universal instantiation becomes derivable once the biconditional Redundancy principle is used and permutation (which Kit Fine (1983) had shown to be independent of Kripke's (1963) axiomatization of free quantified modal logic) is derivable in Stalnaker's system with the help of his principles about identity.

[3] Obviously, it makes no difference if $x = t_i$ in Existence and $s = t_i$ in Identity are replaced by $t_i = x$ and $t_i = s$ respectively. However, if the latter of these replacements is made without the former, then the symmetry principle $s = t \to t = s$ is underivable, because a deviant semantics for identity on which $s = t$ is true if and only if either s and t denote the same thing or s fails to denote validates the axiomatization while invalidating symmetry (analogously if the former replacement is made without the latter). The problem arose for an earlier, unpublished version of the system. Delia Graff and Gabriel Uzquiano prompted the correction; Stalnaker supplied the independence argument. Thanks to all three (p.cs.) for this information.

modal language, delete instead Abstraction, Quantification, Redundancy, Existence, Identity and Universal Generalization.

Now consider the following schemas, where E is the existence predicate $\hat{x}\exists\hat{y}\, x = y$ (thus $v^s_w(E) = \mathrm{D}_w$):[4]

CBF	$\Box\forall\hat{x}\,\varphi \rightarrow \forall\hat{x}\,\Box\varphi$
QCBF	$\Box\forall\hat{x}\,\varphi \rightarrow \forall\hat{x}\Box(Ex \rightarrow \varphi)$
EI	$\forall\hat{x}\forall\hat{y}\,(x = y \rightarrow \Box(Ex \rightarrow x = y))$
NEI	$\forall\hat{x}\forall\hat{y}\Box(x = y \rightarrow \Box(Ex \rightarrow x = y))$
ND	$\forall\hat{x}\forall\hat{y}\,(\sim x = y \rightarrow \Box \sim x = y)$

Each of these principles except CBF is valid—it (or all its instances, in the case of QCBF) is true on all assignments at all worlds in all models—on Stalnaker's semantics.

CBF is the converse Ibn-Sina-Barcan schema, usually known just as the converse Barcan formula.[5] Informally, it says that that if necessarily everything meets a certain condition, then everything is such that necessarily it meets that condition. CBF is highly controversial; for instance, it implies that if necessarily everything exists then everything has necessary existence. Many philosophers regard it as a trivial necessary truth that everything exists (everything is something), but an obvious falsehood that everything has necessary existence. Since Stalnaker's semantics permits contingent existence, it invalidates CBF. QCBF is a weak consequence of CBF; it says that if necessarily everything meets a certain condition, then everything is such that necessarily it exists only if it meets that condition. The existence qualification in QCBF finesses the usual objections to CBF; the relevant instance says only that if necessarily everything exists then everything is such that necessarily it exists only if it exists, which is clearly harmless. On Stalnaker's semantics, if the antecedent of QCBF is true, then in every accessible world everything in the domain satisfies $\hat{x}\,\varphi$, so everything in the domain of the original world is such that in every accessible world if it is in the domain of that world then it satisfies $\hat{x}\,\varphi$, so the consequent is true; thus QCBF is valid.

EI is the essentiality of identity. Informally, it says that if things are identical then necessarily one of them exists only if they are identical. NEI is a strengthened

[4] Stalnaker defines E as $\exists\hat{y}\, x = y$ (p. 151). Since the latter is an open sentence, not a predicate, as he makes the grammatical distinctions, he intends $\hat{x}\exists\hat{y}\, x = y$. But Stalnaker notes (p.c.) that it is essential that the Existence principle be stated as it is, rather than with $\hat{y}\,(\exists\hat{x}\, x = y)t_i$ in the consequent, since otherwise $\hat{y}\,(\exists\hat{x}\, x = y)s \leftrightarrow \exists\hat{y}\, y = s$ will be underivable. One can prove this by considering a deviant semantics on which all predications and universal quantifications count as true (adding the symmetry principle from the previous footnote does not help).

[5] I am grateful to Zia Movahed for the information that the first known discussion of the Barcan and converse Barcan principles was by Ibn Sina (Avicenna, 980–1037); see Movahed 2004. Of course the credit for independently rediscovering and initiating modern discussion of them remains with Ruth Barcan Marcus. I follow Movahed's obvious proposal for renaming. For clarity, I continue to use Stalnaker's acronyms.

form of EI. It says that all things are such that necessarily if they are identical then necessarily one of them exists only if they are identical. Both EI and NEI are valid on Stalnaker's semantics because variables are rigid designators: they are assigned values absolutely, not relative to worlds. If they have the same value at any world, they have the same value at every world. The existence qualification in EI and NEI is needed on Stalnaker's semantics, for the extension of '=' at a world is restricted to the domain of that world.

ND is the necessity of distinctness. Informally, it says that if things are not identical then they could not have been identical. It too is valid on Stalnaker's semantics because variables are rigid designators. If they have different values at any world, they have different values at every world. No existence qualification is needed, for the identity claim is automatically false at any world whose domain does not contain the values of both variables.

EI is derivable in Stalnaker's axiomatization. CBF is of course underivable, since it is invalid on his semantics, for which the axiomatization is sound. However, although QCBF, NEI and ND are all valid on his standard semantics, they are all underivable in his axiomatization. Stalnaker proves their underivability by providing non-standard semantics on which the axioms are valid, the rules preserve validity but the formula in question is invalid. For QCBF and NEI, the non-standard semantics deviates by treating variables as non-rigid: although they are initially assigned a world-independent value, they are subsequently evaluated as denoting a counterpart of the initial value with respect to a given world. The way in which QCBF and NEI fail on this semantics is quite subtle; the details can be checked in Stalnaker's paper and are not crucial for the argument to come. Stalnaker shows that even if one adds QCBF as an axiom schema (which permits the derivation of NEI too), ND remains underivable. He provides another deviant semantics on which variables behave rigidly but '=' is interpreted as indiscernibility rather than identity. His original axioms and QCBF are valid on this semantics, and the rules preserve validity, but ND is invalid: discernibles could have been indiscernible.

2. What philosophical import does Stalnaker attribute to his independence results? Concerning QCBF (and NEI), he writes:

> The variant [counterpart] semantics brings to the surface an assumption about the relation between the modal properties of an individual in different possible worlds—an assumption implicit in the standard semantics that is not grounded in the nonmodal logic of predication, or in the modal logic of propositions, or in their combination. (p. 153).

Concerning ND, he says:

> the combination of identity theory with modality provides the resources to distinguish identity from a weaker relation [indiscernibility] that cannot be

> distinguished from it in a nonmodal context. Perhaps this shows that there is in some sense something modal about the concept of identity. (p. 157).

Are these claims justified?

Let us first ask what methodology is implicit in Stalnaker's form of argument. In schematic terms, he considers a language L1 with an axiom system A1 that is sound and complete with respect to a semantics S1 (all and only formulas valid on S1 are derivable in A1), and a language L2 with an axiom system A2 that is sound and complete with respect to a semantics S2. He combines L1 and L2 into a joint language L1+L2, S1 and S2 into a joint semantics S1+S2 for L1+L2 and A1 and A2 into a joint axiom system A1+A2 for L1+L2. He then treats A1+A2 as exhausting what the logics of L1 and L2 tell us, when combined, about the logic of L1+L2. But what do 'combine', 'joint' and '+' mean here?

First, it is not always clear what it would be to combine two languages. For example, what is English+Japanese? In the particular case at issue, another way of combining the first-order language of predication, quantification and identity with the propositional modal language would have permitted the application of $\Box$ only to closed formulas; such a language would have a lower grade of modal involvement (Quine 1966). However, let us simply take Stalnaker's way of combining the two languages as given. It is standard enough, and philosophically attractive.

Second, it is not always clear what it would be to combine two semantic theories for different languages. For example, in his official combined semantics, Stalnaker makes the assignment of values to variables absolute, not relative to possible worlds, on the plausible grounds that 'open sentences are devices for the formation of complex predicates that express properties of individuals, and not properties of some kind of intension' (p. 150). But one can also give a semantics for first-order modal languages in which the assignment of values to variables is world-relative; the variables may then be said to stand for individual concepts (see Hughes and Cresswell 1996: 334–42 for an introduction to such systems). Thus the form of the semantics for the two original languages underdetermines the form of the semantics for the more inclusive language. For the time being, let us simply take Stalnaker's standard form of the semantics for the first-order modal language as given.

Third, it is not always clear what it would be to combine two axiomatizations for different languages. We have to decide what counts as an instance in one language of an axiom schema or rule of inference originally formulated with respect to another language; more than one extrapolation is generally possible, since we may or may not permit the distinctive vocabulary of one language to occur in instances of an axiom schema or rule inherited from the axiomatization of the other language. In the present case, Stalnaker permits such instantiations; the combined system would otherwise be extremely weak. Given his way of presenting his axiomatizations, his way of combining them is natural enough. We can accept it too as given.

There is a more urgent question. Distinguish *logics* (consequence relations, or sets of theorems) from *axiomatizations* (sets of axioms and inference rules); many different axiomatizations can generate the same logic. But then why should we assume that Stalnaker's combined axiomatization generates all and only the formulas in the first-order modal language to which one is committed by acceptance of his combined logics for the first-order non-modal language and the propositional modal language, even granted that his logics for those languages are sound and complete with respect to their original semantics?

Let us start with the soundness of Stalnaker's combined axiomatization with respect to his combined semantics. It does *not* follow merely from the soundness of his two axiomatizations for the original languages with respect to their original semantics. For some other axiomatizations for those languages are sound with respect to their semantics even though, when one combines them in a way analogous to Stalnaker's, the resulting combined axiomatization is *un*sound with respect to Stalnaker's combined semantics.

Here is an example. Replace the axiom schema Identity above with this axiom schema, a familiar form of Leibniz's Law:

Identity* $\vdash s = t \rightarrow (\varphi \rightarrow \varphi^{s}/_{t})$

Every instance of Identity is an instance of Identity*, so the completeness of the resulting axiomatization for first-order non-modal logic follows from the completeness of Stalnaker's original axiomatization. But instances of Identity* in which φ is not a predication are not instances of Identity. However, as Stalnaker says, every instance of Identity* in the first-order non-modal language is derivable in his original axiomatization (p. 148). Since his axiomatization is sound with respect to his semantics for the non-modal language, every instance of Identity* in that language is valid on his semantics for that language. Consequently, the proof system in which Identity* replaces Identity is sound with respect to Stalnaker's semantics for the first-order non-modal language. Nevertheless, the axiomatization for the first-order modal language that results from combining (in Stalnaker's way) the axiomatization that uses Identity* with Stalnaker's axiomatization for the propositional modal language is not sound with respect to Stalnaker's semantics for the combined language. For, as Stalnaker says, Identity* has instances in the combined language that are invalid on his semantics (p. 148). One such instance is:

(1) $s = t \rightarrow (\Diamond(s = s \wedge \sim s = t) \rightarrow \Diamond(s = s \wedge \sim s = s))$

Here s and t are distinct constants. The reason is that Stalnaker classifies individual constants as descriptive terms and does not require them to be rigid designators; he allows them a flexibility that he does not allow to individual variables (in his standard semantics). Thus s and t may designate the same individual with respect to one world while designating distinct individuals with respect to another world accessible from it.

Although that particular example depends on Stalnaker's questionable treatment of individual constants, the general point does not. A less controversial example uses an extensionality principle for predicates:

Coextensiveness $\vdash \forall \hat{x}(Fx \leftrightarrow Gx) \rightarrow (\varphi \rightarrow \varphi^F/_G)$

For simplicity, F and G here are one-place atomic predicates; $\varphi^F/_G$ is the result of substituting F for G in φ. All instances of Coextensiveness in the first-order non-modal language are valid on any standard semantics, but the same schema has clearly invalid instances in the first-order modal language, such as:

(2) $\forall \hat{x}(Fx \leftrightarrow Gx) \rightarrow (\sim \Box \forall \hat{x}(Fx \leftrightarrow Gx) \rightarrow \sim \Box \forall \hat{x}(Fx \leftrightarrow Fx))$

For F and G may be accidentally coextensive.

The general point is this. The mere soundness and completeness of an axiomatization with respect to the semantics for a language does not entitle one to extrapolate the axiom schemas and rules of inference of the axiomatization in the natural way to a more inclusive language as being what the logic of the logical constants in the original language has to offer the combined logic of the logical constants in the extended language. Different axiomatizations that yield the same set of theorems in the original language may yield different sets of theorems from each other when extrapolated to the extended language. In effect, Stalnaker makes this point himself about the example above: 'The validity of the general schema [Identity*], unlike the validity of the identity axioms [Identity], depends on the expressive limitations of the extensional theory' (p. 148). In axiomatizing the restricted language, one must choose axiom schemas and rules of inference whose appropriateness does not depend on the expressive limitations of the restricted language. Doing that is no straightforward matter, since proving soundness and completeness for the restricted language does not suffice. One must somehow use a conception of how the restricted language may legitimately be extended, in both syntax and semantics.

In the present case, of course, Stalnaker has carefully chosen his axiom schemas and rules of inference for the original languages so that they remain sound when extrapolated to the combined language. But an analogous issue arises about completeness. For one might provide a sound and complete axiomatization for a restricted language that is unnecessarily weak when extrapolated to more inclusive languages.

Here is an example. Stalnaker's propositional modal logic is K. Like most of the familiar systems of propositional modal logic, K is decidable. Thus, instead of using Stalnaker's axiom schemas of Propositional logic and the K schema and rules of Modus Ponens and Necessitation, one can in principle axiomatize K simply by taking all formulas that pass some given decision procedure for K (in effect, all theorems of K) as axioms. Call that axiom family 'Cheap K'. Analogously, Stalnaker's axiom family of Propositional Logic in effect axiomatizes non-modal propositional logic simply by taking all tautologies (in effect, all theorems) as axioms. Cheap K by itself constitutes a sound and complete axiomatization of K. Now combine it in the natural way with Stalnaker's axiomatization of the logic of the first-order non-modal language (Propositional Logic, Abstraction, Quantification, Redundancy, Existence, Identity, Modus Ponens and Universal Generalization) into an axiomatization for the first-order modal language. That axiomatization is manifestly inadequate. The instances of Cheap K are just substitution instances in the first-order modal language of theorems of K. There is no rule of Necessitation. One cannot even derive:

(3) $\Box\forall\hat{x}(Fx \rightarrow Fx)$[6]

But it would not be plausible to conclude that the underivability brings to the surface 'an assumption implicit in the standard semantics that is not grounded in the nonmodal logic of predication, or in the modal logic of propositions, or in their combination' in any serious sense. Rather, its underivability is a mere artefact of the specific way in which the propositional modal logic was axiomatized.

The question now naturally arises: are Stalnaker's underivability results similarly artefacts of the specific ways in which he axiomatized the logics of the propositional modal language and of the first-order non-modal language? If so, they lack the philosophical significance that he claims for them.

Let us start with QCBF. The first point to notice is that in Stalnaker's system we can derive a schema very close to QCBF:

QCBF* $\Box\forall\hat{x}\varphi \rightarrow \forall\hat{x}\Box(Ex \rightarrow \hat{x}\varphi x)$

For where F is an atomic one-place predicate and t is an individual constant, the formula $\forall F \rightarrow (Et \rightarrow Ft)$ is valid on Stalnaker's semantics for the first-order non-modal language, and therefore provable. Consequently, substituting $\hat{x}\varphi$ for F in the proof gives us in his axiomatization for the first-order modal language:

[6] Proof: Consider a deviant semantics in which worlds are divided into *sensible* worlds, in which $\forall$ is interpreted in the usual way as 'all', and *silly* worlds, in which $\forall$ is interpreted as 'not all'; the semantics is exactly like Stalnaker's in all other respects, except that validity is defined as truth in all sensible worlds in all models. On the deviant semantics, all instances of Cheap K are true in every world, sensible or silly, because their truth does not depend on the specific interpretation of $\forall$; *a fortiori*, all instances of Cheap K are valid. Since all instances of Propositional Logic are also instances of Cheap K, all instances of Propositional Logic are valid. Similarly, Modus Ponens preserves validity, because in every world it preserves truth. All instances of Abstraction, Quantification, Redundancy, Existence and Identity are valid, because true in every sensible world, since their truth at a world depends only on the interpretation of the non-modal vocabulary, which is standard in every sensible world. Similarly, Universal Generalization preserves validity. Thus the axiomatization is sound on the deviant semantics. The formula $\forall\hat{x}(Fx \rightarrow Fx)$ is derivable in the usual way; it is valid because it is true in all sensible worlds, even though it is false in all silly worlds. But since a silly world can be accessible from sensible ones, (3) is false in some sensible worlds in some models, and so is invalid on the deviant semantics. Thus (3) is underivable on this axiomatization.

(4) $\forall \hat{x} \varphi \rightarrow (Et \rightarrow \hat{x} \varphi t)$

(Exercise: lay out the proof in full.) By Necessitation and the K schema, we have:

(5) $\Box \forall \hat{x} \varphi \rightarrow \Box (Et \rightarrow \hat{x} \varphi t)$

Universal Generalization yields QCBF* from (5). From QCBF*, we could easily prove QCBF if we had:

(6) $\forall \hat{x} \Box (\hat{x} \varphi x \rightarrow \varphi)$

Although (6) is valid on Stalnaker's semantics for the first-order modal language, it is unprovable in his axiomatization. We cannot derive it from his axiom schema of Abstraction, because the latter does not allow us to slip a modal operator between the outer and inner occurrences of the predicate abstraction operator. His Abstraction schema constrains the effect of the abstraction operator only in a limited range of contexts. But that is an artefact of Stalnaker's version of Abstraction. Consider this rule of inference, designed to constrain the effect of the abstraction operator in a wider range of contexts:

Free Abstraction — If B is the result of replacing some or all occurrences of $\hat{y} \varphi x$ in A by $(Ex \wedge \varphi^x/_y)$, where $\varphi^x/_y$ is as in the original Abstraction principle, and $\vdash$ A, then $\vdash$ B.

The rule is called 'Free Abstraction' because it includes an existence qualification, which is characteristic of free logic. Since x and y are allowed to be the same variable, Free Abstraction has the special case in which occurrences of $\hat{x} \varphi x$ are replaced by $(Ex \wedge \varphi)$. Thus from QCBF*, in the system in which Free Abstraction replaces Stalnaker's Abstraction principle, the rule yields this theorem:

(7) $\Box \forall \hat{x} \varphi \rightarrow \forall \hat{x} \Box (Ex \rightarrow (Ex \wedge \varphi))$

By dropping a conjunct from (7), we easily prove QCBF.

We have still to justify Free Abstraction. To show that it preserves validity on Stalnaker's semantics for the first-order modal language, it suffices to check that in any model, $v^s_w(\hat{y} \varphi x) = v^s_w((Ex \wedge \varphi^x/_y))$, for any assignment s and world w, for then the compositional nature of the semantics ensures that the sentences A and B in the rule always have the same semantic value. The check is easily made. First, suppose that $v^s_w(\hat{y} \varphi x) = 1$. Then $s(x) \varepsilon v^s_w(\hat{y}\varphi) \subseteq D_w$, so $v^s_w(Ex) = 1$ and moreover, by the validity of Stalnaker's own Abstraction schema, $v^s_w(\varphi^x/_y) = v^s_w(\hat{y} \varphi x) = 1$. Thus $v^s_w((Ex \wedge \varphi^x/_y)) = 1$. Conversely, if $v^s_w((Ex \wedge \varphi^x/_y)) = 1$, then $v^s_w(Ex) = 1$ and $v^s_w(\varphi^x/_y) = 1$; but by the former $s(x) \varepsilon D_w$, so by the validity of Stalnaker's own Abstraction schema $v^s_w(\varphi^x/_y) = v^s_w(\hat{y} \varphi x)$. Thus $v^s_w(\hat{y} \varphi x) = 1$. By dropping the world subscript w throughout the argument, we can also show that Free Abstraction preserves validity on Stalnaker's semantics for the first-order non-modal language. The underlying point is that abstracting on an open sentence with respect to a variable and then applying the resultant predicate to that variable makes no difference when the value of the variable is in the relevant domain of quantification but produces falsity otherwise (since that value is excluded from the extension of the predicate), on Stalnaker's standard semantics. He allows abstraction with respect to an individual constant to have more extensive effects, since it may involve replacing a non-rigid by a rigid designator, but both Stalnaker's original Abstraction schema and Free Abstraction concern only abstraction with respect to variables.

We can also check that, in the presence of Free Abstraction, Stalnaker's own Abstraction schema becomes redundant. Using just Propositional Logic, Universal Generalization and Modus Ponens we obtain:

(8) $\forall\hat{x}(\hat{y}\,\varphi x \leftrightarrow \hat{y}\,\varphi x)$

Consequently, by Free Abstraction:

(9) $\forall\hat{x}(\hat{y}\,\varphi x \leftrightarrow (Ex \wedge \varphi^{x}/_{y}))$

We must eliminate the first conjunct on the right-hand side. Let T be any closed truth-functional tautology. As with (8), we obtain:

(10) $\forall\hat{y}\mathrm{T}$

By an instance of Redundancy (with $\hat{y}\,\mathrm{T}$ for F) we have:

(11) $\forall\hat{x}\hat{y}\mathrm{T}x$

Applying Free Abstraction to (11) gives:

(12) $\forall\hat{x}\,(Ex \wedge \mathrm{T}))$

From (9) and (12), Propositional Logic, Universal Generalization, Quantification and Modus Ponens yield Stalnaker's Abstraction schema. Thus we lose no theorems by dropping his Abstraction schema and employing Free Abstraction instead.

In the first-order non-modal logic, Free Abstraction makes no difference. Stalnaker's original axiomatization is sound and complete; as we have seen, the new axiomatization is sound and extends Stalnaker's, so is also complete; thus the set of theorems is the same. In the first-order modal logic, the new system is still sound, but properly extends Stalnaker's, because QCBF is derivable only in the new system. As Stalnaker notes, NEI is derivable given QCBF (p. 155). NEI is therefore another theorem of the axiomatization with Free Abstraction in place of Stalnaker's Abstraction schema. Thus the underivability of QCBF and NEI is no deep fact about the relation between the logic of quantification and the logic of modality. It merely reflects Stalnaker's unforced choice amongst ways of formulating the logic of quantification that are equivalent in a non-modal setting but not in a modal setting. As we have seen, such choices can leave utterly innocuous truths of quantified modal logic unprovable. Thus his result casts no metaphysical doubt on QCBF and NEI. The new Abstraction rule is both formally correct and informally plausible: (x is such that... it...) is equivalent to (x exists and... x...). QCBF and NEI are straightforwardly valid in quantified modal logic.

Could Stalnaker reply that the invalidity of QCBF and NEI on his deviant counterpart semantics shows that Free Abstraction is metaphysically contentious? Indeed, the derivability of QCBF and NEI from Free Abstraction implies that the latter is invalid on Stalnaker's deviant counterpart semantics. For example, that semantics allows an object d not in the domain of a world w to satisfy $\hat{y} \Diamond Gyx$ with respect to w (because the counterpart of d in w has a counterpart in a world w^* accessible from w that belongs to the extension of G in w^*) even though d does not satisfy $\Diamond Gx$ with respect to w (because no counterpart of d in any world belongs to the extension of G in that world). However, for Stalnaker to object to Free Abstraction on that basis would be to argue in a circle. For his original reason for taking the counterpart semantics seriously was precisely that it validated all the principles of first-order non-modal logic and of propositional modal logic. But now that turns out to be so only on an unjustifiedly narrow view of the principles of first-order non-modal logics. For any theorem of first-order modal logic, one can cook up an unintended formal semantics on which it is invalid, and a metaphysical fairy tale to add colour to the semantics. Such a methodology is a recipe for shallowness and confusion. But it was not Stalnaker's methodology. His original argument laudably relied on the constraints of first-order non-modal logic and propositional modal logic; it failed only because it made unjustified claims about the limits of those constraints. That is no reason to throw the constraints away altogether, as the direct appeal to the counterpart semantics would do.[7]

Of course, counterpart semantics in its own right still finds some defenders, who are willing to put up with its ugly complications for the sake of the freedom that it delivers from constraints to which they object on metaphysical grounds. But we have already seen that the contentiousness amongst metaphysicians of a principle is compatible with its being a valid law of logic. Stalnaker's argument promised to do something more interesting: to introduce an objective procedure for determining how far logic constrains modal metaphysics. The trouble is that the putatively objective procedure is over-sensitive to the way in which a given logic is axiomatized. Metaphysicians who start from the idea that counterpart theory must be logically coherent and then tailor their logic to suit are not even attempting to do the more interesting thing.[8]

[7] The basic objection was formulated in Williamson 1996b.

[8] Some defenders of counterpart theory (I exclude Stalnaker) think that they are not really rejecting classical logic (for example, the classical logic of identity) because sentences of quantified modal logic do not really have the logical form that they superficially appear to have, but rather one given by a counterpart-theoretic translation. However, such claims must then be assessed by the

3. Free Abstraction is not the only means by which one can argue for QCBF and NEI within something like Stalnaker's framework. In particular, suppose that we are granted the stronger identity principle Identity*. Of course, Stalnaker regards Identity* as invalid in a modal context, because he allows non-rigid individual constants. However, that decision has no obvious bearing on the logical status of QCBF. Individual constants occur neither in the schema itself nor in the instance of it that Stalnaker shows to be invalid on the counterpart semantics (p. 153). One could instead declare some or all individual constants rigid without undermining the rationale for other aspects of Stalnaker's combined system. Indeed, since he axiomatizes the logic of quantification using only closed formulas (p. 147), individual constants are pressed into playing a double role, as both descriptive terms and the analogue of free variables ('arbitrary names') in proofs. In effect, the rule of Universal Generalization exploits them in the latter capacity. By contrast, their non-rigidity is justified only by their descriptive content. For since Stalnaker's official semantics treats variables as rigid, the role of closed terms in acting like free variables in the corresponding logic of quantification is best served by rigid designators. To mark the difference between these contrasting functions for individual constants, we could divide them into two categories: rigid arbitrary names and possibly non-rigid descriptive terms. Identity* would then be valid for the arbitrary names but not for the descriptive terms. Moreover, as already noted, Identity* has only valid instances on Stalnaker's semantics for first-order non-modal logic, so it is in any case unclear with what right he rejects Identity* as a constraint on the semantics for first-order modal logic, given his methodology elsewhere in the paper.

Consider an extension of Stalnaker's axiomatization by Identity* for arbitrary names. Thus for any formula φ in which only the variable x is free, and distinct arbitrary names s and t that do not occur in φ, we have the theorem:

(13) $\sim\varphi^{s}/_{x} \rightarrow (\varphi^{t}/_{x} \rightarrow \sim s = t)$

Then Universal Generalization gives:

(14) $\sim\varphi^{s}/_{x} \rightarrow \forall\hat{x}\,(\varphi \rightarrow \sim s = x)$

Hence by Quantification:

(15) $\sim\varphi^{s}/_{x} \rightarrow (\forall\hat{x}\varphi \rightarrow \forall\hat{x} \sim s = x)$

But in Stalnaker's system we can already prove:

(16) $Es \rightarrow\sim\forall\hat{x} \sim s = x$

standard methods for assessing claims about logical form in semantics. It is doubtful that they can withstand such assessment. If defenders of counterpart theory drop the claims about logical form and protest that it is just a theory about the metaphysical truthmakers for sentences of quantified modal logic, then they cannot reconcile it with classical logic in the way just envisaged. Unclarity on this point has made counterpart theory look more defensible than it really is. For a more specific critique along related lines see Fara and Williamson 2005.

Propositional reasoning from (15) and (16) yields:

(17) $\forall\hat{x}\varphi \rightarrow (Es \rightarrow \varphi^{s}/_{x})$

Hence Necessitation and the K schema give:

(18) $\Box\forall\hat{x}\varphi \rightarrow \Box(Es \rightarrow \varphi^{s}/_{x})$

But from (18) Universal Generalization yields QCBF. As before, NEI can then be derived as a corollary of QCBF.

Thus all that blocks this alternative derivation of QCBF and NEI are Stalnaker's decisions to allow non-rigid individual constants and have no separate category of rigid arbitrary names (to make the logic as free as possible, one could still permit constants in the latter category to be world-independently empty). Those decisions are in no way compelled by first-order non-modal logic. This reinforces the conclusion that the underivability of QCBF and NEI in Stalnaker's axiomatization is an artefact of its detailed workings and does not undermine their status as logical truths.

4. We turn to the case of ND, the necessity of distinctness. As Stalnaker notes, it is underivable even when QCBF is added to his axiomatization. More generally, it is underivable when the Free Abstraction rule replaces his Abstraction schema. For Free Abstraction preserves validity on Stalnaker's other deviant semantics, on which '=' is interpreted to mean indiscernibility but everything else is standard, while ND is invalid. ND would remain underivable even if we were to add the schema Identity* to the axiomatization, because it remains invalid even when we validate Identity* by requiring individual constants to be rigid designators on that deviant semantics. Is ND a principle that really cannot be settled by the combination of first-order non-modal logic with identity and propositional modal logic?

Stalnaker himself notes a reason for qualifying his claim that ND cannot be so settled (p. 156). The deviant semantics equates the extension of '=' at a world with the set of ordered pairs of members of the domain of that world that are mutually indiscernible, in the sense that they are in the extension of the same one-place predicates (open or closed, simple or complex, but not containing '=' itself) at that world. Stalnaker's counter-model to ND on the indiscernibility semantics requires two individuals *a* and *b* in the domain of a world *w* that are discernible in *w* but indiscernible in some world *w** accessible from *w*. This can happen only if *w* is not accessible from *w**, for otherwise the discernibility of *a* and *b* in *w* makes them discernible in *w** too by modal predicates: hence, *a* but not *b* is in the extension of $\hat{x}\Box(x = x \rightarrow x = y)$ at *w** on an assignment that maps the variable *y* to *a*. Thus the counter-model depends on an underlying propositional modal logic in which non-symmetric accessibility relations are permitted. Stalnaker opts for the weakest normal propositional modal logic K, which imposes no constraints whatever on accessibility (since it can be non-reflexive, necessity does not even entail truth in K). But many philosophers take the propositional logic of metaphysical modality to be S5, the logic of the class of models in which every world is accessible from every world, and also of the wider class of models in which accessibility is an equivalence relation. In such models, accessibility is symmetric. The same holds of weaker propositional modal logics with the Brouwerian schema:

B $\vdash \Diamond\Box\varphi \rightarrow \varphi$

This schema (which is of course derivable in S5) corresponds to the symmetry of accessibility but not to its reflexivity or transitivity. Indeed, even axioms that are weaker than B in the presence of the reflexivity axiom T ($\vdash \Box\varphi \rightarrow \varphi$) will suffice to rule out Stalnaker's counter-model to ND, for it has no counter-model on the deviant interpretation of '=' in which, whenever w^* is accessible from w, w can be reached from w^* in finitely many steps of accessibility.

Syntactically, we can derive ND once we strengthen Stalnaker's axiomatization by Free Abstraction on the first-order non-modal side and by the B schema on the propositional modal side. For we can derive NEI using the Abstraction rule, from which it is routine to derive:

(19) $\forall\hat{x}\forall\hat{y}(\Diamond x = y \rightarrow \Diamond\Box(Ex \rightarrow x = y))$

But from B we can also derive:

(20) $\forall\hat{x}\forall\hat{y}(\Diamond\Box(Ex \rightarrow x = y) \rightarrow (Ex \rightarrow x = y))$

From (19) and (20) we have:

(21) $\forall\hat{x}\forall\hat{y}(\Diamond x = y \rightarrow (Ex \rightarrow x = y))$

Using (12) above (everything exists), we easily obtain ND from (21) by contraposition.[9] Thus the underivability of ND in Stalnaker's axiomatization depends on the separate weaknesses of his abstraction principle and his propositional modal logic.

Suppose, however, that we are working with an interpretation of $\Box$ for which we do not wish to impose any constraints on the accessibility relation. Nevertheless, Stalnaker's semantic framework enables us meaningfully to expand the language by introducing a second necessity-like operator $\blacksquare$ by the semantic clause:

> If φ is a sentence, $v^s_w(\blacksquare\varphi) = 1$ if $v^s_u(\varphi) = 1$ for all $w\varepsilon W$; otherwise $v^s_w(\blacksquare\varphi) = 0$.

The dual operator $\blacklozenge$ is of course defined as $\neg\blacksquare\neg$. The semantic clause obviously validates the principles of S5 for $\blacksquare$:

[9] The main idea of the proof is from Prior 1955: 206–7.

K schema for $\blacksquare$	$\vdash \blacksquare(\varphi \rightarrow \psi) \rightarrow (\blacksquare\varphi \rightarrow \blacksquare\psi)$
T schema for $\blacksquare$	$\vdash \blacksquare\varphi \rightarrow \varphi$
E schema for $\blacksquare$	$\vdash \blacklozenge\varphi \rightarrow \blacksquare\blacklozenge\varphi$
Necessitation for $\blacksquare$	If $\vdash \varphi$ then $\vdash \blacksquare\varphi$

It is well known that those principles for $\blacksquare$ enable one to derive:

B schema for $\blacksquare$	$\blacklozenge\blacksquare\varphi \rightarrow \varphi$
4 schema for $\blacksquare$	$\blacksquare\varphi \rightarrow \blacksquare\blacksquare\varphi$

We also add a valid schema linking the two box operators:

Bridge schema $\qquad \vdash \blacksquare\varphi \rightarrow \Box\varphi$

For if φ is true in all worlds whatsoever, then *a fortiori* it is true in all accessible worlds. Of course, the Bridge schema is essentially a bimodal principle: we obviously cannot hope to derive it from the separate logics of $\blacksquare$ and $\Box$. But it is not clear that Stalnaker can object to that. For, analogously, he does not attempt to derive his non-modal logic of quantification with identity from separate non-modal logics of quantification without identity and identity without quantification.

Informally, we can think of $\blacksquare$ and $\blacklozenge$ as, respectively, metaphysical necessity and metaphysical possibility, and of $\Box$ and $\Diamond$ as restricted modalities of some sort; Stalnaker himself sometimes speaks of metaphysical necessity in just such terms (2003: pp. 202–3). Thus we have a bimodal logic. In the presence of Necessitation for $\blacksquare$ and the bridge schema, Necessitation for $\Box$ is of course redundant, and so may be dropped from the axiomatization, although we still need the K schema for $\Box$ in addition to that for $\blacksquare$. We can now employ a strategy of first using the S5 principles to prove a result for $\blacksquare$ and then using the Bridge schema to deduce a corresponding result for $\Box$.

As an instance of the strategy, we start by proving ND in the system that results from replacing Stalnaker's Abstraction schema by Free Abstraction in his combined axiomatization and adding the K, T and E schemas and Necessitation for $\blacksquare$ and the Bridge schema. In the way already sketched, we use the B schema for $\blacksquare$ to derive:

ND for $\blacksquare \qquad \forall\hat{x}\forall\hat{y}(\sim x = y \rightarrow \blacksquare \sim x = y)$

From the Bridge schema and Universal Generalization we prove:

(22) $\qquad \forall\hat{x}\forall\hat{y}(\blacksquare \sim x = y \rightarrow \Box \sim x = y)$

ND for $\blacksquare$ and (22) yield ND for $\Box$ by non-modal reasoning.

We can generalize the result to the following, for any natural numbers j and k, where $\Box^j$ is a sequence of j occurrences of $\Box$:

ND+: $\qquad \Box^j\forall\hat{x}\forall\hat{y}(\sim x = y \rightarrow \Box^k \sim x = y)$

To see this, note that we can prove the following for any j:[10]

Extended Bridge schema: $\blacksquare\varphi \rightarrow \Box^j\varphi$

Consequently, we can strengthen (22) to:

(23) $$\forall\hat{x}\forall\hat{y}(\blacksquare \sim x = y \rightarrow \Box^j \sim x = y)$$

Just as ND for $\blacksquare$ and (22) yield ND for $\Box$, so ND for $\blacksquare$ and (26) yield:

(24) $$\forall\hat{x}\forall\hat{y}(\sim x = y \rightarrow \Box^j \sim x = y)$$

To complete the argument for ND+, subject (24) to Necessitation for $\blacksquare$ and then apply the Extended Bridge schema.[11] Many similar results are derivable by such means.

According to Stalnaker, 'the necessity (or essentiality) of identity is more central to the logic of identity than the necessity of distinctness' (p. 161). That may well be so in the sense that natural systems of first-order modal logic with identity require significantly richer resources to prove the necessity of distinctness than they require to prove the necessity (or essentiality) of identity: more axioms or rules of inference and, in some cases, greater expressive powers. But proofs of the necessity of distinctness and strengthenings of it such as ND+ need not employ principles that derive neither from first-order non-modal logic with identity nor from propositional modal (bimodal) logic. The proofs for $\Box$ above used only principles taken from first-order non-modal logic (including Free Abstraction) with identity and the bimodal logic of $\Box$ and $\blacksquare$. No distinctively modal principles concerning '=' were assumed. Thus Stalnaker's suggestion that there may be 'in some sense something modal about the concept of identity' is not supported when one examines the issue in a wider range of logical settings.

5. The result of the preceding discussion is that the underivability in Stalnaker's axiomatization of the principles QCBF, NEI and ND, which are valid on his semantics, casts no serious doubt on their status as logical truths. His treatment of individual constants as non-rigid was also queried in passing. The remainder of the paper raises some more radical questions about his system. Nothing in the critique of Stalnaker's argument above depends on what follows.

[10] Proof: By mathematical induction on j. For $j = 0$, the Extended Bridge schema reduces to the T schema. Suppose that the Extended Bridge schema holds for j. By Necessitation and the K schema for $\blacksquare$ we prove $\blacksquare\blacksquare\varphi \rightarrow \blacksquare\Box^j\varphi$. Thence the 4 schema for $\blacksquare$ yields $\blacksquare\varphi \rightarrow \blacksquare\Box^j\varphi$. An instance of the original Bridge schema is $\blacksquare\Box^j\varphi \rightarrow \Box^{j+1}\varphi$. Together, these yield the Extended Bridge schema for $j+1$.

[11] Somewhat similar results are established in Williamson 1996a using the logic of an 'actually' operator rather than $\blacksquare$. Stalnaker (2003: 159–61) comments on those results in a postscript added in 2002 to the reprinting of Stalnaker 1994. He objects that ND+ remains unprovable (although valid) even when the QCBF schema and a complete logic for 'actually' are added to his axiomatization, although ND itself is provable. The present result shows that to be a specific feature of the logic of 'actually', not a more general phenomenon. For discussion of related issues see Karmo 1983 and Humberstone 1983.

We may start by reflecting on Abstraction principles. Free Abstraction preserves validity on Stalnaker's semantics. It has the extra complexity of the existential conjunct *Ex*. Is that extra conjunct really wanted? If we delete it, we obtain this simpler rule:

Simple Abstraction rule: If B is the result of replacing some or all occurrences of $\hat{y}\varphi x$ in A by $\varphi^x/_y$, where $\varphi^x/_y$ is as in the original Abstraction principle, and $\vdash$A, then $\vdash$ B

On Stalnaker's first-order non-modal semantics, Simple Abstraction preserves validity (and, indeed, truth in a model). Admittedly, in any model $v^s(\hat{y}\,\varphi x)$ and $v^s(\varphi^x/_y)$ will differ for some assignments s and formulas φ, for Stalnaker treats assignments as partial functions from individual variables to members of the domain; if the domain is empty, there are no such total functions (p. 158). For example, if s assigns no value to x, $v^s(\hat{y} \sim Eyx) = 0$ (because $\langle v^s(x)\rangle \varepsilon v^s(\hat{y} \sim Ey)$ only if v^s is defined on x); but $v^s(\sim Ex) = 1$ (because $v^s(Ex) = 0$). However, for any member d of D, $v^{s[d/x]}(\hat{y}\varphi x) = v^{s[d/x]}(\varphi^x/_y)$. Now Stalnaker restricts his logic to closed formulas, and in his language a variable x is bound only by the predicate-forming operator $\hat{x}$, which forms a predicate whose semantic value relative to s depends only on the semantic values of the formula to which $\hat{x}$ was applied relative to assignments $s[d/x]$ that assign x a member d of D. Consequently, if B is the result of replacing some or all occurrences of $\hat{y}\,\varphi x$ in a closed formula A by $\varphi^x/_y$, then $v^s(\text{A}) = v^s(\text{B})$ for all assignments s, including those undefined on x. For example, $v^s(\forall\hat{x} \sim \hat{y} \sim Eyx) = v^s(\forall\hat{x} \sim\sim Ex) = 1$. Thus Simple Abstraction preserves validity in the non-modal system.

The corresponding argument fails on Stalnaker's semantics for the first-order modal language. For example, $v^s_w(\forall\hat{x}\Box \sim \hat{y} \sim Eyx) = 1$ since $v^{s*}_{w*}(\hat{y} \sim Ey)$ is empty for any $s*$ and w^*; but $v^s_w(\forall\hat{x}\Box \sim\sim Ex) = 0$ whenever some member of the domain of w is absent from the domain of some world accessible from w. In general, although $v^s_w(\hat{y}\varphi x) = v^s_w(\varphi^x/_y)$ whenever $s(x)\varepsilon\, \text{D}_w$, a modal operator may intervene between those formulas and the occurrence of $\hat{x}$ that binds occurrences of the variable x in them, so that the truth-values of A and B in Simple Abstraction can be sensitive to differences between $v^s_w(\hat{y}\varphi x)$ and $v^s_w(\varphi^x/_y)$ for assignments s such that $s(x)$ belongs only to the domains of some worlds other than w. But that depends on Stalnaker's decision to relativize domains to worlds. For models in which all worlds have the same domain, the earlier argument for the non-modal case can easily be adapted to show that Simple Abstraction preserves truth. One can combine the propositional modal semantics with the first-order non-modal semantics without relativizing domains to worlds, just as Stalnaker himself deliberately refrains from relativizing assignments of values to variables to worlds.

Simple Abstraction gives a smoother account of the effect of abstraction than Free Abstraction does, since the content of $\hat{y}\,\varphi x$ is unpacked wholly in terms of the abstracted formula φ and the variable x, without the introduction of extraneous elements such as *Ex*. The smoother account *might* be unsatisfactory if the variable had been assigned no value, but, as before, since the logic is confined to closed formulas, in which the variable x cannot occur unbound by $\hat{x}$, the relevant assignments all assign it a value. The truth-condition of the sentence φ is equivalent to a condition on the individual assigned to y, and the truth-condition of the sentence $\varphi^x/_y$ is equivalent to the condition on the individual assigned to x, irrespective of whether it belongs to the domain of the current world, that it meets the former condition.

Formally, the rationale for Simple Abstraction is strong. But once Stalnaker's system is extended to include the rule, highly controversial theorems are forthcoming. Note first that his Existence schema yields:

(25) $\hat{y} \sim Eyt \rightarrow Et$

From (25), Necessitation and then Universal Generalization give:

(26) $\forall \hat{x} \Box (\hat{y} \sim Eyx \rightarrow Ex)$

Applying Simple Abstraction to (26), we have:

(27) $\forall \hat{x} \Box (\sim Ex \rightarrow Ex)$

Since $(\sim p \rightarrow p) \rightarrow p$ is a truth-functional tautology, by standard reasoning we can derive from (27):

NE $\forall \hat{x} \Box Ex$

Everything has necessary existence (the necessity of existence). Indeed, by applying Necessitation j times before Universal Generalization and i times afterwards, we can prove:

NE+ $\Box^i \forall \hat{x} \Box^j Ex$

But, it might be thought, NE is quite bad enough already. Is it not obvious that many actually existing things, including ourselves, exist only contingently? If so, NE constitutes a reductio ad absurdum of the system in which it was derived. On that basis, one might reject Simple Abstraction, or perhaps keep it and reject Stalnaker's Existence principle instead.

That reaction would be too quick. What is obvious enough is that many things that exist in space and time, such as ourselves, could have failed to exist in space and time. But we should not assume that existing in space and time is the only way of existing. For example, if pure sets exist, as they arguably do, they presumably do it without existing in space and time. To say that pure sets exist is just to say that there are pure sets. To say that the null set exists is just to say that there is one and only one set with no members. Something in space and time is a counterexample to NE only if it could have failed to exist *at all*; it is insufficient that it could have failed to exist in space and time. Of course, if we had existed without existing in space and time, we would not then have been persons, let alone sets. Rather, we would have been merely possible persons: non-persons that could have been persons. Being a merely possible person is not a way of not existing; it is a way of existing in the only sense of the term of special interest to logic, that is, of being something or other. Elsewhere, I have defended such a conception of modal metaphysics in more detail (Williamson 1990, 1998, 2000a, 2000b, 2002).

One can validate Simple Abstraction by modifying the semantics to have a single domain D, unrelativized to a world. For the general plan of combining the semantics of the propositional modal language with the semantics of the first-order non-modal language simply leaves it open whether the domain of the latter should be relativized to the worlds of the former or not.

Given Simple Abstraction, one can derive CBF, the unqualified converse Ibn-Sina-Barcan schema. For Simple Abstraction is equivalent to Free Abstraction in the presence of NE+; the existence conjunct becomes redundant. We have already seen how to derive QCBF by Free Abstraction. We can therefore derive QCBF by Simple Abstraction, and then obtain CBF from QCBF by NE. It is well known that when domains are world-relative, CBF corresponds to the semantic condition that whenever a world x is accessible from a world w, the domain of w is a subset of the domain of x. Similarly, consider the Ibn-Sina-Barcan schema itself:

BF $\quad \forall\hat{x}\Box\varphi \rightarrow \Box\forall\hat{x}\varphi$

BF corresponds to the condition that whenever x is accessible from w, the domain of x is a subset of the domain of w. Together, the two conditions are equivalent to the constancy of the domain across chains of accessibility; variation in domain between worlds not linked by a chain of accessibility by itself makes no difference to the truth of any sentence. The semantics with a constant domain validates both CBF and BF. However, BF differs from CBF in being underivable even when Stalnaker's system is expanded by Simple Abstraction. For, as already noted, that rule is derivable from Free Abstraction together with NE+. But Stalnaker's system, Free Abstraction and NE+ are all validated by a semantics like Stalnaker's with world-relative domains but subject to the semantic condition corresponding to CBF, whereas BF remains invalid on that semantics.

One way to expand the system to permit the derivation of BF is by adding a modal operator $\Box^{-1}$ whose accessibility relation S is required to be the converse of the accessibility relation R for $\Box$ in all models: $w\text{S}x$ if and only if $x\text{R}w$. Naturally, $\Diamond^{-1}$ is $\neg\Box^{-1}\neg$. Thus $\Diamond$ and $\Diamond^{-1}$ are related like past and future tense operators in tense logic.[12] These new modalities make as good sense as $\Box$ and $\Diamond$ do within Stalnaker's semantics. Their interrelationship automatically validates two axiom schemas for their propositional bimodal logic:

[12] See again Karmo 1983 and Humberstone 1983.

Converse1 $\vdash \varphi \rightarrow \Box\Diamond^{-1}\varphi$

Converse2 $\vdash \varphi \rightarrow \Box^{-1}\Diamond\varphi$

Of course, the K schema for $\Box^{-1}$ is also valid, and Necessitation for $\Box^{-1}$ preserves validity. Thus we consider the extension of Stalnaker's system with Converse1, Converse2 and the K schema for $\Box^{-1}$ as additional axiom schemas, Necessitation for $\Box^{-1}$ as an additional rule and Simple Abstraction in place of his Abstraction schema. We can now derive CBF for $\Box^{-1}$ just as we derived it for $\Box$, using the extra power of Simple Abstraction in the expanded language. To derive BF for $\Box$ we proceed thus. As already noted, $\forall F \rightarrow (Et \rightarrow Ft)$ is derivable in Stalnaker's system, so it is derivable in this extension. By substituting $\hat{x}\Box\varphi$ for F in the proof and then using Universal Generalization, the proof that everything exists and Simple Abstraction we prove:

(28) $\forall\hat{x}(\forall\hat{x}\Box\varphi \rightarrow \Box\varphi)$

Applying Necessitation for $\Box^{-1}$ gives:

(29) $\Box^{-1}\forall\hat{x}(\forall\hat{x}\Box\varphi \rightarrow \Box\varphi)$

As noted, we can derive CBF for $\Box^{-1}$, so (29) yields:

(30) $\forall\hat{x}\Box^{-1}(\forall\hat{x}\Box\varphi \rightarrow \Box\varphi)$

By standard manipulations on (30) we obtain:

(31) $\forall\hat{x}\Diamond^{-1}\forall\hat{x}\Box\varphi \rightarrow \forall\hat{x}\Diamond^{-1}\Box\varphi$

We can remove the outer quantifier in the antecedent by Universal Generalization and the modalities in the consequent by Converse2 (as contraposed), so we have:

(32) $\Diamond^{-1}\forall\hat{x}\Box\varphi \rightarrow \forall\hat{x}\varphi$

By Necessitation and the K schema for $\Box$ we have:

(33) $\Box\Diamond^{-1}\forall\hat{x}\Box\varphi \rightarrow \Box\forall\hat{x}\varphi$

Finally, Converse1 allows us to remove the outer two modalities in the antecedent, thereby obtaining BF for $\Box$. We obtain BF for $\Box^{-1}$ in exactly parallel fashion. The proofs involve only principles from the propositional bimodal logic of $\Box$ and $\Box^{-1}$ and first-order non-modal logic; they do not require any extra assumptions about the interaction of modal operators with quantification and identity.

In the special case in which the accessibility relation for $\Box$ is assumed to be symmetric, $\Box^{-1}$ reduces to $\Box$, Converse1 and Converse2 both reduce to the B schema, and the proof of BF reduces to the usual proof for monomodal systems with CBF and B.

6. How should we decide between variable domains and constant domains versions of possible worlds semantics? Given the overall approach, both versions seem to confer truth-conditions on sentences of the formal object-language in a coherent way. Of course, the model theory by itself does not completely fix the meaning of those sentences. For example, it is compatible with many different readings of $\Box$, depending on how the accessibility relation R is to be understood. In fixing the intended interpretation of the formal language, we also need to decide how any domains are to be understood: that may enable us to decide whether they should be variable or constant.

However, a prior issue arises. For whether domains are variable or constant, their role is to restrict the universal quantifier of the meta-language that is used in the semantic clause for $\forall$ to state its contribution to the truth-conditions of sentences of the object-language in which it occurs. On this view, for $\forall F$ to be true at a world, it suffices that the predicate F applies at that world to all members of the relevant domain, even if it fails to apply to some things outside the domain. In effect, $\forall$ is being interpreted as a restricted quantifier. The Ibn-Sina-Barcan schema and its converse can undoubtedly fail on a restricted reading of the quantifier, even when the modality is read as metaphysical. For example, call a number *popular* if and only if it is many people's favourite number. 'Every popular natural number is necessarily prime' does not entail 'Necessarily every popular natural number is prime', since the only actually popular ones may be prime, and therefore necessarily prime, even though some composite natural numbers could have been popular. Conversely, the obviously true 'Necessarily every popular natural number is popular' does not entail the obviously false 'Every popular natural number is necessarily popular'. But for both metaphysics and logic the most interesting reading of $\forall$ is as totally unrestricted, ranging over everything whatsoever. The restricted readings can then be recovered as complex quantifiers constructed out of the unrestricted one and a restricting condition; the basic reading is the unrestricted one. On that unrestricted reading, $\forall F$ is true at a world if and only F applies to everything whatsoever at that world; domains do not come into it. Such a domain-free semantics automatically validates BF, CBF and NE (Williamson 2000a). On this view, the existence and identity of individuals (being something and being the same thing) are entirely non-contingent matters.

It might be objected that the appearance of domains in the standard semantics is no real restriction on the quantifiers, because in a given world there is nothing except what exists there to quantify over, and on the intended interpretation the domain of a world by definition contains whatever exists there. But that objection fails to take the possible worlds semantics seriously. It treats the semantic clause for $\forall$ as though it were a misleading approximate translation of a more fundamental clause in which an unrestricted universal quantifier of a more fundamental meta-language occurred within the scope of a modal operator. For present purposes, we take the standard meta-language for the semantics of quantified modal logic seriously, as Stalnaker does; the discussion of a more homophonic form of semantics for quantified modal logic must be postponed for another occasion. Once the standard semantic clauses for ∀ are taken at face value, the restriction to the domain of a world must be understood as genuinely imposing a restricted reading on the object-language quantifiers; so the objection fails.

The logic of absolutely unrestricted quantification is highly controversial in at least two ways. First, it is controversial whether absolutely unrestricted quantifiers even make sense, especially given the threat of set-theoretic paradox (see Cartwright 1994 and Williamson 2003 for arguments that they do). Second, even granted that they do make sense, it is controversial whether they should count as logical constants (see Williamson 2000a and Rayo and Williamson 2003 for an account on which they do). If they do so count, then the formalization of the claim that there are at least n things is a logical truth for each natural number n, in the Tarskian sense that it is true under all interpretations of the non-logical vocabulary, for it is true (after all, the formula itself contains at least n variables) and contains no non-logical vocabulary; it cannot be invalidated by a restricted domain of fewer than n things because its semantics involves no restriction to a domain. Thus logic has substantive existential commitments, just as it has on Frege's logicist conception. Those controversies about absolutely unrestricted quantification cannot be resolved here.

The case of absolutely unrestricted quantification illustrates in an extreme way the potential of logic for metaphysical controversy. Less extreme illustrations are provided by the formulas that are invalid on Stalnaker's deviant counterpart semantics but nevertheless valid on the intended semantics, and provable in an appropriate axiomatization. But no other science is bound by the constraint that its laws must be uncontroversial. Why should logic be any different? Of course, when logic is controversial it cannot easily act as arbiter of fair play in extralogical disputes; but we have already observed that to define logic by that role would be to condemn it to extreme unsystematicity. There is no science of fair play. A better proposal is that the primary function of logic is to investigate logical consequence, that is, truth-preservation from premises to conclusion however the argument is interpreted, given its logical form. After all, we need *some* science to investigate that, and logic is by far the best candidate. Logic is not defined by its dialectical or epistemological status. But if we carry out the investigation well enough, the generality of its results can still carry an authority sufficient for our needs.

REFERENCES

Cartwright, R. 1994. 'Speaking of everything', *Noûs* **28**: 1–20.

Fara, M., and Williamson, T. 2005. 'Counterparts and actuality', *Mind* **114**: 1–30.

Fine, K. 1983. 'The permutation principle in quantificational logic', *Journal of Philosophical Logic* **12**: 33–7.

Hughes, G. E., and Cresswell, M. J. 1996. *A New Introduction to Modal Logic*. London: Routledge.
Humberstone, I. L. 1983. 'Karmo on contingent non-identity', *Australasian Journal of Philosophy* **61**: 188–91.
Karmo, T. 1983. 'Contingent non-identity', *Australasian Journal of Philosophy* **61**: 185–7.
Kripke, S. 1963. 'Semantical considerations on modal logic', *Acta Philosophica Fennica* **16**: 83–94.
Movahed, Z. 2004. 'Ibn-Sina's anticipation of Buridan and Barcan formulas', to appear in A. Enayat, I. Kalantari and M. Moniri (eds.), *Proceedings of the Workshop and Conference on Logic, Algebra and Arithmetic, Tehran 2003* (in the Lecture Notes in Logic series of the Association for Symbolic Logic). Natick, Mass.: A.K. Peters.
Prior, A. N. 1955. *Formal Logic*. Oxford: Clarendon Press.
Quine, W. V. 1953. 'Three grades of modal involvement', in *Proceedings of the XIth International Congress of Philosophy*, Brussels, 1953, vol. 14. Amsterdam: North-Holland. Reprinted in Quine 1966.
____ 1966. *The Ways of Paradox and Other Essays*. New York: Random House.
Rayo, A., and Williamson, T. 2003. 'A completeness theorem for unrestricted first-order languages', in J.C Beall (ed.), *Liars and Heaps: New Essays on Paradox*. Oxford: Clarendon Press.
Sinnott-Armstrong, W., Raffman, D., and Asher, N. (eds.) 1994. *Modality, Morality and Belief: Essays in honor of Ruth Barcan Marcus*. Cambridge: Cambridge University Press.
Stalnaker, R. C. 1994. 'The interaction of modality with quantification and identity', in Sinnott-Armstrong, Raffman and Asher 1994. Reprinted with revisions and a postscript in Stalnaker 2003, to which page numbers refer.
____ 2003. *Ways a World Might Be: Metaphysical and Anti-Metaphysical Essays*. Oxford: Clarendon Press.
Williamson, T. 1990. 'Necessary identity and necessary existence', in R. Haller and J. Brandl (eds.), *Wittgenstein—Towards a Re-Evaluation: Proceedings of the 14th International Wittgenstein-Symposium*, vol. I. Vienna: Holder-Pichler-Tempsky.
____ 1996a. 'The necessity and determinacy of distinctness', in S. Lovibond and S. Williams (eds.), *Essays for David Wiggins: Identity, Truth and Value*. Oxford: Blackwell.
____ 1996b. Review of Sinnott-Armstrong, Raffman and Asher 1994. *Philosophy* **71**: 167–72.
____ 1998. 'Bare possibilia', *Erkenntnis* **48**: 257–73.
____ 2000a. 'Existence and contingency', *Aristotelian Society* sup. vol **100**: 181–203.
____ 2000b. 'The necessary framework of objects', *Topoi* **19**: 201–8.
____ 2002. 'Necessary existents', in A. O'Hear (ed.), *Logic, Thought and Language*. Cambridge: Cambridge University Press.
____ 2003. 'Everything', *Philosophical Perspectives* **17**: 415–65.

7

Conditional-Assertion Theories of Conditionals

William G. Lycan

Some clever ass has said that 'if' is the biggest word in the language, but I say it's the most useless.

Sir Harry Flashman[1]

Now under what circumstances is a conditional true? Even to raise this question is to depart from everyday attitudes. An affirmation of the form 'if *p* then *q*' is commonly felt less as an affirmation of a conditional than as a conditional affirmation of the consequent.... If, after we have made such an affirmation, the antecedent turns out true, then we consider ourselves committed to the consequent, and are ready to acknowledge error if it proves false. If on the other hand the antecedent turns out to have been false, our conditional affirmation is as if it had never been made.
Departing from this usual attitude, however, let us think of conditionals simply as compound statements which, like conjunctions and alternations, admit as wholes of truth and falsity.

W. V. Quine

Quine attributes the view here called the "usual attitude" to Dr Philip Rhinelander.[2] His casual departure from it is now (half a century on) a bit controversial. While many theories have been offered according to which conditionals, subjunctive and indicative alike, are compound statements that have truth-values as wholes,[3] descendants of the conditional-assertion theory that he calls the "usual

[1] George MacDonald Fraser, *Flashman and the Tiger*. New York: Alfred A. Knopf, 2000, 198.

[2] Quine (1950), 12. The conditional-affirmation view of conditionals may be Professor (later Dean) Rhinelander's best-known contribution to philosophy, but his influence extends outside the field and into the military and political history of the United States; see J. B. Stockdale, "Master of My Fate—A Stoic Philosopher in a Hanoi Prison," *The World and I* (May, 1995), reprinted at http://www.geocities.com/stoicvoice/journal/0600/js0600a1.htm.

[3] The recipient and honoree of this volume led the field, in Stalnaker (1968), closely followed by Lewis (1973); see also Stalnaker (1975). Others who have offered truth-conditional accounts of

attitude" have recently made a strong comeback as competitors of the standard truth-conditional theories of indicative conditional sentences. My purpose in this paper is to assess their prospects in that role.

I. THE QUINE–RHINELANDER THEORY

What Quine and Rhinelander termed "affirmation" is now more commonly called "assertion," and affirming and asserting are illocutionary acts. The parallel is with common conditional speech acts such as conditional requests ("If you're going out anyway, could you please pick up some Dos Equis?") or conditional bets ("If Kerry gets the nomination, I bet you $100 he'll win"); should the antecedents prove false, no request has been issued and the bet is off. (Or so it is widely supposed; the latter claims are not obvious.) Call the conditional-assertion view thus interpreted the "Simple Illocutionary theory".

But to understand the conditional-assertion view in that light is to diminish its force as a competitor of the standard theories of conditionals, for those are semantic theories applying to sentences, or to sentences in contexts of utterance, not to acts of asserting or their social products. I suppose that is one reason why Quine did his departing; as is well known, there is no simple relation between speech acts of asserting and the semantic properties of the sentences uttered.[4] (No doubt a more fundamental reason is that he was bent on what he called the "regimentation" of factual discourse, not on either commonsensical views of asserting or, for that matter, on the ordinary meanings of conditional sentences in natural language, whatever those might be if they existed at all.)[5]

It is curious, though, that Quine took the Simple Illocutionary (SI) theory to be the "usual attitude" or common sense, or the ordinary person's interpretation of conditional speech. For that is a very strange imputation.[6] I strongly doubt that any ordinary person, hearing a speaker utter an ordinary conditional

whole conditionals include: Davis (1979); Jackson (1979, 1987); Lycan (1984, 2001); and Kratzer (1986).

My own semantic ideas about conditionals have been most heavily influenced by Bob Stalnaker's, and they remain very close to his. He does not entirely welcome my company in this area, but I wholeheartedly thank him for his kind and critical conversation over the years.

[4] Only some of a sentence's entailments are asserted when the sentence is uttered; in particular, there is a contrast between assertion and presupposition (though it is vexed). Explicit performatives strongly motivate a distinction between the truth of such a sentence and the truth of the assertion made in uttering it (consider the sentence, "Mindful that I am under oath, I hereby state that I have never met or had any contact with the defendant"). According to some recent Relevance theorists, one can unequivocally and actionably assert propositions that are not entailed by the sentence one utters (e.g., Carston 2002). And so far as speaker-meaning figures in the analysis of "what is asserted," a mismatch between speaker-meaning and sentence meaning may result in a truth-value difference.

[5] On regimentation and the death of meaning, see respectively chs. V and II of Quine (1960).

[6] His famous footnote to p. 12 does not say whether Professor Rhinelander himself offered the conditional-assertion view merely as representing common sense, or as correct.

sentence, would suspend judgment on *whether any assertion had been made* until it had been established whether the antecedent was actually true. My neighbor says to her daughter, "If you don't finish mowing the yard this afternoon, you can't go to the mall after dinner." I think the daughter, at least, would perceive that something fairly substantial had been asserted, right then. And it is odd to suppose that by mowing the rest of the yard, she could see to it that her mother had asserted nothing at all.

Setting aside the question of common sense or ordinary usage, the SI theory is implausible in its own right. Some other examples to show that: "The element will burn out if you throw that switch while the red light is on"; "That figurine will break if dropped"; "She won't pass unless she scores at least 75 on the final exam"; "If they get measles they'll break out in spots." In none of these cases do we feel that nothing at all has been asserted unless the antecedent happens to be true. (Call this the "Initial Implausibility" objection.) The late Richard Jeffrey (1963) objected that if Quine and Rhinelander are right, "The hearer of a conditional whose antecedent will turn out to have been false loses nothing if he fails to hear the consequent" (p. 42). Jeffrey offered a useful analogy. According to SI, to utter a conditional is as if to hand one's hearer a sealed envelope. In our example, the mother would hand the daughter an envelope marked, "To be opened in case you do not finish mowing the yard this afternoon, but otherwise to be destroyed unopened."[7] Only if the daughter does not finish mowing the yard would she then open the envelope and find within it a piece of paper on which is written, "You can't go to the mall after dinner." And similarly for the further examples.

(The analogy is not perfect, because the SI theory at least allows that the hearer knows *what will have been* asserted should the antecedent prove true; she gets a peek at the paper inside the envelope. But the fact remains that SI allows for no more definite actual illocutionary or cognitive event; what would have been asserted is irrelevant if the antecedent is false.)

Notice, incidentally, that the opening argument from analogy to other conditional speech acts is flawed. There is a special reason why the conditional request and conditional bet examples work as they do: In each case, the locutionary content in question itself presupposes the truth of the antecedent. It is a logical truth that you cannot do anything for me while you are out if you are not going to be out; it is at least a constitutive institutional fact that to win the election one must first have been nominated. Absent the truth of their respective

[7] Notice that that instruction is itself the conjunction of two conditionals. So, strictly, there would have to be duplicate envelopes, each containing the piece of paper with "You can't go to the mall . . ." and each inside a bigger envelope also containing a cover sheet. One of the bigger envelopes would say, "To be opened in case you do not finish mowing the yard this afternoon, but otherwise to be destroyed unopened," and the cover sheet inside would say, "Open the accompanying envelope." The other bigger envelope would say, "To be opened in case you do finish mowing the yard this afternoon, but otherwise to be destroyed unopened," and the cover sheet inside it would say, "Destroy the accompanying envelope unopened." But of course the instructions on the bigger envelopes would be conjunctions of conditionals . . . .

antecedents, the request and bet would be so grievously defective as speech acts that they could hardly count as request or bet at all. But ordinary conditional declaratives exhibit no such presuppositional phenomenon. To assert to someone that she cannot go to the mall after dinner presupposes nothing whatever about mowing yards. Notice that there are declarative conditionals whose consequents presuppose their antecedents: "If Richard Nixon stole money, someone somewhere knows he did"; "If Sheila owns a heavy overcoat, it's green." These are plausibly taken to be conditional assertions, precisely because of the presuppositional feature but only because of it.

By the same token, there are interrogative and imperative conditionals that are not plausibly taken as vehicles of merely conditional speech acts. In our mother–daughter situation, "Is it true that if she doesn't finish mowing the yard, she can't go to the mall?" would unequivocally be used to ask a question (attempting to confirm the mother's threat), whether or not the daughter will later finish the mowing. "Make it so that the mine will explode if anyone touches it even lightly" is an unmistakable command whether or not anyone is ever going to touch the mine; if the mine is not already so rigged and the subordinate does nothing, s/he has disobeyed the command.

II. FURTHER CRITICISM OF QUINE–RHINELANDER

The SI theory faces further problems. One is that contraposition is utterly ruled out from the beginning. If a conditional is used to assert something true, that is because its antecedent is true and so is its consequent. But then its contrapositive has a false antecedent, and so cannot be used to assert anything whatever.[8] Of course, given the frequent dislocation between semantic properties and (illocutionary) assertion properties, this is not in itself surprising, but if a speaker utters a conditional with assertive intent, we should expect that speaker to be ready, willing and able to use its contrapositive to make an assertion as well.

Yet another problem is that of nested conditionals. Conditionals with conditional consequents are perhaps manageable. On the SI theory, to utter "If we have a tomato, then if I can get some bacon I'll make a BLT" cannot be to assert that if I can get some bacon I will make a BLT on the condition that we have a tomato, because one cannot assert a distinctively conditional content. The SI theorist must say that the uttering is a conditional *conditional* assertion: If we do have a tomato, then I have issued the (merely) conditional assertion, "If I can get some bacon I'll make a BLT," and if also I can get some bacon, I will have

[8] Stalnaker (1968) and others have questioned the validity of contraposition for English conditionals. But counterexamples to contraposition have been limited to what Goodman (1947) called "semifactuals" and Wayne Davis' (1983) "weak" conditionals. Most conditionals still do contrapose. (See Lycan (2001), 31–6.)

succeeded in asserting that I will make a BLT.[9] But conditionals with conditional antecedents are harder to handle. According to SI, one who utters "If this vase will break if I drop it on the driveway, then I will be careful not to drop it on the driveway" asserts, on the condition that the vase will break if s/he drops it on the driveway, that s/he will be careful not to drop the vase on the driveway. But also according to SI, the "condition" specified is not a matter of fact, and so the antecedent cannot be true, and so nothing can have been asserted at all. (It would not help to let the condition be, rather, that the speaker does drop the vase and it does break, because that would make the original consequent, hence the assertion, automatically false.) "If you will fail unless you study hard, I suggest you hit the books." Another nice example (from Edgington (1995), 284) is "If John should be punished if he took the money, then Mary should be punished if she took the money."

And what about disjunctions with conditional disjuncts?: "Either Geoff will go to the dance, or if Laura allows him in, he will go to her party." Are we to suppose that if Laura does not allow Geoff in, the speaker will have asserted nothing? No, because if Geoff does go to the dance, the speaker will have asserted something true. I suppose the best position for the SI theorist to take is that if either Geoff goes to the dance or Laura allows him in, the speaker will have succeeded in asserting that either Geoff will go to the dance or he will go to Laura's party, but if Geoff does not go to the dance and Laura does not allow him in, nothing has been asserted; this seems ad hoc at best.[10]

Finally, it is hard to see how to extend the SI theory to the propositional attitudes, because it is a purely illocutionary theory, about the performing of public, convention-governed speech acts. (By the same token exactly, one may call this criticism unfair, since the SI theorist is a philosopher of language, not a philosopher of mind, SI being a thesis in linguistic pragmatics. But remember that we are considering conditional-assertion theories in their recent role as competitors of truth-conditional semantic theories of conditional sentences; and truth-conditional theories of conditional sentences extend very naturally to the corresponding propositional attitudes.)

It is easy enough to carry over the SI model to belief or judgment: As E.W. Adams and others do,[11] we can let belief having conditional form be belief of the conditional's consequent conditionally upon supposition of the antecedent, modeling this as the subject's subjective conditional probability of the consequent given the antecedent. But this will run into trouble over complex belief sentences.

[9] Someone might appeal to the principle that a nested conditional A > (B > C) is equivalent to (A & B) > C (Gibbard (1981)); but I think that principle is fairly easily refuted (Lycan (2001), 82).

[10] Barker (1995) offers an ingenious discussion of such disjunctions on the conditional-assertion theorist's behalf (pp. 202–5). He considers the sentence "Either Fred will marry Jane, if she asks him, or we are just completely wrong about Fred and this marriage business"—very much a case in point—but I do not see how his remarks on it are responsive (though they are true).

[11] Adams (1965), and most prominently, Dorothy Edgington (1986).

How would the Adams adaptation of SI handle something like, "Angie believes that Bob believes that Cindy will attend if she does, and Dave dislikes Cindy so much that he thinks if it's true that she will attend if Angie does, he will try to persuade Angie that if Cindy does attend, Angie should try to convince her that if she gets anywhere near him she'll catch something"?

Also, what of nondoxastic attitudes?[12] "Dave is afraid that if Cindy looks at him his left cheek will tic visibly"; "Angie is embarrassed that if Dave sees her he will send her what he thinks are subtle secret signals"; "Bob hopes that if Cindy and Dave both attend, no one will sing 'You Must Have Been a Beautiful Baby'"; "Cindy is sad that Bob will leave the room if she sings anything at all"; "Bob is ashamed that he will run crying from the room if someone does sing 'You Must Have Been a Beautiful Baby'." I do not see how such attitudes can be explicated in terms of conditional probability, unless possibly by some convolute analysis of embarrassment, hope *et al.* in terms of belief.

I believe, then, that the trick for the conditional-assertion theorist is to move beyond the SI theory, avoiding the four objections raised in these two sections.[13]

Notice, incidentally, that there is a close affinity between the conditional-assertion view even in its SI form and the thesis defended by Adams (1965) and others[14] that indicative conditionals lack truth-conditions and truth-values. (Call that thesis "NTV"; we shall return to it.) Suppose a conditional has a truth-condition, hence a truth-value given totality of fact. Then if a speaker utters that sentence in assertive mode, presumably s/he would (unconditionally) assert of its truth-condition that that condition obtains; if the conditional sentence is true constitutively iff P, then the speaker has asserted that P. But for the SI theorist, no conditional sentence can be used to assert a conditional content. Therefore, the SI theorist would find it hard to explain how a conditional sentence could have a truth-condition or a truth-value.

[12] This point is made by Michael Kremer (1987).

[13] G. H. von Wright (1957) offers a type of conditional-assertion account, and he emphasizes that by "assertion" he too means the speech act of asserting, or rather a slightly idealized version of that act. But his view differs crucially from the SI theory. He does hold that one who utters an indicative conditional "If p, then q" "licenses others to take him as having asserted q, *if* the condition p is found to be, in fact, fulfilled" (p. 130), but he does not hold that if the antecedent p is false, nothing has been asserted. Whether the antecedent is true or false, von Wright maintains, the assertion pattern is as follows: The speaker asserts the material conditional $p \supset q$ (*and* any logical equivalent of it such as $\sim(p \,\&\, \sim q)$), but neither asserts nor denies p and neither asserts nor denies q (p. 134).

Nonetheless von Wright insists that the conditional sentence does not express any proposition. (He gives a compressed and bilocated argument (pp. 131, 135): If a sentence expresses a proposition, then that proposition has a negation. But when one asserts a conditional sentence, the only proposition asserted is the corresponding material conditional, and that is only part of the assertive act—the rest of it being the four *refrainings* listed above—so there is no proposition that is the negation of what is asserted. He does not say why $\sim(p \supset \sim q)$ does not qualify.) Notice that on this view, either a proposition that is centrally and literally asserted by utterance of a sentence may not be entailed by the sentence, or a sentence that does not express a proposition may entail a proposition.

[14] Gibbard (1981); Appiah (1985); Edgington (1986, 1995); Bennett (2003).

III. SEMANTICIZING QUINE-RHINELANDER

Rhinelander's immediate followers in the matter of conditional assertion happily abandoned strictly illocutionary notions and gave the issue a purely semantic turn.

Jeffrey (1963) explored an extensional, truth-functional treatment of conditionals in three-valued logic, adding a value "indeterminate" to "true" and "false." He argued that if certain disputable assumptions are made, we get a logic for the three-valued conditional that preserves the core conditional inferences such as *modus ponens*, *modus tollens*, contraposition and conditional proof. However, he also pointed out that if we dispute the disputable assumptions, we start to lose the core inferences. For example, if we hold that the negation of an indeterminate sentence is itself indeterminate rather than false, we cannot preserve contraposition. And if we give up the strong thesis that all indeterminate sentences are "assertable" (meaning roughly that for any one of them, there could always be some point in asserting it), undesirable indeterminacies would result.

Jeffrey further points out a damaging feature of his truth-functional version of the conditional-assertion theory, one that does not depend on any of the disputable assumptions. Obviously, according to the SI theory, whenever someone assertively utters a conditional whose antecedent and consequent are both true, that person has made a true assertion, and that feature carries over into Jeffrey's truth table. But suppose a conditional consequent is true only by fluke. His example (p. 41): The dentist says, "If you don't undergo this treatment, you'll lose all your teeth." I turn down the treatment. Then I am in an auto accident that knocks out all my teeth, but the dental surgeons who treat me report that in fact all my teeth had been perfectly sound. The dentist's conditional seems to have been false; it would be no defense for her/him to point out that I did, after all, lose all my teeth.

Better yet, suppose antecedent and consequent are mutually irrelevant, *and* the consequent is true only by fluke: "If you do not finish mowing the yard this afternoon, then in 2017 there will be violent solar flares"; "If I fail to learn Swedish within five years, then there will have been a mouse in the cellar this afternoon"; "If you cough sharply twice, then the winning lottery number will be 275489." That these would be severely defective assertions does not prove that they would not be true ones, but it is hard to hear them as true ones. I join Jeffrey in rejecting such sentences, indeed holding them to be false.[15] (Call this the "TT" objection.)

[15] If the "then" is removed from each of those sentences, they acquire readings on which they are arguably true: semifactual or "weak" readings in the sense of Davis (1983). Contrast, e.g., "If you open the refrigerator, it won't explode" with "If you open the refrigerator, then it won't explode." I was careful to put in the "then"'s precisely to block the weak readings.

Nuel Belnap (1970) offered an intensional treatment, saying that he "believe[d Jeffrey's analysis] does not yield a structure rich enough to do justice to the underlying idea of conditional . . . [assertion]" (p. 4). He stipulates (pp. 4–6): Every atomic sentence "asserts" a proposition (conceived as a set of worlds), but not every compound sentence does. A negation asserts a proposition at a world iff its negand does there, and the proposition it asserts is the negation of the proposition asserted by the negand. A conjunction asserts a proposition at a world iff at least one of its conjuncts does there, and the proposition it asserts is the conjunction of the proposition(s) asserted by its conjuncts; similarly for disjunction. But the rule for the conditional is more complicated, and follows Quine: A > B asserts a proposition at a world iff A is true at that world; if A is true there, the proposition A > B asserts is the proposition B asserts there.

Now, how does this intensional treatment improve on Jeffrey's truth-functional one? Once Belnap has gone on to develop similar semantic rules for quantifiers, he applies the resulting apparatus to the analysis of Aristotelian A-, I-, E- and O-forms; it also affords a partial solution to the Raven Paradox. But if we assess it according to the concerns raised against Jeffrey, there is no visible improvement. Contraposition fails for Belnap's conditional. And Belnap is subject to the TT objection as well; on his semantics, a conditional whose antecedent and consequent are both true is automatically true. That may not be obvious, precisely because Belnap's system is intensional rather than truth-functional; but recall that if the conditional's antecedent is true, the proposition then asserted by the whole conditional is exactly the proposition asserted by its consequent, and by hypothesis that proposition is true.

Belnap's account faces still a further problem, distinctive to it. He makes it very clear that his term "asserts," applying to sentences, is only the thin semantic shadow of the SI theory's core illocutionary notion of asserting. But then, what is the relation between a sentence's "asserting" a proposition and the sentence's *expressing* one? A dilemma arises: Suppose that for a sentence to "assert" a proposition is just for it to express that proposition. (Belnap says nothing to discourage this reading.) Then a conditional with a false antecedent expresses no proposition at all. That result would in itself be fine with the proponents of NTV (mentioned in the previous section), because they hold that indicative conditionals are not in the business of expressing propositions in the first place. But the failure here is selective, for on the present interpretation of Belnap, conditionals with true antecedents do express propositions, while conditionals with false antecedents do not. That seems too radical a difference to be made in a sentence's semantic status by the (possibly chance) obtaining or not of a contingent fact.

Suppose then, that asserting is a stronger notion than that of expressing (I pass over the strange possibility that asserting is weaker than expressing). But, once "asserting" has been drained of its action-theoretic and illocutionary character, what could the difference be? On this second horn of the dilemma, a sentence's failing to assert a proposition is compatible with its expressing one. If it expresses a proposition and the proposition is true, then the sentence is true; but on Belnap's view, a sentence that fails to assert a proposition lacks truth-value. Here we have an apparent contradiction and no help in resolving it.

IV. FURTHER OBJECTIONS TO THE SEMANTICIZED ACCOUNTS

Let us revisit our other objections to the SI theory and see if they cause trouble for the semanticized views. *Contraposition* has already been seen to carry over.

What about *Initial Implausibility*? Recall my neighbor's sentence, "If you don't finish mowing the yard this afternoon, you can't go to the mall after dinner," and her daughter's likely reception of it. Belnap holds that if the daughter does mow the yard, the sentence has asserted no proposition.[16] As before, that would be all right with and for NTVists, but Belnap's claim is selective and would not apply if the daughter had not mowed the yard. From the daughter's point of view, at least, the selective claim is implausible. The same can be said in regard to our other sample sentences from section I.

Jeffrey's truth-table account[17] perhaps does better against the implausibility charge, for it says, not that the mother's sentence asserts no proposition, but only that it lacks truth-value. Some will see that as a distinction without a difference; others, having a more generous notion of a proposition, will grant the improvement, however small the improvement may be.

Nesting: I believe that for either semanticized account, the problem of nesting is only technical. If Jeffrey's three-valued tables work at all, they should work for nested conditionals, though I have not tried to verify that. And I would never doubt that someone of Belnap's ingenuity and technical skill could accommodate nested constructions. What made nesting a grave difficulty for the SI theory was that SI deals not just with formal semantics of sentences but with speech acts that are individuated according to their actual illocutionary force.

Propositional attitudes: Here too, the semanticizing turns out to help, because what made the SI theory unattractive for nondoxastic attitudes was primarily its official focus on the speech act of asserting. The semanticized views are merely about the semantics of sentences, and all propositional attitudes—not just the doxastic ones—have sentence-like properties. In particular, being embarrassed that P, hoping that P, being sorry that P, etc., have representational and other semantical properties, even if those properties are not the most important things about those attitudes. Nonetheless, an analogue of *Initial Implausibility* remains: Are the respective fears, embarrassments, hopes and sadnesses of Angie *et al.* from

[16] N.B., Belnap would never say that the sentence is *meaningless* in the context; he is careful to detach his notion of asserting-a-proposition from that of meaningfulness. A nonassertive sentence "continues to have determinate semantic relations, etc." (p. 2).

[17] Remember that he did not himself accept it, though I shall continue to call it that.

section 2 *contentless* if their antecedent conditions do not obtain? For example, the extension of Belnap's view to propositional attitudes entails that if in fact Dave will not see Angie, then either she is *not* after all embarrassed that if Dave sees her he will send her secret signals, or she is embarrassed but (contrary to its description) her embarrassment lacks propositional content.

Now, here are three further objections to the semanticized accounts. First, it is widely agreed that many if not all conditionals are equivalent to the corresponding disjunctions. (Since like most conditional-assertion theorists I reject the material or horseshoe theory of indicatives, I deny that conditionals are equivalent to *truth-functional* disjunctions;[18] I believe the relevant disjunctions are intensional.) If A > B fails to assert any proposition, or at least lacks truth-value, when A is false, is the same true of ~A-or-B? (The Disjunction objection.)

Second, what if a conditional is entailed by a law of logic, a law of nature, or just a true universal generalization, or some other categorical fact? Every piece of iron heated to 200°C glows red; it follows that if this piece of iron is heated to 200°C, it will glow red. What, then, do we say if this piece of iron is not going to be heated? That the true generalization logically entails an untrue sentence, or that the generalization is itself not true after all? Chapel Hill is in North Carolina; it follows that you do not live in Chapel Hill unless you live in North Carolina. And so on. (The Entailment objection.)

Third, Stalnaker (1984, 111–12) has emphasized the many linguistic parallels between indicative and subjunctive conditionals. On the assumption that subjunctive conditionals have truth-conditions and express propositions, our semanticized views imply that when an indicative's antecedent fails, the sentence differs from its corresponding subjunctive in a very fundamental semantical way—for example, "If your piece of iron is now at 200°C, it is glowing red" would get a Belnap conditional-assertion semantics and asserts a proposition at all only if the iron is in fact at 200°C, asserting only its consequent even if its antecedent is true, but "If your piece of iron were now at 200°C, it would be glowing red" would be assigned a straightforward truth-condition of quite a different kind, say in terms of closeness of possible worlds.[19] But that consequence makes nonsense of the glaring parallels and analogies between indicatives and subjunctives. Indicatives and subjunctives are expressed by the same lexemes, not only in English but

[18] Lycan (2001), 27). (In that discussion I offered two other examples of clearly nonconditional sentences that are allegedly equivalent to conditional ones, but David Sanford and Jonathan Bennett have persuaded me that the equivalences hold only for the material conditional and not even for my own; see especially Bennett (2003), 354–5. I am duly ashamed.)

[19] Most conditional-assertion theorists and NTVists do grant the assumption that subjunctives have normal truth-conditions. However, shining exceptions are Dorothy Edgington (1986, 1995) and Stephen Barker (1995); so the present objection does not apply to them. Von Wright (1957) is an intermediate case; he gives an initially conditional-assertion account of subjunctives, but he then argues that subjunctives have truth-conditions and truth-values (p. 162), for "[i]t is not part of what we do, when we assert *q* on the counterfactual condition *p*, that we leave certain propositions unasserted."

in most other languages. Indicatives and subjunctives have the same syntax but for their distinctive tense and aspect differences, including their modification by "only" and "even." More importantly, they have virtually the same logic. And they admit almost all the same paraphrases. All this would be surprising, to say the least, if the two kinds of sentences differ so greatly in their semantics. (The Subjunctive Parallel objection.)

V. NTV AND CONDITIONAL ASSERTION

Dorothy Edgington (1986, 1995), Stephen Barker (1995) and Michael Woods (1997) defend versions of NTV, but they also describe their theories as conditional-assertion views. The relation is not immediately clear. Indeed, one reason it is positively unclear is that according to either of the semanticized conditional-assertion views we have seen, an indicative with a true antecedent and a true consequent is true, which is incompatible with NTV; more obviously, an indicative with true antecedent and false consequent is false. (Of course, Edgington's, Barker's and Woods' theories are competitors of standard truth-conditional theories, if only because they deny that indicative conditionals have truth-conditions at all.) I shall discuss primarily Edgington's version, because it is the one I think I understand the best, but I shall note any significant points on which Barker's or Woods' views differ.

Edgington holds that a conditional belief is (roughly) the believer's corresponding subjective conditional probability. It is not belief of a proposition, not a belief *that* anything. If this seems odd—a belief that is not a belief that P despite being described using the usual sort of 'that'-complement—so be it; call it "acceptance" or "endorsement" or some such. And of course it has no truth-value.

There are several objections to this in its own right (before we get to the conditional-assertion theory of indicative sentences). First, there is the problem of nondoxastic attitudes. Edgington offers an account of conditional desire (1995, 288): One desires that if A then B iff one prefers A & B to A & ~B. But what about fear, embarrassment, sadness, and shame?

Second, there is the probability version of the TT objection: If I firmly believe A and B, my conditional probability of B given A will be high; but sometimes I will not and should not believe that if A then B. (Edgington is well aware of this objection and stiffarms it; she thinks the intuition can be explained away (268–9).)

The Disjunction and Entailment objections carry over to belief as well.

Moreover, though I do not expect to convert many souls here, I reject in the first place Adams' widely accepted contention (which Edgington (p. 263) calls "the Thesis") that one's degree of belief in a conditional matches the relevant conditional probability. I have what I maintain are counterexamples. Here is one:

I believe each of two propositions that are not only probabilistically independent of each other but mutually irrelevant in topic. I believe the first only tentatively. Right now I am inclined to think that my copy of Stalnaker's *Inquiry* is lying on a shelf in my office near my desk, though I may well be wrong because I am always absent-mindedly picking up things and putting them somewhere else without realizing I have done so. I also believe, but pretty firmly, that the Australians will win their next test against Sri Lanka. I am confident to degree. 9 that Australia will win. Since the two propositions are probabilistically independent, my subjective conditional probability for the second given the first is equal to its own subjective probability, .9. According to Adams' generalization, then, I should believe to degree .9 that if my copy of Stalnaker's *Inquiry* is lying on a shelf in my office near my desk, then the Australians will win their next test against Sri Lanka. But I do not have the latter belief at all; I think it is false.[20]

Edgington's account of conditional belief is not itself analogous to a conditional-assertion theory of conditional sentences, for it does not say that if the antecedent is false, nothing is believed. (As before, she holds that no *proposition* is believed, but that holds regardless of the antecedent's truth-value, so she does not join in the conditional-assertion-analogue view that whether a proposition is expressed depends on the truth of the antecedent.) Indeed, she explicitly acknowledges that when one assigns a high subjective conditional probability to B given A, and one learns that A is true, one does not necessarily affirm B; one might be obliged to withdraw the conditional (1995, 270). But when she turns to the topic of speech acts, she does defend a conditional-assertion theory. It has two distinctive features.

First (p. 289), she rejects Jeffrey's envelope model. Of course it is not true that, should the antecedent prove false, nothing whatever has been said and no speech act has occurred. Nothing has been *asserted*, but something has been said and something has been conditionally-asserted. (She seems to regard conditional assertion as a speech-act type of its own.)[21]

Second, of course conditional assertions—unlike conditional beliefs and conditional sentences—have truth-values when their antecedents are true, because they then constitute assertions of their consequents. So one can make a true or

[20] Fairness compels me to record that Edgington's view has a great advantage over my own theory (Lycan 2001). My theory assigns a vanishingly low probability to conditionals that are epistemically uncertain, as in, "How likely is it that if I undergo this treatment I will be cured?," when intuitively the probability of such a conditional is or is close to the conditional probability of its consequent given its antecedent. See Edgington (2003).

[21] As does Barker. But although Barker declares allegiance to Rhinelander (p. 186), his text does not make it clear whether he agrees that the utterer of a conditional has actually and flatly asserted its consequent when its antecedent is in fact true.

Barker thinks of the conditional assertion of A > B as part of a pretense: We are affecting to believe A, and while doing so we assert B. The usual sincerity condition on assertion is suspended, as it would be for a stage actor. "If" is a particle of conventional implicature, or what Lycan (1984) calls "lexical presumption"; it serves simply to signal that the conditional consequent B is asserted only conditionally upon A. I think this is a neat and well-motivated view.

false assertion by uttering a truth-valueless sentence. (This distinguishes Edgington's view from the semanticized accounts, according to which a conditional sentence is itself a true or a false one when its antecedent is true.)

Each of those two points is correct. That a speech act is not one of assertion does not entail that there is no related illocutionary category such as that of conditional-assertion. And we have already noted the common mismatch between truth-value of assertion made and that of sentence uttered. The conditional-assertion theory gives us yet another example of such a mismatch: one may utter a truth-valueless sentence and thereby make a true or false assertion, because when the sentence is a conditional with a true antecedent, one has asserted the consequent alone.[22]

But on the first point, I would reply as before that in our mother–daughter example from section 1, an assertion flatly has been made. It is not just that the mother has said something which may or may not turn out to have been an assertion. I contend that in this case, conditional-assertion vs. assertion is a distinction without a difference, and the objection applies as before. (I do not expect Edgington to concede this, since she obviously does see a difference between conditional-assertion and assertion. She makes a related claim about conditional commands (pp. 289–90).

> A child is told 'If you go out, wear your coat'. If he cannot find his coat, he stays in, in order to comply with the command. On my interpretation, if the child can't find his coat, he has a choice between disobeying the command, and behaving in such a way that no categorical command has been made (not: behaving as though nothing had been said). If he wishes not to disobey, he must stay in [in which case no categorical command has been issued, though a conditional command has been].

I join Dummett (1973) in denying the alleged distinction between not disobeying the command and obeying it. Edgington says that "other examples make this implausible" (p. 290), but the one example she gives is tendentious.)[23]

[22] Moreover, Edgington offers a new argument in favor of the conditional-assertion view—roughly, that (a) the treatment of assertions of indicative conditionals should parallel that of conditional imperative and conditional interrogative speech acts, but (b) no standard truth-conditional account of the latter is adequate. Barker independently makes similar arguments, in considerable detail.

DeRose and Grandy (1999) offer several further considerations. I have not the space to consider the positive arguments here, but I would self-servingly note that DeRose and Grandy's main argument fails. It is that the conditional-assertion view can offer a unified view of ordinary indicatives and "biscuit" conditionals (e.g., Austin (1961, 158): "There are biscuits on the sideboard if you want them") considered together, while "the best the leading theories can do by way of 'handling' biscuit conditionals . . . [is] to cordon off the abnormal [biscuit] cases for separate treatment" (p. 407). Ahem: Lycan (2001, 206–10) offers a unified truth-conditional account of ordinary indicatives, biscuits, and a number of cases in between.

[23] "If, in the emergency ward, you're told 'If the patient is still alive in the morning, change the drip', and you smother the patient, you can hardly claim to have merely carried out an order." Of course you cannot. If you smother the patient you have committed murder or at least active

Let us see how Edgington's account fares against our remaining objections to previous theories. TT, Disjunction and Entailment continue to apply. Subjunctive Parallel does not; Edgington doughtily favors an NTV, subjective conditional probability theory of subjunctives themselves (pp. 311 ff.). (I am not sure whether she accepts a conditional-assertion view of subjunctive conditional utterances.)

That leaves Nesting, especially the problem of conditionals with conditional antecedents. Here Edgington makes what has become a standard NTVist move: to insist that the antecedents are anomalous and at best must be reinterpreted ad hoc as expressing propositions that are not (of course) conditional propositions (Gibbard (1981, 234–38); Appiah (1985, 205–10); Woods (1997, 65–6)). Edgington (283–4) suggests that we reinterpret the speaker as asserting the "obvious [categorical] basis" of the conditional belief being expressed, such as the categorical fact that serves as the speaker's immediate evidence for the conditional, or perhaps the existence of a disposition that grounds the conditional.

I have to say that this has always struck me as desperate. The examples I and others have given of conditionals with conditional antecedents are perfectly clear and in need of no reinterpretation at all. True, there are further examples which are odd; but that is due to the absence of any natural context for them.[24] The Nesting objection remains severe.

To his credit, Barker (1995) does not take the standard line on conditional antecedents, but insists that they are (often) fine as they stand. His position on them (p. 206) is that while the "if" in a simple conditional signals that the antecedent is a supposition which initiates a pretense, the "if" in a conditional with a conditional antecedent signals that that antecedent is "stipulated [merely] to be assertible" (as opposed to the usual case of being stipulated to be true).

I am fairly sure Barker does not mean that the speaker institutes a metalinguistic supposition, about the assertibility of a sentence; that would be to no purpose. But what, then, is supposed? This departure from Barker's existing theory is both hard to parse and ad hoc.

Incidentally, I do not see that any conditional-assertion view of conditional utterances is required by NTV. Indeed, Bennett (2003) has persuasively argued that the NTVist would do well to embrace an expressivist view: that to utter a conditional is to express one's subjective conditional probability rather than to make any statement of fact.[25]

euthanasia. That leaves open the question of whether you did obey the order; and I say you did obey it, however defectively. Perhaps there is a sense in which you did not *carry it out*, but that does not entail that no order was issued.

[24] Edgington herself uses the phrase "ad hoc" (p. 283). She also uses exactly my present sort of argument to impugn "irrelevance" counterexamples to TT (p. 269).

[25] He thereby takes most of the sting out of the first of my ten arguments against NTV (Lycan (2001), 73–4). I had said that NTV makes indicative conditionals a surd in truth-conditional semantics. But Bennett reminds us that many otherwise truth-conditionally minded philosophers are expressivists about moral judgments, aesthetic judgments, modal discourse, . . . , so NTV is no

In fact, I think the expressivist account would fit a bit better with the NTV theory of conditional belief. Utterances generally express beliefs. Even if the "belief" in question is not belief of a proposition, and the corresponding sentence lacks truth-value, the more obvious account of the matter is that a conditional utterance (even when the antecedent is true) genuinely asserts nothing but only expresses the relevant cognitive state.

VI. VERDICT

We have not found that conditional-assertion theories are very plausible in their own right. Nor are they required by NTV, by the Thesis, or by Edgington's theory of conditional belief. At this stage, the conditional-assertion idea seems expendable.

REFERENCES

Adams, E.W. (1965). 'The Logic of Conditionals', *Inquiry* 8: 166–97.

Appiah, A. (1985). *Assertion and Conditionals*. Cambridge: Cambridge University Press.

Austin, J. L. (1961). 'Ifs and Cans', in *Philosophical Papers*. Oxford: Oxford University Press.

Barker, S. J. (1995). 'Towards a Pragmatic Theory of "If"', *Philosophical Studies* 79: 185–211.

Belnap, N. (1970). 'Conditional Assertion and Restricted Quantification', *Noûs* 4: 1–12.

Bennett, J. (2003). *A Philosophical Guide to Conditionals*. Oxford: Oxford University Press.

Blackburn, S. (1995). *Essays in Quasi-Realism*. Oxford: Oxford University Press.

Carston, R. (2002). *Thoughts and Utterances: The Pragmatics of Explicit Communication*. Oxford: Basil Blackwell.

Davis, W. (1979). 'Indicative and Subjunctive Conditionals', *Philosophical Review* 88: 544–64.

—— (1983). 'Weak and Strong Conditionals', *Pacific Philosophical Quarterly* 64: 57–71.

DeRose, K., and R. Grandy (1999). 'Conditional Assertions and "Biscuit" Conditionals', *Noûs* 23: 405–20.

Dummett, M. (1973). *Frege: The Philosophy of Language*. London: Duckworth.

Edgington, D. (1986). 'Do Conditionals Have Truth Conditions?," *Critica* 18: 3–30.

(1995). 'On Conditionals', *Mind* 104: 235–329.

(2003). 'William G. Lycan's *Real Conditionals*: Comments by Dorothy Edgington', <http://www.unc.edu/~ujanel/Edgington_LYCAN.htm> (Delivered at an 'Author Meets Critics' session on Lycan (2001), APA Central Division meetings.)

surd but is in fairly good company. (For highly developed expressivist views of such other sentences, see Gibbard (1992) and Blackburn (1995).) I had missed that, (a) because I myself am a compulsive metaphysical realist about nearly everything and (b) because qua philosopher of language I do not believe that an entirely well-formed sentence can lack a truth-value.

Gibbard, A. (1981). 'Two Recent Theories of Conditionals', in Harper, Stalnaker and Pearce (1981).
____ (1992). *Wise Choices, Apt Feelings*. Cambridge: Harvard University Press.
Goodman, N. (1947). 'The Problem of Counterfactual Conditionals', *Journal of Philosophy* 44: 113–28.
Harper, W., R. Stalnaker, and G. Pearce (eds.) (1981). *Ifs*. Dordrecht: D. Reidel.
Jackson, F. (1979). 'On Assertion and Indicative Conditionals', *Philosophical Review* 88: 565–89.
____ (1987). *Conditionals*. Oxford: Basil Blackwell.
Jeffrey, R. (1963). 'On Indeterminate Conditionals', *Philosophical Studies* 14: 37–43.
Kratzer, A. (1986). 'Conditionals', in *Papers from the Parasession on Pragmatics and Grammatical Theory, Twenty-Second Regional Meeting of the Chicago Linguistic Society*. Chicago: University of Chicago.
Kremer, M. (1987). '"If" Is Unambiguous', *Noûs* 21: 199–217.
Lewis, D. K. (1973). *Counterfactuals*. Cambridge: Harvard University Press.
Lycan, W. G. (1984a). 'A Syntactically Motivated Theory of Conditionals', in P. French, T. E. Uehling and H. Wettstein (eds.), *Midwest Studies in Philosophy, vol. IX: Causation and Causal Theories*. Minneapolis: University of Minnesota Press.
____ (1984*b*). *Logical Form in Natural Language*. Cambridge: Bradford Books/MIT Press.
____ (2001). *Real Conditionals*. Oxford: Oxford University Press.
Quine, W. V. (1950). *Methods of Logic*. New York: Holt.
____ (1960). *Word and Object*. Cambridge: MIT Press.
Stalnaker, R. (1968). 'A Theory of Conditionals', in N. Rescher (ed.), *Studies in Logical Theory* (*American Philosophical Quarterly* Monograph, 2, Oxford: Basil Blackwell).
____ (1975). 'Indicative Conditionals', *Philosophia* 5: 269–86; reprinted in Harper, Stalnaker and Pearce (1981).
____ (1984). *Inquiry*. Cambridge: Bradford Books/MIT Press.
von Wright, G. H. (1957). 'On Conditionals', in *Logical Studies*. London: Routledge and Kegan Paul.
Woods, M. (1997). *Conditionals*, ed. David Wiggins and with a Commentary by Dorothy Edgington. Oxford: Oxford University Press.

8

Non-Catastrophic Presupposition Failure[1]

Stephen Yablo

1. BACKGROUND

I will be talking in this paper about the problem of presupposition failure. The claim will be (exaggerating some for effect) that there is no such problem—more like an *opportunity* of which natural language takes extensive advantage.

The last two sentences are a case in point. The first was, "I am going to talk about the F"; the second was, "there is no F." If the second sentence is true—there is no F—then the first sentence, which presupposes that there is an F, suffers from presupposition failure. In theory, then, it should strike us as somehow compromised or undermined. Yet it doesn't. So here is one case at least where presupposition failure is not a problem.

The title is meant to be understood compositionally. *Presuppositions* are propositions assumed to be true when a sentence is uttered, against the background of which the sentence is to be understood. Presupposition *failure* occurs when the proposition assumed to be true is in fact false.[2] Failure is *catastrophic* if it prevents a thing from performing its primary task, in this case making an (evaluable) claim. Non-catastrophic presupposition failure then becomes the phenomenon of a sentence still making an evaluable claim despite presupposing a falsehood.

I said that presuppositions were propositions taken for granted when a sentence is uttered, against the background of which the sentence is to be

[1] Papers sort of like this one were presented at Indiana, UC Davis, UC Berkeley, UC San Diego, Yale, Brown, Penn, Kentucky, Oxford (as the 2005 Gareth Evans lecture), ANU, Monash, the Chapel Hill Colloquium, an APA session on Metaontology, and graduate student conferences at Pittsburgh and Boulder. I am grateful to Richard Holton, Sally Haslanger, Agustin Rayo, Caspar Hare, Kai von Fintel, Danny Fox, Irene Heim, Karen Bennett (who commented at the APA), Anne Bezuidenhout (who commented at Chapel Hill), Larry Horn, Sarah Moss, John Hawthorne, Lloyd Humberstone, and especially Bob Stalnaker for questions and advice. I learned too late of the literature on "logical subtraction" (see Humberstone 2000 and references there); it holds out hope of a different and perhaps more straightforward route to incremental content.

[2] Really I should say "untrue" rather than "false," to allow for presuppositions that lack truth-value because they themselves suffer from presupposition failure. Looking ahead to the Donnellan examples, "The man drinking a martini is *that* guy" is (so it seems) not false but undefined if no one is drinking a martini.

understood.[3] It would be good to have some tests for this. Here are three, loosely adapted from the paper that got me thinking about these issues (von Fintel 2004—don't miss it!).[4]

One is the "hey, wait a minute" test.[5] If π is presupposed by *S*, then it makes sense for an audience previously unaware of π to respond to an utterance of *S* by saying "hey, wait a minute, I didn't know that π." If π were asserted, that response would be silly; of course you didn't know, the point of uttering *S* was to tell you. Suppose you say, "I'm picking my guru up at the airport." I can reply, "Hey, I didn't know you had a guru," but not, "hey, I didn't know you were going to the airport." This suggests that your having a guru was presupposed while your going to the airport was asserted. A likelier response to what is asserted is, "is that so, thanks for telling me."[6] [7]

Second is the attitude attributed when we say that someone denies that *S*, or hopes or regrets that *S*; the presupposition π is exempted from the content of that attitude. Hoping you will pick up your guru at the airport may be in part hoping your guru will be picked up, but it is not hoping that you have a guru in the first place. Denying that you are going to pick up your guru at the airport is not denying the conjunction of *you have a guru* with *you are going to pick your*

[3] Are we to think of presupposition as a relation that *sentences* bear to propositions (Strawson), or a relation that *speakers* bear to propositions (Stalnaker)? There may be less of a difference here than meets the eye. The first relatum for Strawson is *utterances* or tokens of *S*, from which it is a short step to speakers presupposing this or that *in uttering S*. Stalnaker for his part appreciates that certain sentences *S* should not be uttered unless this or that is (or will be as a result of the utterance) pragmatically presupposed. It does little violence to either's position to treat "*S* presupposes π" as short for "All (or most, or contextually salient) utterances of *S* presuppose π," and that in turn as short for "Speakers in making those utterances always (often, etc.) presuppose that π." (Von Fintel ms and Simons 2003 are illuminating discussions.) Semantic presupposition would be the special case of this where *S* presupposes π as a matter of meaning, that is, *S*-users presuppose π not for conversational reasons but because semantic rules require it.

[4] Strawson noticed that while some King-of-France sentences strike us as unevaluable ("The KoF is bald"), others seem false ("The KoF visited the Exhibition yesterday"). Von Fintel criticizes earlier accounts of this contrast (by Strawson and Peter Lasersohn) and proposes an interesting new account. He does not address himself to a third possibility noted by Strawson, that a sentence with false presuppositions should strike us as true. This paper agrees with von Fintel's basic idea: some KoF-sentences "are rejected as false... because they misdescribe the world in two ways: their presupposition is false, but in addition there is another untruth, which is *independent* of the failure of their presupposition" (2004, 325). But it implements the idea differently.

[5] Taken apparently from Shanon 1976.

[6] This test seems to work best for semantic presuppositions (see note 3). Looking ahead a bit, "The man drinking a martini is a philosopher" does not invite the reply, "Hey, I didn't know *that* guy was the one drinking a martini." One can, however, say, "Hey, I didn't know that guy was drinking a martini." So perhaps a version or variant of the test applies to (some) non-semantic presuppositions as well.

[7] Von Fintel attributes to Percus a test that is in some ways similar. "*R*, and what's more, *S*" sounds fine if *S* asserts more than *R*, but wrong if *S* only presupposes more. So, "John thinks Judy is a chiropractor" can be followed by "And what's more, he is right to think Judy is a chiropractor," but not "And what's more, he realizes Judy is a chiropractor." This seems to indicate that "He realizes that BLAH" presupposes what "He is right to think that BLAH" asserts, viz. that BLAH, and asserts what it presupposes, viz. that he believes that BLAH.

guru up at the airport.[8] So a second mark of presuppositions is that π does not figure in what you hope or deny or regret in hoping or denying or regretting that *S* (Stalnaker 1999, 39).

A third test is that presuppositions within limits *project*, that is, π continues to be presupposed by more complex sentences with *S* as a part. If you say, "I don't have to pick up my guru after all," or, "it could be I will have to pick my guru up," these statements still intuitively take it for granted that you have a guru. Our earlier tests confirm this intuition. One can still reply, "hold on a minute, you have a guru?" And to hope that you don't have to pick your guru up is not to hope that you have a guru.[9]

Note that one test sometimes used to identify presuppositions is missing from this list: π is presupposed iff unless π holds, *S says nothing true or false*. That test is useless in the present context because it makes NCPF impossible; π is not classified as a presupposition unless its failure would be catastrophic.

A sentence suffers from catastrophic presupposition failure only if, as Strawson puts it, "the whole assertive enterprise is wrecked by the failure of [*S*'s] presupposition" (1964, 84). There is also the phenomenon of what might be called *disruptive* presupposition failure. This occurs when π's failure does not wreck the assertive enterprise so much as reveal it to have been ill advised. It could be, for instance, that π was an important part of the speaker's *evidence* for *S*. It could be that π was part of what made *S relevant* to the rest of the conversation. It could even be that *S entails* π so that π's falsity guarantees that *S* is false too.[10]

Disruption is bad, but it is not (in our sense) a catastrophe. On the contrary, a remark is implausible or irrelevant or false because of what it says, and that something was said suggests that the assertive enterprise has not been wrecked after all. I mention this because Stalnaker, who has written the most about these topics, is addressing himself more often to the disruptive/non-disruptive distinction

[8] This observation goes essentially back to Frege. Frege considers the sentence "whoever discovered the elliptic form of the planetary orbits died in misery." He notes that its negation is "whoever etc. did not die in misery" rather than "Either whoever discovered the elliptic form of the planetary orbits did not die in misery or there was nobody who discovered the elliptic form of the planetary orbits" (1872, 162–3). If we assume (as he did) that denial is assertion of the negation, this amounts to the claim that "somebody discovered the elliptic form of the planetary orbits" is no part of what is denied when we deny that "whoever etc. died in misery."

[9] Related to this, presuppositions fail to project in certain contexts, such as conditionals with π as antecedent. "I don't remember if I have a guru, but if I do, it could be I am supposed to pick my guru up at the airport" does not presuppose that I have a guru.

[10] This relates to a passage in "Pragmatic Presuppositions": "using the pragmatic account [of presupposition], one may say that sometimes when a presupposition is required by the making of a statement, what is presupposed is also entailed, and sometimes it is not. One can say that 'Sam realizes that *P*' entails that *P*—the claim is false unless *P* is true. 'Sam does not realize that *P*,' however, does not entail that *P*. That proposition may be true even when *P* is false. All this is compatible with the claim that one is required to presuppose that *P* whenever one asserts or denies that Sam realizes it" (1999, 54).

than the catastrophic/non-catastrophic distinction.[11] This paper is meant to be entirely about the latter.

2. RELEVANCE TO PHILOSOPHY

Why should we care about non-catastrophic presupposition failure? There are reasons from the philosophy of language, from epistemology, and from metaphysics.

The philosophy of language reason is simple. All of the best-known theories of presupposition (among philosophers, anyway) suggest that failures are or ought to be catastrophic. This is clearest for Frege's and Strawson's theories—for those theories more or less *define* a sentence's presuppositions as preconditions of its making an evaluable claim. Assuming as before that a sentence's primary task is to offer a true or false account of how things are, presuppositions on Frege's and Strawson's theories are *automatically* propositions whose failure has catastrophic effects.

Next consider Stalnaker's theory of presupposition. Stalnaker-presupposition is in the first instance a relation between speakers and propositions; one presupposes π in uttering S if one thinks that π is (or will be, as a result of the utterance) common ground between relevant parties. A *sentence* presupposes π only to the extent that S is not appropriately uttered unless the speaker presupposes that π.

Why on this account should presupposition failure be problematic? Well, the point of uttering S is to draw a line through the set of worlds still in play at a particular point in the conversation—one is saying that *our* world is on the S-true side of the line rather than the side where S is false. Since the worlds still in play are the ones satisfying all operative presuppositions, the speaker by presupposing π is arranging things so that her remark draws a line through the π-worlds only.

But then what happens when π is false? Because the actual world is outside the region through which the line is drawn, it is hard to see how in drawing this line the speaker is saying anything about actuality. It's as though I tried to locate Sicily for you by saying that *as between North and South Dakota*, it's in the North,

[11] For instance here:

Where [presuppositions] turn out to be false, sometimes the whole point of the inquiry, deliberation, lecture, debate, command, or promise is destroyed, but at other times it does not matter much at all... Suppose... we are discussing whether we ought to vote for Daniels or O'Leary for President, presupposing that they are the Democratic and Republican candidates respectively. If our real interest is in coming to a decision about who to vote for..., then the debate will seem a waste of time when we discover that in reality, the candidates are Nixon and Muskie. However if our real concern is with the relative merits of the character and executive ability of Daniels and O'Leary, then our false presupposition makes little difference (1999, 39).

although truth be told it's not in either Dakota. Similarly it is not clear how I can locate actuality for you by saying that as between the π-worlds where S is true and the ones where it is false, it's in the first group, although truth be told it's not a π-world at all.[12]

That was the philosophy of language reason for caring about NCPF; the standard theories seem to rule it out. A much briefer word now on the epistemological and metaphysical reasons.

The epistemological reason has to do with testimony, or learning from others. Someone who utters a sentence S with truth-conditions C (S is true if and only if C obtains) might seem to be telling us that C *does* obtain. But if we bear in mind that π is one of the conditions of S's truth, we see that that cannot be right. For it makes two false predictions about the phenomenon of NCPF. The first is that *all presupposition failure is non-catastrophic*; if π is false, then the speaker is *telling* us something false, hence the assertive enterprise has not been wrecked. The second is that what the speaker is telling us *can never be true*. The fact is that some presupposition failure is catastrophic and some isn't; and the claim made can be either true or false. To suppose that speakers are saying inter alia that π in uttering S collapses the first two categories—catastrophic, non-catastrophically true—into the last—non-catastrophically false.

So here is the epistemological relevance of NCPF. It reminds us that speakers are not in general vouching for *everything* the truth of their sentence requires; they vouch for the asserted part but not (in general) for the presupposed part.

This leads to the metaphysical reason for caring about NCPF. Quine famously argues like so: "scientists tell us that the number of planets is 9; that can't be true unless there are numbers; so scientists tell us inter alia that there are numbers; so unless we consider ourselves smarter than scientists, we should believe in numbers." This assumes that speakers are vouching for *all* the truth-conditions of the sentences coming out of their mouths. But there being a thing that numbers the planets is no part of what Clyde Tombaugh (the discoverer of Pluto) was telling us—no part of what he was giving his professional opinion about—when he spoke the words, "the number of planets is 9." A different metaphysical upshot will be mentioned briefly at the end.

3. FREGE AND STRAWSON

I said that the best-known theories suggest that all presupposition failure ought to be catastrophic, and that the suggestion is implausible. I did not say that the best-known theorists are unaware of this problem. Well, Frege might have been

[12] See Beaver 2001 for theories of the kind favored by many linguists. These seem at least as unaccommodating of NCPF as the ones philosophers like, for a reason noted by Simons: "Dynamic theories of presupposition claim that presupposition failure results in undefinedness of the context update function—the dynamic correlate of truth valuelessness" (2003, 273).

unaware of it. Even he, though, gives an example that might be taken as a case in point: "Somebody using the sentence 'Alfred has still not come' actually says 'Alfred has not come,' and at the same time hints—but only hints—that Alfred's arrival is expected. Nobody can say: 'since Alfred's arrival is not expected, the sense of the sentence is false' " (1918, 331).

Frege's use of *hint* makes it sound as though we are dealing with an implicature. But "still" is by the usual tests a presupposition trigger. ("Hang on, I didn't know Alfred was supposed to be here!") Suppose for argument's sake that the tests are right.

Frege says that the thought is not automatically false if Alfred was unexpected. By this he presumably means that the thought's truth-value depends not on how expected Alfred was but on whether he has indeed come. Even if the presupposition fails—he was *not* expected—a claim is still made that can be evaluated as true or false.

So the Alfred example looks like a case of non-catastrophic presupposition failure. Of course, Frege would not see it that way, because the presuppositions that he (and later Strawson) has mainly in mind are *existential* presuppositions: "If anything is asserted there is always an obvious presupposition that the simple or compound proper names used have reference" (1872, 162).

The sentence "Whoever discovered the elliptic from of the planetary orbits died in misery" is said to lack truth-value unless someone did indeed make the indicated discovery (1872, 162). Strawson in similar fashion says that if someone produced the words "The King of France is bald," we would be apt to say that "the question of whether his statement was true or false *simply did not arise*, because there was no such person as the King of France" (1950, 12).

But, and here he goes beyond Frege, Strawson notices that failure even of a sentence's existential presuppositions does not prevent it from making an evaluable claim:

> Suppose, for example, that I am trying to sell something and say to a prospective purchaser *The lodger next door has offered me twice that sum*, when there is no lodger next door and I know this. It would seem perfectly correct for the prospective purchaser to reply *That's false*, and to give as his reason that there was no lodger next door. And it would indeed be a lame defense for me to say, *Well, it's not actually false, because, you see, since there's no such person, the question of truth and falsity doesn't arise* (1954, 225).

This is an example of what Strawson calls "radical failure of the existence presupposition" (1964, 81), radical in that "there just is no such particular item at all" as the speaker purports to be talking about. It shows that for the existence presupposition to fail radically is not necessarily for it to fail catastrophically.

Now, if the existence presupposition can fail radically—there is no such item as the speaker purports to be talking about—one expects that the uniqueness presupposition could fail radically too—there are *several* items of the type the speaker purports to be talking about. Consider another example of Strawson's:

if, in Oxford, I declared, "The Waynflete Professor of Logic is older than I am" it would be natural to describe the situation by saying that I had confused the titles of two Oxford professors [Waynflete Professor of Metaphysics and Wykeham Professor of Logic], but whichever one I meant, what I said about him was true (1954, 227).

This becomes *radical* failure of the uniqueness presupposition if we suppose that in confusing the titles Strawson had confused the individuals too, so that his remark was no more directed at the one than the other. Does the failure thus reconstrued remain *non-catastrophic*? I think it does. The remark strikes us as false if the Waynflete and Wykeham Professors are both younger than Strawson, and true (or anyway truer) if he is younger than them.[13]

What about *non*-radical failure of the existential and uniqueness presuppositions? By a *non*-radical failure I mean that although the description used is not uniquely satisfied, the subject *does* have a particular item in mind as the intended referent. The uniqueness presupposition fails non-radically when one says, "The square root of N is irrational," meaning to refer to the positive square root, forgetting or ignoring that N has a negative root too. This kind of remark does not court catastrophe since it strikes us as correct if both roots are irrational, and incorrect if both are rational, and no other outcome is possible.

That was my example of non-radical failure of the uniqueness presupposition, not Strawson's; his would be the Oxford mix-up, assuming that the intended referent was, say, Gilbert Ryle, then Waynflete Professor of Metaphysics. Strawson also gives an example where it is the existential presupposition that non-radically fails:

perhaps, if I say, "The United States Chamber of Deputies contains representatives of two major parties," I shall be allowed to have said something true even if I have used the wrong title, a title, in fact, which applies to nothing. (1954, 227)[14]

[13] Suppose Strawson had said, "The Philosophy Professor at St Andrews is older than me," not realizing that St Andrews had two professors. Such a statement again seems correct if both are older and incorrect if both are younger—indeed (arguably) if either is younger. Stalnaker in conversation suggests treating this as a case of pragmatic ambiguity; the utterance seems true when it is true on both disambiguations, false when it is false on both (or perhaps false on either). I do not see how to extend this treatment to superficially similar cases. "All eight solar planets are inhabited" seems false, but it is presumably not ambiguous between nine attributions of inhabitedness, each to all solar planets but one.

[14] This example is important in Strawson's debate with Russell. Some empty-description sentences strike us as false, as Russell's semantics predicts. But others are such that "if forced to choose between calling what was said true or false, we shall be more inclined to call it true" (Strawson 1954, 227). Russell cannot claim too much credit for plugging truth-value gaps, if he sometimes plugs in the wrong value. (I ignore the wide-scope negation strategy as irrelevant to the examples Strawson is concerned with here.)

So although Strawson doesn't put it this way, his discussion suggests a four-fold classification along the following lines:[15]

	uniqueness presupposition	*existential* presupposition
radical failure of the	Waynflete Prof of Logic[16]	lodger next door
non-radical failure of the	square root of N	Chamber of Deputies

The fourth of Strawson's categories—non-radical failure of the existential presupposition—proved the most influential, as we shall see.

4. DONNELLAN AND STALNAKER

Strawson appreciates, of course, that the judgments just noted seem at odds with his official theory, particularly with the principle that "If someone asserts that the ϕ is ψ he has not made a true or false statement if there is no ϕ" (Donnellan 1966, 294). Donnellan's famous counterexample to that principle would thus not have come as a surprise to him:

> Suppose one is at a cocktail party and, seeing an interesting-looking person holding a martini glass, one asks, "Who is the man drinking a martini?" If it should turn out that there is only water in the glass, one has nevertheless asked a question about a particular person, a question it is possible for someone to answer (1966, 287).

Given that "Strawson admits that we do not always refuse to ascribe truth to what a person says when the definite description he uses fails to fit anything (or fits more than one thing)" (1966, 294), what does Donnellan think he is adding to Strawson's own self-criticism? Donnellan is not very explicit about this but here is my best guess as to his reply.

What Strawson admits is that the person has said *something* true. He does not (according to Donnellan) admit that the statement *originally at issue*, viz. "the man drinking a martini is a famous philosopher" is true. One might wonder, of course, what we are doing if not "awarding a truth value. . . . to the original statement." The answer is that we "amend the statement in accordance with [the speaker's] guessed intentions and assess the amended statement for truth or falsity" (Strawson 1954, 230). The statement Strawson is willing to call true, then, is not the one suffering from presupposition failure, and the one suffering from presupposition failure he is not willing to call true. (Elsewhere Strawson says the original statement is true only in a *secondary* sense.) Donnellan is bolder: he

[15] This classification is not meant to be exhaustive; perhaps, e.g., the description applies to exactly one thing, but that thing is not the intended referent.

[16] Understood so that the speaker is thinking confusedly of both professors at once.

thinks that the *un*amended, original statement "The ϕ is ψ" can be true in the absence of ϕs, if the description is used referentially.

A second difference between Donnellan and Strawson is this. Strawson paints a mixed picture featuring on the one hand a *presupposition* that the description is uniquely satisfied, and on the other hand an *intention to refer* with that description to a certain object. Donnellan simplifies matters by turning the referential intention into an additional presupposition:

> [W]hen a definite description is used referentially, not only is there in some sense a presupposition... that someone or something fits the description,... but there is also a quite different presupposition; the speaker presupposes of some *particular* someone or something that he or it fits the description. In asking, for example, "Who is the man drinking a martini?" where we mean to ask a question about that man over there, we are presupposing that that man over there is drinking a martini—not just that *someone* is a man drinking a martini (1966, 289).

This may not seem like progress; before we had one failed presupposition to deal with, now we have two. But, and this is the third difference between Donnellan and Strawson, the "new" failed presupposition, rather than being an obstacle to evaluation, is what *enables* evaluation, by pointing the way to an evaluable hypothesis: that man is a famous philosopher.

Stalnaker attempts to put all this on a firmer theoretical foundation. Imagine O'Leary saying, "The man in the purple turtleneck is bald," where it is understood that the man in question is *that* man (Daniels). The propositional content of O'Leary's statement is that Daniels is bald. The fixation of content here is along lines more or less familiar from Kaplan. Just as the character of an expression like "you" determines its denotation as a function of context, "there are relatively systematic rules for matching up [referential] definite descriptions with their denotations in a context" (1999, 41). The rule for "you" is that it contributes the addressee; the rule for a referential description is that it contributes "the one and only one member of the appropriate domain who is presupposed to have the property expressed in the description" (1999, 41). Crucially from our perspective,

> it makes no difference whether that presupposition is true or false. The presupposition helps to determine the proposition expressed, but once that proposition is determined, it can stand alone. The fact that Daniels is bald in no way depends on the color of his shirt (1999, 43).

So we see that Stalnaker does have an account to offer of *some* cases of NCPF. NCPF occurs (in these cases) for basically Kaplanian reasons. A conventional meaning is given by a systematic character function mapping contexts (=sets of worlds) to propositions. And there is nothing to stop a set of worlds from being mapped to a proposition defined on worlds outside of the set.

This is fine as far as it goes. But NCPF is ubiquitous, and character as Kaplan understands it is reserved to a few special terms. Stalnaker knows this better than anyone, of course; he was one of the first to charge two-dimensionalists with an undue optimism about the project of extending Kaplan-style semantics from demonstratives to the larger language. Some NCPF may be a matter of characters mapping contexts to propositions defined outside those contexts, but not much. An example of Kripke's brings out the extent of the difficulty:

Two people see Smith in the distance and mistake him for Jones. They have a brief colloquy: "What is Jones doing?" "Raking the leaves." "Jones," in the common language of both, is a name of Jones; it *never* names Smith. Yet, in some sense, on this occasion, clearly both participants in this dialogue have referred to Smith, and the second participant has said something true about the man he referred to if and only if Smith was raking the leaves. (Kripke 1977, 14)

Assuming Smith was raking the leaves, the second participant says something true with the words, "Jones is raking the leaves," despite (or because of) the false presupposition that it is Jones they see off in the distance. The example has a Donnellan-like flavor, but the explanation will have to be different; a proper name like "Jones" does not have a reading on which it denotes whoever is presupposed to be Jones in the relevant context. This is why I say there is no general account of NCPF in Stalnaker.[17] I will be suggesting, however, that he does provide the materials for such an account.

So, to review. A sentence's presuppositions are (generally) no part of what it says. Presuppositions can however function as *determinants* of what is said. The suggestion is that they can influence what is said equally well even if false. It remains to explain how exactly the trick is pulled off. Explaining this will be difficult without an account of the mechanism by which presuppositions exert their influence. Because we are really asking about that mechanism. Does it ever in the course of its π-induced operations find itself wondering whether π is true?

There are hints in the literature of three strategies for making π (not a part of but) a guide to asserted content. The first tries to get at what S says by *ignoring* the possibility that π fails. The second tries to get at what S says by *restoring* π when it does fail. The third tries to get at what S says by asking what *more* than π needs to be true for S to be true. I will be arguing against IGNORE and RESTORE and defending SAY-MORE.

[17] This is not to say he doesn't have particular explanations to offer in particular cases. Often he appeals to a device like Strawson's (see above). The original statement—"Jones is raking the leaves"—suffers from presupposition failure, so is not evaluable. Had the speaker been better informed, she would have made a statement—"Smith is raking the leaves"—whose presuppositions are true. Our evaluation of the second statement is then projected back onto the first.

5. IGNORE

Asserted content as conceived by the first strategy addresses itself only to π-worlds. It just ignores worlds where π fails. Thinking of contents as functions from worlds to truth-values, ignoring a world is being undefined on that world. *S*'s asserted content is thus a partial function mapping π&*S*-worlds to truth, π&∼*S*-worlds to falsity, and worlds where π fails to nothing at all.

> [1] *S*'s asserted content *S* is the proposition that is true (false) in a π-world *w* iff *S* is true (false) in *w*, and is otherwise undefined.[18]

There might seem to be support for this in a passage from Stalnaker:

> [I]n a context where we both know that my neighbor is an adult male, I say, "My neighbor is a bachelor," which, let us suppose, entails he is adult and a male. I might just as well have said "my neighbor is unmarried." The same information would have been conveyed. (1999, 49)

The same information would indeed have been conveyed if by "information conveyed" we have in mind assertive content in the sense of [1] above, for (ignoring worlds where my neighbor fails to be an adult male), my neighbor is a bachelor if and only if he is unmarried.[19]

Never mind whether the IGNORE strategy can be attributed to Stalnaker; does it succeed in making π not a part of *S*'s asserted content but a determinant of that content? It does. π influences what *S* says by marking out the set of worlds on which *S* is defined. But π is not a part of what *S* says, for [1] makes *S* undefined in worlds where π is false, and it would be false in those worlds if *S* said in part that π.

The IGNORE proposition has some of the features we wanted. But what we mainly wanted was an *S* that could still be evaluated in worlds where π failed. And here [1] does not deliver at all. "The King of France is sitting in this chair" sounds to most people just false. But there is nothing in the IGNORE proposition to support this judgment, for the IGNORE proposition is undefined on worlds where France lacks a king.

[18] So far this says nothing about *S*'s truth-value in worlds where π fails. Let *S* be "The KoF is so and so." Russellians will call *S* false in worlds where France lacks a king. Strawsonians will say it is undefined. They agree, however, on *S*'s truth-value in worlds where France has a unique king, and those are the only worlds that [1] cares about. Later I will be stipulating that *S*'s truth-value in a world goes with the truth-value of the IGNORE proposition, the one defined by [1].

[19] Stalnaker would not identify what is said with a proposition defined only on π-worlds. Such an identification would make nonsense of passages like the following: "To make an assertion is to reduce the context set in a particular way... all of the possible situations incompatible with what is said are eliminated" (1999, 86). It is not clear to me how closely his notion of what is said—he sometimes calls it "the proposition expressed"—lines up with my assertive content, but certainly the correspondence is not exact.

Methodological digression: I said that "The KoF is sitting in this chair" sounds to most people just false. Why not go further and declare that it really *is* false? Strawson for his part is reluctant to take this further step. "The KoF is sitting in this chair" is *not* false in what he considers the term's primary sense: "*sometimes* [however] the word 'false' may acquire a *secondary* use, which collides with the primary one" (1954, 230)

One option is to follow Strawson in calling sentences like "The KoF is sitting in this chair" false only on a secondary use of that term, and sentences like "The US Chamber of Deputies has representatives from two major parties" true only on a secondary use of "true." The task is then to explain why some gappy sentences *count* as false, while others count as true. Another option would be to follow Russell and call both of the above sentences false in the *primary* sense. The task would then be to explain why some primarily false sentences ("The man with the martini is a philosopher") count as true, while others ("The King of France is bald") count as neither true nor false.

Given that both theories (Russell's and Strawson's) need an analogous sort of supplement to deal with intuitive appearances of truth and falsity, either could serve as our jumping-off point; the choice is really between two styles of theoretical bookkeeping. That having been said, let's consider ourselves Strawsonians for purposes of this paper. *S*'s semantic content—what in context it *means*—will be a proposition defined only on π-worlds; it is semantic content that determines *S*'s truth-value.[20] Truth-value intuitions are driven not by what a sentence means, however, but by what it says: its asserted content.

So, "The KoF sits in this chair" strikes us as false because it says in part that someone sits in this chair. "The US Chamber of Deputies has representatives from two major parties" strikes us as true because it says that the House of Representatives has representatives from two major parties. Both of our remaining strategies are aimed at carving out a notion of asserted content that predicts truth-value intuitions in a way that semantic content is prevented from doing by the fact that it is undefined on worlds where π fails.

6. RESTORE

Let *S* be "The KoF is sitting in this chair." Even if we agree with Strawson that *S* is lacking in truth-value, there is still the feeling that it escapes on a technicality. The chair's emptiness is all set to falsify it, if France's lack of a king would just get out of the way. One response to this obstructionism is to say, fine, let's

[20] Von Fintel 2004 and Beaver and Krahmer (ms) also take this option. Because sentences and their semantic contents have the same truth-value (if any) in all worlds, we can be casual (sloppy) about the distinction between them. So, for instance, it makes no difference to an argument's validity whether we think of it as made up of (i) sentences, (ii) the propositions that are those sentences' semantic contents, or (iii) sentences and propositions combined.

give France a king; then *S*'s deserved truth-value will shine through. This is the idea behind RESTORE. Instead of *ignoring* worlds where π fails, we attempt to *rehabilitate* them, in the sense of bringing them back into line with π. Of course one can't literally turn a non-π world into a π-world, so in practice this means looking at *S*'s truth-value in the closest π-worlds to w.

Now, for S to be true (false) in the π-worlds closest to w is, on standard theories of conditionals, precisely what it takes for a conditional $\pi \rightarrow S$ to be true (false) in w. So we can let the idea be this:

[2] S is true (false) in w iff $\pi \rightarrow S$ is true (false) in w.[21]

Why does "The KoF is sitting in this chair" strike us as false? Even if France is supplied with a king, still he is not to be found in this chair. Why does "The KoF is bald" strike us as lacking in truth-value? Supplying France with a king leaves the issue still unresolved; in some closest worlds the added king is bald, in others not.[22] So the RESTORE strategy has prima facie a lot going for it.

I don't doubt that for *some* similarity relation and *some* associated similarity-based conditional, [2] gives the right results. But if we confine ourselves to the conditionals we know of and have intuitions about—the indicative and the subjunctive—the strategy fails. Let me give some examples before attempting a diagnosis.

Bertrand Russell, invited to imagine what he could possibly say to God if his atheism proved incorrect, replied (not an exact quote), "I would ask him why he did not provide more evidence of his existence." I infer from this that Russell accepted a certain indicative conditional

G. If God exists, he is doing a good job of hiding it.

Now the consequent of this conditional presupposes what its antecedent affirms; so *G* is of the form $\pi \rightarrow S$, read as *if it is the case that* π, *then it is the case that S*. This according to [2-ind] is the condition under which what *S* says is true. But then it would seem that *S* ought to count for Russell as true, given that he accepts *G*. And something tells me that it does *not* strike Russell as true that God is doing a good job of hiding his existence.

So this remark of Russell's shows that [2] in its indicative version does not give a correct account of asserted content. Now consider a different Russell remark: "If there were a God, I think it very unlikely that he would have such an uneasy vanity as to be offended by those who doubt his existence." From this it seems that Russell would have accepted

H. If there were a God, he would be generous to doubters.

[21] I assume that $\pi \rightarrow S$ is false iff $\pi \rightarrow \sim S$ is true.

[22] See Lasersohn 1993 and von Fintel 2004.

H is of the form $\pi \rightarrow S$, read as *if it were the case that* π, *it would be the case that* S. This according to [2-sub] is the condition under which what S says is true. So it would seem that S ought to count for Russell as true, given that he accepts H. But Russell is not at all inclined to think that God is generous to doubters.

Non-theological example: Is the King of France at this moment somewhere in Europe, Africa, Australasia, or the Americas? Not a chance. But the corresponding conditionals are plausibly correct; those are the places he would be, if he existed, and the places he is, if he exists.

[2-sub] does get "The King of France sitting in this chair" right, for the King if he existed would not be in this chair. But imagine for a moment that this chair is the long lost French throne; the King of France *would* (let's say) be sitting in this chair if France had a king. [2-sub] predicts that our intuitions should shift. But it does not make it any more plausible to suppose that the King of France *is* sitting in this chair to be told that he *would* be sitting in it if France had a king. Imagine now that this chair is the long lost French throne *and* French kings if any are master illusionists; if France has a king, he *is* sitting in this chair. This does not affect our truth-value intuitions at all. It is enough for them that the chair is empty.

The problem we are finding with [2] (I will focus for simplicity on [2-sub]) is an instance of what used to be called the "conditional fallacy." According to Shope (1978, 402—I have taken some liberties), the conditional fallacy is

> A mistake one makes in analyzing a statement *p* by presenting its truth as dependent upon the truth of a conditional of the form: 'If *a* were to occur, then *b* would occur', when one has overlooked the fact that although statement *p* is actually true, if *a* were to occur, it would undermine *p* and so make *b* fail to occur.

Philosophers have tried, for instance, to analyze dispositions in counterfactual terms:

x is fragile = if x were to be struck, it would shatter.

But x would not shatter if the molecular properties M making it fragile go away the moment that x is struck. What we meant to say, it seems, is that

x is fragile = if x were struck and retained M, it would shatter.

[2-sub] tries to analyze false-seemingness in counterfactual terms:

S counts as false = if π, S would be false.[23]

[23] Perhaps the fallacy comes in an indicative version too. One is tempted to analyze "Jones is totally reliable" as: if Jones says X, X is true. But if Jones says $0 = 1$, that means not that $0 = 1$ but that Jones is unreliable.

But suppose S counts as false in virtue of certain facts F, and restoring S's presuppositions chases those facts away. (Europe would not have been King-of-France-free if France had had a king.) What we should have said, it seems, is that

S counts as false = if $\pi \& F$, S would be false

This is *essentially* what we do say in the next few sections. I mention this now because the motivation to be offered below is different, and we won't be stopping to connect the dots.

7. SAY-MORE

A passage discussed earlier deserves a second look. Stalnaker had us choosing between "my neighbor is a bachelor" and "my neighbor is unmarried," it being understood that my neighbor is an adult male. He says that the same information would be conveyed whichever sentence we chose. But in a part of the passage we didn't get to, he puts the word "increment" before "information": "the *increment of information*, or of content, conveyed by the first statement is the same as that conveyed by the second" (1999, 49).

The word *increment* suggests that we are to ask what *more* it takes for S to be true, supposing the requirement of π's truth is waived or assumed to be already met. This is the idea behind SAY-MORE. What S *says*, its assertive content, is identified with what *more* S asks of a world than that it should verify π.[24]

Determining these additional requirements may sound like a tricky business; but it is not so different from something we do every day, when we look for the missing premises in an enthymematic argument. To ask what further conditions (beyond π) a world has to meet to be S is essentially to ask what premises should be added to π to obtain a valid argument for S:

$$\frac{\begin{array}{l}\pi \\ ???\end{array}}{S}$$

So we can put the SAY-MORE strategy like this:

[3] S is whatever bridges the logical gap between π and S.

Of course, the gap might be bridgeable in more than one way. I propose to finesse this issue for now by letting S be the result of lumping all otherwise qualified gap-bridgers together. So, for instance,

[24] Suppose we use prop(π) for the properties a world needs to verify π, prop(S) for the properties a world needs to verify S, and prop($S\backslash\pi$) for the additional properties π-worlds must have to be worlds where S is true. It is not in general the case that prop($S\backslash\pi$) = prop(S)-prop(π). An analogy might be this. A rich man can get into heaven only by giving millions to charity; so prop(heaven-goers\rich) includes giving millions to charity. But giving millions away is not in prop(heaven-goers)\prop(rich), because lots of people who don't give millions away still get into heaven, e.g., the deserving poor.

France has exactly one king.
???
The King of France is sitting in this chair.

becomes valid if for ??? we put either "Some French king is sitting in this chair" or "All French kings are sitting in this chair." Assuming both statements bridge the gap equally well (see below), the assertive content is "Some and all French kings are sitting in this chair."

A lot more needs to be said, obviously, and some of it will be said in the next section. Right now though I want to try [3] out on a series of examples, one from Strawson, two adapted from Strawson, one from Donnellan, one from Kripke, and one from Langendoen.

A The lodger next door offered me twice that sum.
B The author of *Principia Mathematica* also wrote *Principia Ethica*.
C All ten solar planets are inhabited.
D The man drinking a martini is a philosopher.
E Jones is burning the leaves.
F My cousin is not a boy anymore.

All six sentences are meant to strike us as false—the first because there is no lodger next door; the second because neither *PM* author wrote *PE*; the third because most solar planets (they number nine, not ten) are uninhabited; the fourth because Daniels (who is in fact drinking water) is not a philosopher but an engineer; the fifth because that man (it's really Smith) is not burning but raking the leaves; and the sixth because my cousin (whether a boy or not) is only eight years old.

How would Stalnaker explain the appearance of falsity in these cases? This is to some extent speculative, but here is what I suspect he would say. *A* seems false for Russellian reasons: it is equivalent to a conjunction one of whose conjuncts is "There is a lodger next door." *B* seems false for supervaluational reasons: it is false on all admissible disambiguations. *C* and *E* seem false for the sort of reason Strawson offered (section 4): we amend them to "All nine solar planets are inhabited" and "Smith is burning the leaves" before assigning a truth-value. *D* seems false for Kaplanian reasons: its character applied to the context of utterance issues in a falsehood, viz. Daniels is a philosopher. *F* seems false because I use it to make an assertion not about my cousin's sex (that's presupposed) but my cousin's age.

The hope is that we can replace these various explanations with one, perhaps closest in spirit to Stalnaker's undeveloped proposal about *F*: the sentences seem false because what they assert is false. [3] tells us how to find the propositions asserted; we ask what assumptions have to be added to

π_A There is exactly one lodger next door.
π_B *Principia Mathematica* has exactly one author.

π_C There are exactly ten solar planets.
π_D That man [pointing] is the man drinking a martini.
π_E That man [pointing] is Jones.
π_F My cousin is a male human being.

for it to follow that

A The lodger next door offered me twice that sum.
B The author of *Principia Mathematica* also wrote *Principia Ethica*.
C All ten solar planets are inhabited.
D The man drinking a martini is a philosopher.
E Jones is burning the leaves.
F My cousin is not a boy anymore.

The needed assumptions would seem to be

A Some and all lodgers next door offered me twice that sum.
B Some and all *Principia Mathematica* authors also wrote *Principia Ethica*.
C All solar planets are inhabited.
D That man is a philosopher.
E That man is burning the leaves
F My cousin is an adult.

A–F, the asserted contents of *A–F*, really are what *A–F* only appear to be, namely false.[25] The suggestion (once again) is that this is not a coincidence. *A—F* appear false because of the genuine falsity of what they assert or say.

I have been stressing the role asserted content plays in explaining felt truth-value, but it is also relevant to judgments about what is said, contributing in this second way even where truth-value intuitions are lacking. This is the application that matters to Stalnaker:

> it is possible for... presuppositions to vary from context to context, or with changes in stress or shifts in word order, without those changes requiring variation in the semantic interpretation of what is said. This should make possible a simpler semantic theory... (1999, 53)

There is that much less need to multiply meanings if "one [can] use the same sentence" against the background of different assumptions to assert different things.

[25] The term "asserted content" might be in some cases misleading, since one does not hear "The lodger next door offered me twice that sum" as *asserting* that some and all lodgers next door offered me twice that sum. Other terms sometimes used are "allegational," "proffered," or "at-issue" content.

(Grice of course makes similar claims on behalf of implicature.[26]) [3] shows why asserted content would fluctuate in this way: the shape of the logical gap between π and S is clearly going to depend in part on π. To illustrate with the Donnellan case, the gap between π_D and D is filled by a proposition about Daniels because that is who we presume to be drinking a martini; if we decide it is really O'Leary then the gap-filler changes accordingly. Or consider Stalnaker's elaboration of Langendoen's example:

> normally, if one said "my cousin isn't a boy anymore" he would be asserting that his cousin had grown up, presupposing that he is male. But one might, in a less common context, use the same sentence to assert that one's cousin had changed sexes, presupposing that she is young. (1999, 53–4)

The first proposition he mentions (my cousin has grown up) corresponds to the missing premise in

My cousin is and always has been a male human being.
???
———————————————
My cousin is not a boy any more.

while the second (my cousin has changed sexes) is the premise one needs to make a valid argument of

My cousin is and has always been a human child.
???
———————————————
My cousin is not a boy any more.

A final example concerns the cognitive content of proper names. How are we to reconcile Mill's idea that names mean their referents with Frege's observation that a name's contribution to cognitive content is not predictable from its meaning so defined? Stalnaker's candidate for the role of (unpredicted) cognitive content is the diagonal proposition, but assertive content can be helpful in this regard as well. Suppose that "*n*" is a name and that *n* is presumed to be the so and so. Then "*n* is F" asserts that the so and so is F (that is what it takes to get from the stated presumption to the conclusion that *n* is F). This is why "A meteor is about to destroy the Earth" is experienced as saying that a meteor is about to destroy this very planet.

8. DEFINITIONS

There is more to bridging the gap between π and S than combining with π to imply S. It is crucial that S not be a "bridge too far" at either end. Let me explain what I mean by that.

[26] See "Further Notes on Logic and Conversation" in Grice (1989), esp. the discussion of "modified Occam's razor" on 47–9.

X is a bridge too far at the S end if it combines with π to imply *more* than S. So, to take again the Donnellan example, the proposition we want is *That man is a philosopher,* not *That man is a philosopher & snow is white.* The latter proposition combines with π to yield "The man with the martini is a philosopher & snow is white," which is a stronger conclusion than we were aiming for.

X would be a bridge too far at the π end if it made for redundancy in the premises—if it repeated material already present in π. The proposition we want is *That man is a philosopher,* not *That man is a martini-drinking philosopher*. It is stated already in π that he is drinking a martini, and there is no need to repeat what is already stated.

How do we enforce the requirement of not being a bridge too far at the S end? Suppose X and π imply a stronger statement than S = a statement that S does not imply. Then S does not imply $X\&\pi$ (or it would imply the stronger statement). Turning this around, if we stipulate that S *does* imply $X\&\pi$, that will prevent X from being a bridge too far at the S end.[27]

How do we enforce the requirement of not being a bridge too far at the π end? Suppose we have our hands on the reason why X is true (false)—its truth-maker or falsity-maker. X has a trace of π in it iff X could not be true (false) *for that reason* unless π too were true (false). So

> [4] X is π*-free* iff X is true (false) and could be true (false) for the same reason, that is, with the same truth-(falsity-)maker, even if π were false (true).

I will leave to an appendix my attempt at a theory of truth- and falsity-makers. But the idea is this. Truth-preservation across all worlds w can be called global implication.[28] Local implication (in some particular w) is truth-preservation across all situations in w. A fact in w is a proposition true in w. Given all this,

> [5] A truth-maker for X in w is a fact that implies X globally and is implied by X locally.[29]

27 There is no question of S not implying π—we have stipulated that S has truth-value only if π is true—so the requirement is really just that S should imply X.

28 Implication should preserve definedness as well as truth. The reason is this. "The KoF is bald" should not π-free imply the falsehood, "Among the bald people is a French king." The latter is false because there are no French kings among the bald people, which is compatible with France's having a unique king. This *would* be a case of π-free implication, if it were a case of implication. But it is not a case of implication, because although truth is preserved, definedness is not. (There could be a world lacking in bald people where France had a unique king.) On an intuitive level, requiring X to be defined wherever S is defined is requiring that X's presuppositions be no stronger than S's. This fits with the idea that S counts as false because it implies something whose weaker presuppositions allow it to be false where S is undefined.

29 Falsity-makers are similar. It is not assumed that X has only one truth- or falsity-maker. A disjunction might be made true either via its left disjunct or its right. See Appendix.

Now we can explain what is involved in bridging the gap between π and S. Certainly X should combine with π to imply S. Also though S should return the favor; it should imply $X\&\pi$. So much is basically to say that although X may well be defined on additional worlds, X restricted to the π-worlds is true/false in the very same worlds as S. Since X extends S beyond the π-worlds, let's call it an extension of S. Finally X should be π-free. All in all, then,

[6] X bridges the gap between π and S iff X is a π-free extension of S.

Asserted content was to be the sum or conjunction of gap-bridgers, so

[7] S is the conjunction of S's π-free extensions.

This can be simplified, however. Extensions are just maximum-strength implications; they are implications false in as many π-worlds as possible, compatibly with still being implied by S. Like any maximum-strength implications, they are conjunctions of regular implications. But then conjoining extensions are conjoining conjunctions of implications, which by the associative law for conjunction is just conjoining implications. So we can get (roughly) the same results as [7] in a more digestible form if we switch to

[8] S is the conjunction of S's π-free implications.[30]

This way of putting it further clarifies why an S that is undefined due to presupposition failure might nevertheless strike us as false. S does not seem false merely because it implies a falsehood, for all the sentences we are talking about do that much, just by virtue of implying π. (E.g., "The KoF is bald" implies that France has a unique king.) S seems false because it implies a falsehood, the reasons for whose truth-value have nothing to do with π.

"The KoF is sitting in this chair" implies "Someone is in sitting in this chair." "Someone is sitting in this chair" doesn't suffer from presupposition failure, so we can evaluate it; it is false. Moreover, and this is crucial, "Someone is sitting in this chair" is false for a reason that could still obtain even if France had a unique king, viz. that the chair is empty.

Compare "The KoF is bald." It implies "France has a king." "France has a king" doesn't suffer from presupposition failure, so we can evaluate it; it is false. So "The KoF is bald" counts as false, if "France has a king" is π-free. Could "France has a bald king" have been false *for the same reason* even if France had a

[30] This should be understood to mean "natural, intelligible implications" (and above, "natural, intelligible extensions"). Otherwise asserted content becomes hard to distinguish from semantic content, as the intersection of *all* π-free propositions implied is liable to be defined only on the π-worlds and the actual world.

king? No, it could not. The reason "France has a bald king" is false is that *France has no king*. Clearly it could not have been false *for that reason* in a world where π was true—a world where France had a unique king.

9. CLAIMS

A theory of NCPF should address itself to three questions. First, how is it possible, that is, how can presupposition failure ever fail to be catastrophic? Second, why does catastrophe strike in some cases but not others? Third, limiting ourselves to the "good" cases, why do some of these sentences strike us as true and others as false?

Stalnaker gave the outlines of an answer to our first question: how is NCPF so much as possible? π helps to determine the proposition expressed; π's truth-value is not directly relevant to its role as proposition-determiner; and the proposition determined is not limited to worlds where π is true. I call this the outline of an answer because Stalnaker doesn't really explain how false πs *can* play the same determinative role as true ones.[31]

Now that we have some idea of the mechanism by which π exerts its influence, we can see why truth-value doesn't come into it. Those of us who use stacks of books as bookends know that false books perform just as well in this capacity as true ones. It's the same with presuppositions. The shape of the logical gap between π and S is defined by π's content; whether that content obtains doesn't much matter.[32]

So that's our explanation of how NCPF is possible. It may seem that we have succeeded too well. If π's falsity is irrelevant to its role in determining asserted content, why is presupposition failure ever catastrophic?

This would be a good question if catastrophic presupposition failure had been characterized as presupposition failure resulting in the loss of *asserted content*. But that is not how we explained it. Presupposition failure is catastrophic, we said, if it has the result that S makes no *claim*. And having an asserted content does not suffice for making a claim.

The reason is this. Part of what is involved in S's making a claim is that for matters to stand as S says is for them NOT to stand as $\sim S$ says. It can happen that S is so tainted by its association with π that S and $\sim S$ cease to disagree when π fails. A sentence whose negation is (counts as) just as true as itself takes no risks and cannot be used to convey real information.

[9] *S makes a claim* iff: if S is true, then $\sim S$ is false.
S makes no claim: S is true and $\sim S$ is also true

[31] Leaving aside special cases like the referential use of definite descriptions..

[32] It can matter a little. If π is true, then any false implication X of S is automatically π-free; X's falsity-maker coexists with π in this world, so obviously the two are compatible. Likewise if π is false then S's true implications are automatically π-free.

Now we can define the central notion of this paper:

[10] *S* is a case of *catastrophic presupposition failure* iff π's falsity has the result that *S* makes no claim.

The poster child here is "The KoF is bald." To find its assertive content we ask, what beyond France's having a unique king is required for the KoF to be bald? The candidates are, let's suppose, (i) *France has a bald king*, and (ii) *any French kings are bald*. We have already seen that (i) is not π-free, because it is false for a reason incompatible with π, namely France's lack of a king. What about (ii)? Its truth-maker (again, France's lack of a king) is consistent with π being false, so (ii) is π-free. It appears that there is nothing for "The KoF is bald" to say but *Any French kings are bald*, and nothing for its negation to say but *Any French kings are non-bald*, that is, *France has no bald kings*. Both propositions are true; so according to [9], "The KoF is bald" makes no claim, and according to [10], "The KoF is bald" suffers from catastrophic presupposition failure.

10. "TRUE" AND "FALSE"

This leaves the question of what distinguishes the presuppositionally challenged sentences that count as true from the ones that count as false. First a stipulation: the claim *S* makes—when it makes a claim—is its asserted content. A sentence's felt truth-value goes with the truth-value of the claim it makes:

[11] *S counts as true (false)* iff it makes a true (false) claim.

Take "The KoF is sitting in this chair." What π-free implications does it have? One obvious implication is *The chair contains a French king*. This is false because *No one is sitting in the chair*, or perhaps because *No king is...*, or *No French king is....* [33] All of these are compatible with France's having a unique king. It looks

[33] Note that *France lacks a king* does not make "This chair contains a French king" false, since it does not imply the latter's presupposition that there is such a thing as this chair. What about *France lacks a king and there is such a thing as this chair*? This is formally eligible but a better—more proportional—candidate for the role is *This chair is empty* (see Yablo 2003 and the Appendix). Objection: If "The KoF sits in this chair" π-free implies the false "A French king sits in this chair," shouldn't "The KoF is heavier than this chair" π-free imply the false "A French king is heavier than this chair"? Yet "The KoF weighs more than this chair" does *not* strike us as false. I reply that "A French king is heavier than this chair" is not π-free, because in *this* case there is no better candidate for falsity-maker than *France lacks a king and there is such a thing as this chair*. No simple fact about the chair itself falsifies "A French king is heavier than this chair" as the chair's emptiness falsifies "A French king sits in this chair." More generally, "The KoF bears R to *x*" does not strike us as false when R is an "internal relation" like taller-than or heavier-than—a relation that obtains in virtue of intrinsic properties of the relata. The proposed explanation is that facts purely about *x* do not suffice for the falsity of "A French king bears R to *x*"; France's lack of a king has to be brought in, which makes the implication no longer π-free. (See in this connection Donnellan 1981 and von Fintel 2004.)

then like "The KoF is sitting in this chair" claims in part that *The chair contains a French king*. That claim is false, so the sentence that makes the claim ("The KoF is sitting in this chair") counts as false.

A couple of examples finally of counting as *true* despite presupposition failure. The first will use a concrete term, the second, to move us back to the ontological relevance of NCPF, will use a term that's abstract.

A long time ago there were two popes, one in Rome, the other in Avignon. Imagine that we check into a monastery one fine night, and scrawled on the bedpost we read, "The pope slept here." A bit of research reveals that both popes in fact slept in the bed in question. Then it seems to me that the inscription strikes us as true; for although it was making a stronger claim than it knew, this stronger claim was correct. To check this against the theory, we must hunt around for π-free implications of "The pope slept here." One such implication is that *some* pope slept here; another is that *all* popes slept here. So far then it looks like the assertive content is that some and all popes slept here. It is because this is true in the imagined circumstance that "The pope slept here" counts for us as true.

Looking into my back yard last Monday, I saw one cat; looking into the yard last Tuesday, I saw two cats, and so on. I now form the hypothesis that this pattern will continue forever[34]—to state the hypothesis more explicitly,

> "For all n, the number of cats in my yard on the nth day $= n$."

I presuppose in saying this that no matter what day it is, there is a unique thing that numbers the cats in my yard on that day (and more generally that whenever there are finitely many Fs, there is a unique thing that numbers them). Now consider the statements on this list:

> on the first day there is one cat in my yard
> on the second day there are two cats in my yard
> on the third day there are three cats etc.
> etc.

All of them are implied by my hypothesis, and it is easy to see that each is a π-free implication. A typical falsity-maker is the fact that on the third day there are no cats in my yard. That there are no cats in my yard can happen just as easily in platonistic worlds (where π holds) as in wholly concrete worlds. A typical truthmaker is the fact that on the third day the cats in my yard are Zora, Teasel, and Yossele. For the cats in my yard to be Zora, Teasel, and Yossele can happen just as easily in concrete worlds (where π fails) as in platonistic worlds.

Now I haven't argued that these are *all* the π-free implications. But if they are, then what my hypothesis *says* is that on the first day there is one cat, on the

[34] This example is from Burgess and Rosen (ms).

second there are two, and so on. This fits with our intuitive sense that the hypothesis counts as true or false according to how many cats my yard contains on which days; the existence of numbers plays no role whatsoever. Note the analogy with the King of France. Just as "The KoF is sitting in this chair" counts as false because of the chair's material contents—nothing to do with French royalty—"The number of cats on the *n*th day = *n*" counts as true, if it does, because of my yard's material contents—nothing to do with numbers.

11. PARTING THOUGHTS ON ABSTRACT ONTOLOGY

Nominalists maintain that abstract terms do not refer. They will find it suggestive, then, that (supposedly empty) abstract terms make in some cases the same sort of contribution to felt truth-value as definitely empty concrete terms do.

Platonists will complain that empty concrete terms make a *negative* contribution; simple King-of-France sentences almost all count as false, to the extent that they make a claim at all. One would expect the emptiness of abstract terms, if they were empty, to manifest itself the same way. But the sentences we construct with abstract terms very often strike us as true.

I agree that the failure of a *concrete* term to refer prevents it from exercising positive semantic influence. But there is a reason for this. "The King of France"'s semantic contribution goes way beyond our notions of what a French king would have to be like. He is (or would be, if he existed) an original source of information of the type that makes simple King-of-France sentences count as true. Numbers by contrast are not (would not be) an original source of information on any topic of interest; their contribution is exhausted by what they are *supposed* to be like. This makes the presupposition that numbers exist "fail-safe" in the sense that its failure makes (or would make) no difference whatever to which applied arithmetical sentences count as true. I am tempted to conclude that nothing in the felt truth-values of those sentences has any bearing on the issue of whether numbers exist.

APPENDIX

Situations are parts of worlds; worlds are maximal situations; truth-in-a-world is a special case of truth-in-a-situation. Suppose that A is a sentence and situation s is part of world w. Then s verifies (falsifies) A only if it contains everything potentially relevant to A's truth-value in w. Thus for s to verify "All swans are white," it is not enough that all the swans in s are white; s should also contain all of w's swans. The formal upshot is a condition called *persistence*: if A is true (false) in s and s is part of s^*, then A is true (false) in s^*.[35] Now we introduce two notions of implication (the first is familiar, the second not):

[35] See Kratzer for an enlightening discussion of persistence.

A implies B globally iff for all worlds w, A is true in w only if B is true in w.[36]

A implies B locally in w iff for all $s \leq w$, A is true in s only if B is true in s.[37]

Here is our first stab at a definition of truth-maker. A truth-maker for X in w is a T that implies X across all possible worlds, and that holds in every w-situation where X holds:

T makes X true in w iff
(a) T is true in w,
(b) T implies X globally,
(c) X implies T locally in w.

(F makes X false in w iff it makes $\sim X$ true.) This runs into a problem, however.[38]

It could happen that X's truth in w is overdetermined by T_1 and T_2 (e.g., X might be a disjunction of unrelated disjuncts both of which are true). Then X may well lack a truth-maker in the above sense. It won't imply T_1 in w because there are w-situations where X holds thanks instead to T_2, and it won't imply T_2 because there are w-situations where X holds thanks instead to T_1. Still, there ought to be a $v_1 \leq w$ such that X implies T_1 in v_1 and a $v_2 \leq w$ such that X implies T_2 in v_2. So rather than asking X to imply T in w, we should ask it to imply T in some subsituation of w.

T makes X true in w iff for some $v \leq w$
(a) T is true in v,
(b) T implies X,
(c) X implies T in v,

Now that we are explicitly contemplating multiple truth- and falsity-makers, we need to adjust the definition of π-freedom:

> X is *π-free* in w iff X is true in w and it has a truth-maker that holds also in worlds where π is false, or X is false in w and it has a falsity-maker that holds also in worlds where π is true.

This runs into a different problem. "France has a bald king" had better not be a π-free consequence of "The King of France is bald," or the latter will count as false which it shouldn't. The obvious falsity-maker is *France has no king*, which is indeed not compatible with π. But another technically eligible falsity-maker is *France has no bald kings*. This is compatible with France's having a unique king, and since our definition requires only that *some* falsity-maker be transportable to π-worlds, it seems we are sunk. The solution I suggest is to impose a proportionality requirement along roughly the lines of Yablo 2003. *France has no bald kings*

[36] I ignore that implication should preserve definedness as well. See n. 28.
[37] Compare the definition of lumping in Kratzer (1989).
[38] Originally raised by Heim against Kratzer.

is needlessly complicated in that a strictly simpler condition (*France lacks a king*) still satisfies conditions (a)–(c). Thus we should think of (a)–(c) as defining *candidacy* for the role of truth- (falsity-) maker, and add that the successful candidate should not involve gratuitous complications in whose absence (a)–(c) would still be satisfied.

REFERENCES

Beaney, Michael. 1997. *The Frege Reader* (Oxford: Blackwell)

Beaver, David. 2001. *Presupposition and Assertion in Dynamic Semantics* (Palo Alto: CSLI Press)

—— and Krahmer, E., "Presupposition and Partiality" (ms)

Bezuidenhout, A. and Reimer, M. 2004. *Descriptions and Beyond* (Oxford: OUP)

Burgess, John, and Gideon Rosen. ms. "Nominalism Reconsidered," to appear in Stewart Shapiro (ed.), *Handbook of Philosophy of Mathematics and Logic*

Burton-Roberts, Noel. 1989. *The Limits to Debate: A Revised Theory of Semantic Presupposition* (Cambridge: Cambridge University Press)

Cohen, Ariel. 2000. "The King of France Is, In Fact, Bald," *Natural Language Semantics* 8: 291–5

Cole, P. (ed.). 1981. *Radical Pragmatics* (New York: Academic Press),

Donnellan, Keith. 1966. "Reference and Definite Descriptions," *Philosophical Review* 75: 281–304

—— 1981. "Intuitions and Presuppositions," in Cole (1981, 129–42)

von Fintel, Kai. "What is Presupposition Accommodation?" (ms)

—— 2004. "Would You Believe It? The King of France is Back," in Bezuidenhout and Reimer 2004, 315–41

Frege, Gottlob. 1872. "On Sense and Meaning," page references to Beaney 1997

—— 1918. "The Thought," page references to Beaney 1997

Glanzberg, Michael. "Expression Failure and Presupposition Failure" (ms)

—— 2002. "Context and Discourse," *Mind and Language* 17: 333–75

Grice, H. Paul. 1989. *Studies in the Way of Words* (Cambridge, Mass.: Harvard University Press)

Heim, Irene. 1983. "On the Projection Problem for Presuppositions," *WCCFL* 2

Humberstone, Lloyd. 2000. "Parts and Partitions," *Theoria* 66: 41–82

Kratzer, Angelika. 1989. "An Investigation into the Lumps of Thought," *Linguistics and Philosophy* 12: 607–53

Kripke, Saul. 1977. "Speaker's Reference and Semantic Reference," originally in *Midwest Studies in Philosophy* II, as reprinted in French, Uehling, and Wettstein (eds.), *Contemporary Perspectives in the Philosophy of Language* (University of Minnesota Press: Minneapolis), 6–27

Lasersohn, Peter. 1993. "Existence Presuppositions and Background Knowledge," *Journal of Semantics* 10: 113–22

van der Sandt, Rob. 1989. "Anaphora and Accommodation," in Renate Bartsch, Johann van Benthem and Peter van Emde Boas (eds.), *Semantics and Contextual Expression* (Foris: Dordrecht)

Shanon, Benny. 1976. "On the Two Kinds of Presupposition in Natural Language," *Foundations of Language* 14: 247–9
Shope, Robert. 1978. "The Conditional Fallacy in Contemporary Philosophy," *Journal of Philosophy* 75: 397–413
Simons, Mandy. 2003. "Presupposition and Accommodation: Understanding the Stalnakerian Picture," *Philosophical Studies* 112: 251–78
Stalnaker, Robert. 1970. "Pragmatics," *Synthese* 22, page references to the reprint in Stalnaker 1999
—— 1974. "Pragmatic Presuppositions," in Munitz and Unger (eds.), *Semantics and Philosophy* (New York: NYU Press), page references to the reprint in Stalnaker 1999.
—— 1978. "Assertion," in Cole (ed.), *Syntax and Semantics*, vol. 9, page references to the reprint in Stalnaker 1999
—— 1999. *Content and Context* (Oxford: OUP)
—— 2002. "Common Ground," *Linguistics and Philosophy* 25: 701–21
Strawson, P. F. 1950, "On Referring," *Mind* 59, page references to the reprint in Strawson 1971
—— 1954. "A Reply to Mr. Sellars," *Philosophical Review* 63: 216–31
—— 1964, "Identifying Reference and Truth-Values," *Theoria* 30; page references to the reprint in Strawson 1971
—— 1971. *Logico-Linguistic Papers* (London: Methuen)
Yablo, S. 2003. "Causal Relevance," *Philosophical Issues* 13
Zeevat, Hank. 1992. "Presupposition and Accommodation in Update Semantics," *Journal of Semantics* 9: 379–412

9

The Story of 'Fred'

Frank Jackson

1. Captains of fishing fleets draw closed figures on maps to indicate where they think the fish are. In the terminology of representation, they represent where they think the fish are by dividing a map into a region where the fish may be according to them, and a region where, according to them, there are no fish. They do the same sort of thing using words, as in 'The fish are somewhere inside the rough circle I have just drawn on the map' or 'The fish are in the area between so and so latitudes and such and such longitudes' or 'The fish are five miles ahead'.

Words and sentences represent much as maps, diagrams, semaphore and the like do, and they do so by making divisions among possibilities. In the most general case, they do so by making divisions in logical space. This is why we can and must capture content in the sense of how things are being represented to be with divisions among possible worlds, or, as we'll see later, in some cases, divisions among centred possible worlds or possible individuals. Equivalently, we can talk in terms of functions from possible worlds and centred possible worlds to truth values.

There are many good questions to ask about the representational picture of language. How ontologically seriously should we take the possible worlds? Does the representational picture apply to ethical language and conditionals? Is its notion of content in terms of sets of possible worlds (or functions from possible worlds to truth values)—*mutatis mutandis* for centred worlds and possible individuals—the only or the most fundamental notion of content? And so on. However, I do not think we can seriously doubt two basic ideas that lie behind the remarks of the first two paragraphs: first, much of language is a system of representation that enables us to make public how we represent things to be in thought, and, secondly, to represent is to make divisions among possibilities. Of course some have denied one or both ideas but Robert Stalnaker is not among them, so I can afford, in the context of this volume, to regard both as a datum.[1] My concern in this essay is with a central question that arises for the representationalist picture. I introduce it by noting that there are two very different questions that can

[1] Stalnaker (1984) is a classic articulation of the representationalist picture.

be asked about the content of a sentence or a word (or a thought or a mark on a map if it comes to that).

2. One question about a sentence or a word is, What is its content? What region in logical space does 'There are fish' mark out—that is, at which worlds is the sentence true? Equally, we can ask, which set of things at a possible world is the word 'fish' true of? A quite different question is how a sentence or a word gets to have the content it does. We all know what *three short, three long, three short* represents in Morse Code; how it came to so represent is a different matter altogether. It is discussion of the second question that involves issues to do with what goes on inside the heads of users, how a state of a head gets to be about this or that happening in the world, the adoption of conventions of usage that allow us to make public the contents of thought by means of marks on paper or patterns of sound, the promulgation of agreements to use certain words one way or another in the few cases where the conventions are explicit, and so on. Stalnaker (2003*a*) usefully calls the question of how a word or sentence gets to have the content it does, the meta-semantical question, as opposed to the semantical question.

The question I will be concerned with is whether or not the distinction between what I have called in various places (for example, Jackson 1998, 2004) *A* and *C*-intensions should be thought of as one between two equally correct though different answers, for some sentences and words, to the semantical question, or whether the distinction should be thought of as a product of how the answer to the semantical question may be a function of context, that is, as pertaining to an aspect of the meta-semantical question.

I think the view that some words and sentences have both *A and C*-intensions is a correct view about the answer to the semantical question for those words and sentences.[2] I know from discussions, especially in the States, that many think that this is a (bad) confusion between the semantical and the meta-semantical questions; that what we have are interesting dependencies between context and content, not two contents. Roughly, on this view, the *C*-intension is *the* content, and the *A*-intension is a shadow of the way content is a function of context, of the kind of phenomenon we learn about from Twin Earth and the many arguments for externalism. The same is said about David Chalmers's views about primary and secondary intensions, in, for example, (1996, ch.2, §4). However, although there are very close connections between his notion of a primary intension and my notion of an *A*-intension, and between his notion of a secondary intension and my notion of a *C*-intension, I will focus on *A* and *C*-intensions here, as I can be reasonably confident that I have not misunderstood the intentions of the proponent of the distinction as expressed in those terms. I will, however, say something in passing about a significant difference in our views on the content of *thought*.

[2] See, e.g., Jackson (1998, chs. 2 and 3).

Where does Stalnaker stand on this issue? When I first argued that we need the A versus C-intension distinction to handle various problems in the philosophy of mind and language, I thought of myself as borrowing (gratefully) a picture I owed to a number of philosophers but to him especially. I was disabused by a number of papers but especially by Stalnaker (2001 and 2003*a*). The thrust of (2001) is well caught by the adaptation of an English proverb that heads the essay: 'Hell is paved with primary intensions'. He might as well have said 'Hell is paved with A-intensions', except that it would not have been as much fun. But I confess that I am still not sure exactly where he stands on the question, Do we have two contents, or one content and an account of the dependency of content on context? Perhaps his view is that the answer varies case by case. In any case, this volume is a chance to clear the air. My aim is to state as simply as I can why I think that *sometimes* the answer to the question has to be that we have two contents and, moreover, that we can see this without answering anything as hard as the meta-semantical question. I will start by describing a very simple word game where we have, I will suggest, no alternative but to acknowledge two contents—or at least we do if we like the representationalist approach to language.

3. Narrow and Broad enjoy making up languages for their private use. The languages are not private in any sense that raises serious philosophical issues. They are private but could become public if Narrow and Broad wanted them to. They agree one day to use 'fred' for the shape of the smallest homogenous object (sho). They have no idea what that shape is, where the sho is, and whether it is in existence now. They are though confident that there is exactly one such object. We may suppose that they are right and that, unknown to them, it is round. They have read Kripke (1972), understand the difference between treating 'fred' as a definite description and as name-like and rigid, and are clear that 'fred' is rigid.

Broad argues that their term 'fred' has some important messages for the philosophy of content. We are familiar, he says, with the idea that content is a function of the environment of the speaker and writer but 'fred' teaches us that content is a function of something highly particular, the shape of the sho. Indeed, we can express the way its content is a function of the shape of the sho in a table.

if the sho is round, 'x is fred' iff x is round.

if the sho is square, 'x is fred' iff x is square.

. . .

if the sho is blah-shaped, 'x is fred' iff x is blah-shaped.

Alternatively, Broad notes, we could express the point in terms of a set of functions, one for each possible world (as context), whose value at a possible world (as world of evaluation) is true if *x* in the evaluation world has the shape of the sho in the context world. Broad emphasises that only one of these functions, the one for the context world that is the actual world, gives the content of 'fred'. The role of the other functions is *not* to give the content of 'fred' but to give the content 'fred' might have had and to give the way the content of 'fred' is a function of context. Broad insists that the content of '*x* is fred' is the set of worlds where *x* is blah-shaped, whatever that shape is, and that in no sense of 'content' is the content of '*x* is fred' the set of worlds where *x* has the shape of the sho. He grants, of course, that this set is the set of worlds *w*, with *w* treated as both world of evaluation and context world, where '*x* is fred' is true, but that, in his view, is no reason to regard it as a content of the sentence.

Narrow's view of matters is very different. Narrow says that the idea that content might depend on something as trivial as the shape of the sho is implausible. The key question to ask, Narrow urges, is how much credence we give to the sentence '*x* is fred''s being true, and that surely is exactly the same as the credence we give to *x*'s being the same shape as the sho. In our mouths, the sentence '*x* is fred' is nothing more than a way of saying that *x* is the same shape as the sho, whatever that is. Or consider, says Narrow, what happened when we indulged in the speculations that we expressed in words like 'I wonder what the chances are that fred is a conic section'; we were surely doing nothing more than speculating on the chance that the shape of the sho is a conic section. What else could we be doing, given how the term 'fred' entered our private language? The term was nothing more than an agreed word for the shape of the sho and the fact that we agreed to use it rigidly made no odds to that, or so Narrow argues.

Who is right, Broad or Narrow? We might stipulate a meaning for 'content' that delivers one answer or the other but what would that achieve apart from showing our powers of stipulation? There seems to be only one principled way to tackle the question. It is to ask, first, what role the term 'content' plays in this whole discussion, and then to see whether Broad's or Narrow's answer makes best sense in terms of the identified role.

The role of content comes from the commonplaces we mentioned at the beginning about language, or anyway much of language, being a system of representation. We use it to convey putative information about how we take things to be in much the way we use maps and diagrams and gestures. What putative information about how things are do Broad and Narrow convey by their use of 'fred'? There is only one plausible answer. They use it to say that something is the same shape as the sho. They do not use it to say that something is round despite the fact that, as we are supposing, *x* is sho iff *x* is round. Narrow is right and Broad is wrong.

Broad might object that this misses the radically interesting nature of what we learn from the case of 'fred'. We learn that the belief that *x* is fred is the very same belief as the belief that *x* is round (if that is indeed the shape of the sho), or is the very same belief as the belief that *x* is square (if that is indeed the shape of the sho), or . . . . The situation, Broad might argue, is like that with the belief that Hesperus is Phosphorus. Recent work on reference tells us that this belief is the very same belief as that Hesperus is Hesperus. So what if that was not obvious until the discovery of the necessary *a posteriori* truth that Phosphorus is Hesperus.[3] In the same way, the discovery that the belief that *x* is fred is the belief that *x* is round (or whatever the shape turns out to be) awaits the discovery about the shape of the sho, but the belief is the belief that *x* is round all along. It follows, urges Broad, that that is how Narrow and Broad represent things to be when they use the sentence '*x* is fred'.

Whatever position may be taken on the belief that Hesperus is Phosphorus, Broad's contention that they—Narrow and Broad—are expressing the belief that *x* is round (say) when they say that it is fred, gives to verbal conventions a quite implausible power to make belief. Merely by agreeing to use 'fred' in the way described, they make it the case that they believe that *x* is round whenever they believe that *x* is fred? As Evans (1982, 50) says with reference to Grice (1969, 140), 'We do not produce new thoughts (new beliefs) simply by a "stroke of the pen" (in Grice's phrase)—simply by introducing a name into the language.'

4. We emphasized earlier the distinction between the semantical and the meta-semantical questions. Broad's answer is an application of this distinction. His view is that we have a case where the (single) content of '*x* is fred' is a function of context—the shape of the sho in the world of assertion—and, because I have said nothing about the hard question of what determines content, it might be asked how I can come down on Narrow's side? I agree that the meta-semantical question is hard. There is relatively little controversy about the content of a sentence like '*x* is round' in English, but great controversy about why it has the content it does. All the same, we don't need to solve the meta-semantical question in order to come down on Narrow's side. All Narrow needs is that we know enough to know that the stipulation that gave the meaning to 'fred' in the word game is possible, and we know that. A similar point sometimes comes up in the debate between causal theories of reference for proper names and causal descriptivism. Causal descriptivists accommodate the intuitions that drive the causal theory of reference for proper names by including them in the descriptions associated with the name. Here is how Lewis (1997, fn. 22) puts the strategy:[4]

[3] But note that he had better express his thought in a way which avoids saying that the proposition that Phosphorus is Hesperus has a very different status from that Phosphorus is Phosphorus while being one and the same proposition; see Soames (2002).

[4] Causal descriptivism has had many supporters: e.g., Kroon (1987), and Searle (1983, ch. 9). The view is, I take it, only intended to be an account for some proper names. We can name things outside the light cone. For more on this, see Jackson (2005).

Did not Kripke and his allies refute the description theory of reference, at least for names of people and places?.... I disagree. What was well and truly refuted was a version of descriptivism in which the descriptive senses were supposed to be a matter of famous deeds and other distinctive peculiarities. A better version survives the attack: *causal descriptivism.* The descriptive sense associated with a name might for instance be *the place I have heard of under the name "Taromeo"*, or maybe *the causal source of this token: Taromeo,* and for an account of the relation being invoked here, just consult the writings of causal theorists of reference.

Causal theorists sometimes object that this strategy ducks the important question of how content gets determined; that the important insight behind the theory they like is that causal connections are an important part of the answer to the meta-semantical question. But how can this be an objection to causal descriptivism? Controversial though the meta-semantical question is, no-one thinks it is impossible to use certain words as abbreviations for descriptions that contain *inter alia* causal elements. Causal descriptivism is the contention that, as a matter of fact, that's what some proper names are. There would only be a problem for causal descriptivism from the important distinction between the semantical and meta-semantical questions if the answer to the meta-semantical question implied it was impossible for English to contain words that worked like abbreviations of clusters of descriptions that contain a causal element.

5. So I persist in the view that Narrow is right. But Narrow has a problem. We defended Narrow's answer by reminding about the representational role of language. And, as we said early on, to represent is to divide the possibilities into those in accord with how things are being represented to be and those not in accord with how things are being represented to be. How does 'x is fred' divide the possibilities? The obvious answer is by dividing the possible worlds into those where it is true and those where it is false; equivalently, the sentence determines a function from worlds to truth values that takes the value true whenever the sentence is true at a world. But this delivers as the content of 'x is fred', the set of worlds where x is round—Broad's answer, not Narrow's. Narrow has, I think, only one way to go. It involves three steps.

First, Narrow should urge that 'fred' should be thought of as equivalent to 'the *actual* shape of the sho', where the 'the actual shape of the sho' at any world w refers to the shape of the sho at the actual world; in consequence, which things are sho in w is determined by which things in w are the shape of the sho in the actual world and not by which are the shape of the sho in w. This is more or less implicit in the way that Broad and Narrow added 'fred' to their language; it is not an *ad hoc* addition prompted by a perceived problem.

The second step is to observe that, given the equivalence between 'fred' and 'the actual shape of the sho', we can pick out *two* candidate sets of worlds to be the content of 'x is fred': the set where the sentence is true at w, and the set of worlds *w* where the sentence is true under the supposition that *w* is actual. Call the first the *C*-intension and the second the *A*-intension. The *C*-intension of 'x is fred' is the set of worlds where *x* has the shape the sho actually has and, as we are supposing the sho to be round in the actual world, this will be the set of worlds where *x* is round. The *A*-intension of '*x* is fred' is the set of worlds *w* where *x* has the shape of the sho at the actual world under the supposition that *w* is the actual world—that is, it is the set of worlds where *x* has the shape of the sho at that world. (We can set aside for our purposes what to say about worlds where 'the shape of the sho' is an imperfect description.)

The final step is to observe that both sets can be thought of as contents but it is the second that is plausibly what's being said about how things are by Narrow and Broad in their use of 'fred'; it is the second that is content in the sense tied to the use of words to capture how things are being represented to be. Why? Because, as we noted, the *A*-intension of '*x* is fred' is the set of worlds where *x* is the same shape as the sho, and that's what Narrow and Broad both claim about how things are when they use the sentence.

6. There are two concerns one might have about appealing to a sentence's *A*-intension in elucidating its content. One is the very notion of truth under the supposition of actuality: what's that? The other is why it should be thought to deliver the intended sense of content.

What is meant here by truth under the supposition of actuality is *not* some special way of being true at a world; if you understand truth at a world, you have all you need to understand truth at a world under the supposition of actuality in the intended sense. Truth under the supposition of actuality enters the picture when and only when we are dealing with sentences (*mutatis mutandis* for predicates) containing, explicitly or implicitly, devices that have the effect that the sentences' truth at a world *w* depends, in part or in whole, on how the actual world is. In evaluating under the supposition that *w* is actual, the role of how the actual world is is handed across to how *w* is. The effect of this in practice is most easily seen by looking at some simple examples.

'Actually *p*' is true at *w* iff '*p*' is true at the actual world, whereas 'Actually *p*' is true at *w* under the supposition that *w* is actual iff '*p*' is true at *w*.

'The actual *F* is *G*' is true at *w* iff the *F* at the actual world is *G* at *w*, whereas 'The actual *F* is *G*' is true at *w* under the supposition that *w* is actual iff the *F* is *G* at *w*.

'x is the actual shape of the sho' is true at *w* iff the shape *x* has at *w* is the shape the sho has at the actual world, whereas '*x* is the actual shape of the sho' is true at *w* under the supposition that *w* is actual iff the shape *x* has at *w* is the shape the sho has at *w*.

The crucial point is that, in each case, what appears on the RHS of the second occurrence of 'iff' concerns how things are at *w*, not how things are at *w* under the supposition that *w* is the actual world.

Why should truth at a world under the supposition of actuality deliver the needed sense of content? The answer is that one's opinion about how things are is none other than one's opinion about how the world one is in is, and that's nothing other than one's opinion about which worlds might be the one you are in; that is, might be the actual world. It follows that the representational role of a sentence in making a claim about how assertors take things to be is given by the worlds whose actuality is consistent with the sentences they assert, and that is given by the set of worlds whose actuality is consistent with the sentence, by the sentence's *A*-intension.

7. The distinction between *A* and *C*-intensions for 'fred' is one between two contents for certain words and for sentences containing those words. My thesis is that if you want the content that gives how Narrow and Broad represent things to be when they use 'fred', saying, as it might be, that *x* is fred, you should look to the *A*-intension and not the *C*-intension of '*x* is fred'. The claim is *not* that there are two contents of Narrow and Broad's 'fred' *thoughts*. It couldn't be, as part of the motivation is the conviction that when they assert that *x* is fred, what they think, the thought they are making public, is that *x* is the same shape as the sho. To say that one content of their thought is the *C*-intension of '*x* is fred' would be, from my perspective, to say precisely the wrong thing. For this reason, what I am saying needs to be clearly distinguished from Chalmers's views. We agree about a lot obviously but I dissent from his use of the *A* and *C* distinction—the primary versus secondary one as he usually calls it—as part of an argument that thoughts have both primary and secondary contents.[5] I am a two content theorist about parts of language; I am not a two content theorist about thought.

8. My case for distinguishing two contents in the case of '*x* is fred' was, first, that the representational approach to language requires that we think of content in terms of divisions among possibilities and, secondly, that the right set of possibilities to capture Narrow and Broad's thought when they use the sentence to wonder, affirm, deny, discuss etc. how things are is the sentence's *A*-intension. The set of worlds where the sentence is true, the sentence's *C*-intension, gives the wrong answer.

One might accept the case for distinguishing two contents in the case of the special language invented by Narrow and Broad while denying the need to make the distinction for the natural languages we in fact write and speak. The contention would be that we might have used a language that requires that we make the distinction but in fact we don't. I have argued in, for example, Jackson (2004) that we need to make the distinction for natural kind terms in English (and I don't suppose that the situation is radically different for other natural languages), and I discuss the case for proper names in Jackson (2005 and 2003). But I think a

[5] See, e.g., Chalmers (2002).

similar message comes to us from cases where we have egocentric, perspectival or centred thoughts and seek to express them in language. The rest of this paper will be concerned with this issue and why we need the two content approach when we think about how sentences capture the contents of perspectival thoughts.

9. Cases where our thought is, as it is variously put, perspectival, egocentric, subjective, positioned or centred, are common. A person's take on how things are is typically a take on how they are from their own standpoint: it is raining *now* where *they* are, *they themselves* are tired, the person next to *them* has a bad cough, swans are white in regions near *here* where 'here' is 'their' here, I *myself* am cold, and so on. This is no surprise when we reflect on the fact that perception represents how things are from a centre or location, a location which is typically roughly where the perceiver is (viewing something through a periscope is one of the exceptions). What is perhaps a surprise is that perspectival thought is *irreducibly* so, as we learn from the writings of, for example, Perry (1982), Castañeda (1966), and Lewis (1979). The thought that I myself am cold cannot be reduced to the thought that someone who is thus and so is cold.

Although the conclusion that perspectival thought is irreducibly so is widely accepted, the precise significance of the result is much more controversial. I think it is a mistake to think that we learn that there is something 'extra' about our world, something beyond the ken of science, that has an irreducible perspective or subjectivity built into it. For this reason I think it is best to avoid calling the content in question 'subjective'. I follow Lewis (1979) in holding that what we learn is something about how to model the content of egocentric thought. The key idea is that, as well as thoughts to the effect that one is in such and such a kind of world, there are thoughts to the effect that things are thus and so with regard to oneself. To believe that you yourself are located at a certain place in a shopping centre complex is to believe in part that someone is located at that place, but it is to believe *in addition* that it is you yourself who is located at that place. Likewise, when I believe that I myself am cold, I believe not merely that I am in a world where someone is cold but that I am one of the cold ones in those worlds.

What does this come to in terms of divisions among possibilities? The key point is that to believe that it is I myself who is cold is to believe something more. The more is that as well as excluding worlds where no-one is cold, I exclude worlds where I am not one of the cold ones. That is to say, I make a finer division: I divide among possible individuals as well as among possible worlds. Lewis puts it in terms of self-ascription of properties: I self-ascribe the property of being cold. In the literature it is perhaps more common to put it in terms of centred worlds. The content of my belief that I myself am cold is given by the set of centred worlds that are the right 'cold person' way at their centres. The irreducibility of egocentric content is, on this approach, the product of the fact that a division among centred worlds typically carries information that cannot be captured by a division among worlds.

10. There is a puzzle arising from how we report egocentric or perspectival content using sentences. We report perspectival takes on how things are in sentences like 'I am cold', 'There is a spider crawling up my leg', and 'I am on my way to the dentist'. But the obvious way to connect these sentences to divisions among possibilities to capture content in the way enjoined by representationalism gets things as badly wrong as did *C*-intensions in the case of 'fred'.

One way to bring this out is to suppose that I am seriously confused and deluded about who I am, when it is and where I am. I may still be of the opinion (and justifiably so) that I am cold. A person can wake in the middle of the night in a state of total confusion but still know for sure that they are cold and express this fact saying out loud or to themselves 'I am cold'. But now it is transparent that the wrong way to capture my (token) thought about how things are when I use the (token) sentence 'I am cold' is with the set of worlds where the sentence is true. The sentence will be uttered by FJ, at, say, 1500 on October 29, 2004, in which case the set of worlds where the sentence is true is the set where FJ is cold at 1500 on October 29, 2004. But that is not how I am representing things to be and is not the thought I have, for *ex hypothesi* I have no idea who I am and when it is. But the key point is anyway obvious enough. To say that you yourself are cold is not to say who you are, or when it is, even in cases where you know perfectly well who you are and when it is. It follows that the *C*-intension of 'I am cold', does not capture the content in the sense of how I am representing things to be by my use of the sentence. As with 'fred', we have an example where *C*-intensions don't do the job, and of course the same point could have been made with the other sentences we use to report thoughts with egocentric content; there is nothing special about 'I am cold'.

11. If *C*-intensions don't do the job what does? I think the role of the token sentence 'I am cold' is like the role of the location dot, the 'you are here' dot, on shopping centre maps. Both serve to identify the correct centre for the relevant set of centred worlds. In the case of the sentence, the location is that of being the producer of the token sentence. The token sentence is like a flag that says 'here is where the coldness is'. Think, for example, of the information conveyed when you hear someone in the corridor say 'I am cold'. Your grasp of English tells you that this sentence is designed to inform hearers that a certain causal origin of the token sentence is cold. This means that 'I am cold' is true at a centred world $\langle c, w \rangle$ iff c is the actual producer of 'I am cold' and c is cold in w. (I am thinking of c as a temporal part, or we could add a reference to time.)

It has to be the actual producer because 'I' is rigid. But it would be wrong to conclude that the way things are being represented to be by my use of 'I am cold' is given by the set of centred worlds where the actual producer is cold, as that re-inserts, as part of the content, who I am, and, as we noted, this is not part of what's being asserted. The correct set of centred worlds is instead the set of centred worlds where 'The actual producer of "I am cold"' is true under the supposition that the centre is actual. It is the *A*-intension transposed from sets of worlds to sets of centred worlds.

12. Stalnaker has a different take on the issues of the last few paragraphs. Below is a passage from Stalnaker (2003*b*, 255). The self-locating statements and beliefs in the quotation are the ones we have been discussing and the semantic diagnosis he refers to as uncontroversial is the agreed point that, for example, 'I am cold' said by x at t is true iff x is cold at t.

> The truth-conditions ascribed to self-locating statements and beliefs by the semantic diagnosis are uncontroversial, but I don't think this account is quite right. It seems that what I believe when I believe that I am a philosopher, what I discover when I discover that my pants are on fire, or what I say when I say that I live in Massachusetts, is something about myself, and something different from what someone else might believe, discover, or say about themselves. And it also seems that when TN [Thomas Nagel] expresses his belief about who he is by saying 'I am TN' and his interlocutor understands and believes him, then the information she acquires is the same information that he imparted. But she does not, of course, thereby ascribe the property of being TN to herself. So the account of content as self-ascribed property, or as set of centred possible worlds identifies contents that need to be distinguished, and distinguishes contents that should be identified.

It will help highlight where we differ if I explain why I reject the two contentions he puts forward as data for accommodation by a good theory of content.

Take, first, the identification of contents that Stalnaker thinks should be distinguished. When I come to believe or say that my pants are on fire, what I come to believe and say is the very same thing you come to believe and say when you believe or say that your pants are on fire. When it dawns on you that not only are you in a world where someone's pants are on fire, you are one of the one's whose pants are on fire, all you acquire by way of belief, as far as I can see, is that you come to believe that you belong to a certain class—those unfortunates with their pants on fire. And this is exactly what Fred comes to believe when Fred comes to believe that his pants are on fire. This is why you can move from the belief that you are in 'a pants on fire' world to the belief that you are one of those with pants on fire while having no idea who you are. If one's identity were part of the content of self-ascribing belief, if what you believe and what I believe were different in content because you and I are different people, then an opinion about who one is would be required to have these self-ascribing beliefs, whereas in fact no such opinion is required. To include something in the content to reflect the fact we are different people is to include something that should not be there.

In Stalnaker (1999, 21), there is an argument from disagreement for the contention that the content of x's belief that they themselves are so and so differs in content from y's belief that they themselves are so and so.

Consider the case (discussed by Perry and Lewis) of mad Heimson who believes that he is Hume. All his impersonal beliefs about Hume are correct, let us assume, or better, let us assume that they are the same as Hume's beliefs about Hume. Still, it would not be right to say that Hume and Heimson don't disagree about anything.

The point here, I take it, is that on the kind of view I have been supporting, the contents of Hume and Heimson's beliefs, both personal and impersonal (to use the terms of the quotation) are the same and, runs the argument, disagreement is inconsistent with sameness of contents. But the fact that Hume and Heimson's beliefs have the same contents would not imply that the conditions under which their beliefs are true are the same, and they aren't in the case of their personal beliefs that they are Hume. Heimson's is true if Heimson belongs to the relevant class of individuals, if Heimson has the self-ascribed property of being Hume; whereas Hume's is true if Hume belongs to the relevant class of individuals, if Hume has the self-ascribed property of being Hume. Indeed, in the case to hand, not only are the truth conditions different, Heimson and Hume cannot both believe truly. That would seem to be disagreement enough. (I take it this is the position in Lewis 1979.)

What about Stalnaker's claim that treating perspectival contents as sets of centred worlds 'distinguishes contents that should be identified'? It is certainly true that when Fred believes and affirms that some things are cold and I accept what Fred believes and affirms, there is one content in common to us, namely, that some things are cold, that we are in a world containing cold things. But this is a feature of a case which does not involve egocentric content. The rules are different when we are dealing with egocentric content; that seems to me a lesson from its irreducibility. When Fred affirms and believes that he himself is cold, and I go along with him, what I come to believe is something like that there is someone in such and such a relation to me who is cold. If Fred is standing before me and produces the sentence 'I am cold', and I understand his words and accept that he is correct in what he's claiming, what I will come to believe, as we argued earlier, is something like that there is a person who is a certain kind of causal origin of the sentence token and that that person is cold. I will no doubt acquire some egocentric content but it will be from my perspective. Perhaps it will be that I myself am hearing someone saying that they are cold. If you like, we get a shift in centre from Fred to me.[6]

REFERENCES

Castañeda, Hector-Neri (1966). 'He*: A Study in the Logic of Self-Consciousness', *Ratio*, 8: 130–57.

[6] I have discussed the topics covered in this essay with too many people over too many years for a list to make sense or indeed be possible, but in rereading it I spotted the footprints of David Braddon-Mitchell, Martin Davies and Andy Egan and must express my debt to discussions with them without saddling them with responsibility.

Chalmers, David (1996). *The Conscious Mind* (New York: Oxford University Press).
—— (2002). 'The Components of Content' in *Philosophy of Mind: Classical and Contemporary Readings* (New York: Oxford University Press), ed. David J. Chalmers: 608–33.
Evans, Gareth (1982). *The Varieties of Reference* (Oxford: Oxford University Press).
Grice, H. P. (1969). 'Vacuous Names', in *Words and Objections*, ed. Donald Davidson and Jaakko Hintikka (Dordrecht: Reidel), 118–45.
Jackson, Frank (1998). *From Metaphysics to Ethics* (Oxford: Oxford University Press).
—— (2003). 'Narrow Content and Representationalism—or Twin Earth Revisited', Patrick Romanell Lecture, *Proceedings American Philosophical Association*, 77, 2: 55–71.
—— (2004). 'Why We Need *A*-intensions', *Philosophical Studies*, 118, 1–2: 257–77.
—— (2005). 'What are Proper Names For?', in *Experience and Analysis*, Proc. 27th International Wittgenstein Symposium, 2004, ed. Johann C. Marek and Maria E. Reicher (Vienna: hpt-öbv).
Kripke, Saul (1972). *Naming and Necessity* (Cambridge, Mass.: Harvard University Press).
Kroon, Fred (1987). 'Causal Descriptivism', *The Australasian Journal of Philosophy* 65.
Lewis, David (1979). 'Attitudes *De Dicto* and *De Se*', *The Philosophical Review*, 88: 513–43.
—— (1997). 'Naming the Colours', *The Australasian Journal of Philosophy*, 75: 325–42.
Perry, John (1982). 'The Problem of the Essential Indexical', *Nous*, 13: 3–21.
Searle, John (1983). *Intentionality* (Cambridge: Cambridge University Press).
Soames, Scott (2002). *Beyond Rigidity: The Unfinished Semantic Agenda of Naming and Necessity* (New York: Oxford University Press).
Stalnaker, R. (1984). *Inquiry* (Cambridge, Mass.: MIT Press).
—— (1999). *Context and Content* (Oxford: Oxford University Press).
—— (2001). 'On Considering a Possible World as Actual', *Proceedings of the Aristotelian Society*, 75, Ch. XII; reprinted in Stalnaker (2003: 188–200)
—— (2003). *Ways a World Might Be* (Oxford: Oxford University Press)
—— (2003*a*). 'Conceptual Truth and Metaphysical Necessity' in Stalnaker (2003: 201–15)
—— (2003*b*). 'On Thomas Nagel's Objective Self' in Stalnaker (2003: 253–75)

10

Stalnaker and Indexical Belief

John Perry

§1

Paradigm indexicals include "I", "now", and "here". In this essay I'll also include demonstrative phrases like "this city" and "that lake". Indexical beliefs are those that are appropriately expressed with indexicals, and reported by the person who has the belief with an indexical.

If you asked me why I am going to campus I might say, "I have a class to teach in half an hour." I might add, if your question makes me doubt my own grasp on my schedule, "At least I believe I have a class to teach in half an hour." In the first case I would have expressed my belief, and in the second case reported it. In both cases it would have been decidedly odd to use my name instead of the indexical "I". If I had said, "John Perry has a class to teach," or "I believe John Perry has a class to teach," it would have sounded like there must be some other John Perry who is going to hold a class I want to attend. Or perhaps I would have merely sounded self-important, like President de Gaulle did (with considerable justification) when he used to talk about himself in the third person. Or perhaps one would think I was having an unusually bad senior moment, forgetting that I was John Perry, but wanting to attend his class because I remember how plausible his views have always seemed to me.

§2

Indexical beliefs pose a prima facie problem for the view that belief is simply a matter of a believer believing a certain proposition. This was once the received theory and I'll call it that. There seems to be a real difference in the belief I express with "I have a class to teach," and the one I express with "John Perry has a class to teach." I could have the first without the second if I had a mild case of amnesia, and forgot my name and who I was, but still remembered, like an old milk horse, my basic schedule. I could have the second without the first if I lost even more of my memory, but looked up John Perry's schedule in the Stanford Time Schedule and found that he had a class to teach.

These are pretty bizarre suppositions, but the same point can be made with other indexicals without appealing to anything particularly uncommon. Suppose Elwood has always wanted to see the capital of California. One day, on a long trip up the Central Valley, he drives into Sacramento. He is tired and hasn't been paying much attention, so he doesn't really know where he is. He has the vague impression he is coming into Modesto or Stockton. He would give different answers to the questions, "Do you want to look around this city?" and "Do you want to look around the capital of California?" After he realized he was in Sacramento, he might say, "I didn't realize that this city was Sacramento." That would be the natural and appropriate way for him to get at what he didn't believe. When he realized what he would express with "this city is Sacramento," he seems to have acquired a new belief.

But what belief is it? If belief is simply a matter of believing a proposition, then the new belief must correspond to a new proposition believed. But what proposition is it? Finding that proposition, or amending or supplementing the account of beliefs as relations to propositions, is the problem indexical beliefs pose for those drawn to the received theory.

The problem of indexical belief was advanced by Hector-Neri Castañeda in a number of brilliant papers in the 1960s (Castañeda, 1966, 1967, 1968*a*, 1968*b*). In Perry (1977) I argued that it posed a problem for Frege's theory of Sinn and Bedeutung, and in Perry (1979) I argued that distinguishing between de dicto and de re belief doesn't solve the problem. I advanced a positive thesis, that belief consisted in believing a proposition *by* being in a belief state, whose nature was better characterized by something akin to Kaplan's characters—functions from context to propositions—than by the proposition one believed in virtue of being in the state (Kaplan, 1989). Before he realized where he was, Elwood believed the proposition that Sacramento is the capital of California in virtue of being in one state, the state people get into by reading books about California, and that disposes them to get the right answer when asked which city is the capital of California. But there was another way of believing that very same proposition, a different belief state, that he wasn't in, until he made his discovery. That is the state that disposes people to say, "This city is the capital of California," and to stop and get out of the car and look around if they have a long-standing desire to see the capital of California.

Whatever the merits of my criticisms and positive suggestions, these papers did seem to play a role in helping some outstanding minds, including Robert Stalnaker's, to focus on the problem of indexical belief and develop their own accounts. I would no doubt be wise to be content with having a played a role in this literature and to shut up. Nevertheless in this paper I'll examine Stalnaker's theory in 'Indexical Belief' (1981) and also say a little about how my past and current approaches to semantics fit together with his insights. Page references are to the reprint of the article in Stalnaker (1999).

§3

In (1981) Stalnaker considers the problem of indexical belief. He criticizes my proposal (1977, 1979) and Lewis's (1979) and gives his own account in terms of diagonal propositions.

Stalnaker identified the received doctrine of belief with two theses, which amount to the following although I've reworded them a bit:

1. (a) Belief is a relation between an animate subjects and abstract objects, the objects of belief.
   (b) The objects of belief are propositions.
2. Propositions are objective; they have truth-values, and their truth-values do not vary with time, place, or person. (131)

Stalnaker sees my account as giving up 1(a), Lewis's as giving up 1(b), and his own view as not giving up either. Once we recognize and appreciate diagonal propositions, the received doctrine remains intact.

> We'll start with just an example from Stalnaker:
> Rudolf Lingens is an amnesiac lost in the Stanford Library. He has found and read a biography of himself, and so knows quite a bit about Rudolf Lingens. He knows, for example, that Lingens is a distant cousin of a notorious spy. But he does not know that *he* is Lingens—that *he* is a distant cousin of a notorious spy. No matter how complete the biography, it will not by itself give him the information he lacks. Even if the book includes the fact that Lingens is an amnesiac lost in the Stanford Library, this may not be enough. He can use this information only if he knows, or has reason to believe, that *he* (and no other amnesiac) is lost in the Stanford library (131).

(This example is pleasantly inclusive. Lingens is Frege's character (1918/1967). I had him lost in the Stanford Library (1977). Castañeda had a character who unknowingly not only reads but writes a biography of himself (1967). The spy Lingens has for a cousin turns out to be Quine's character, Bernard J. Ortcutt (1966). Stalnaker's own trademark character, O'Leary, appears elsewhere in the article.)

Lingens does not believe that he* is Lingens, or that he* has a spy for a cousin. ("He*" is Castañeda's device, which he calls a "quasi-indicator," which shows that we are imputing a first-person belief to Lingens. Castañeda thought this was one sense of the pronoun "he".) But he does believe that Lingens has a spy for a distant cousin. What is the proposition he *doesn't* believe? It's the one he would express, if he believed it, with "I have a spy for a cousin." But the sentence "I have a spy for a cousin" doesn't identify the truth conditions of a proposition independently of the person who is uttering it.

§4

On my account in Perry (1979) beliefs have a pair of properties, roughly analogous to the properties of character and content that Kaplan attributes to utterances (modeled as sentences in contexts) in Kaplan (1989). One can believe the same proposition in different ways, just as one can say the same thing in different ways. Lingens can express the singular proposition that Lingens has a spy for a cousin by saying either, "I have a spy for a cousin," or "Lingens has a spy for a cousin." And he can say it in one way, without saying it in the other. Similarly, Lingens can believe that Lingens has a spy for a cousin, by being in one belief state, while not believing it by being in the other. There is third-person way of believing that someone named "Lingens" has a spy for a cousin, and Lingens is in that state. There is a first-person way of believing that someone has a spy for a cousin; Lingens is not in this state. Stalnaker takes this to be a modification of principle 1.

This doesn't seem right to me. Suppose Elwood is at a party and wants to meet Eloise. She is drinking a martini. He thinks that if she spills her drink, he can rush up with another drink and strike up a conversation. He decides to bring it about that Eloise spills her drink. There are three ways he considers for doing this. He could stand next to her and jostle her arm. Or he could stand next to her and quickly hit the bottom of her glass, knocking it up in the air. Or he could pay his confederate Elmo to stand next to her and jostle her arm, while he stands a few feet away, ready to rush up with another martini. He chooses the third method; things go according to plan; Elwood brings it about that Eloise spills her drink.

Now consider this last sentence. It seems the "brings it about" gets at a relation between a person and a proposition. One has the relation to the proposition in virtue of the results of one's acts, which the proposition captures. One might have qualms about having relation to abstract objects; perhaps this isn't quite the right way to think of it. But if "believes" can be thought of as reporting a relation to a proposition, then so can "brings it about".

It is a truism in the philosophy of action that one can bring about the same results in different ways. Elwood could have brought it about that Eloise spilled her drink in at least the three ways he considered. That is, one can be in the same relation (bringing it about) to the same proposition in virtue of different combinations of bodily movements and circumstances. He could have brought the result about by jostling Eloise, by hitting her glass, or by paying Elmo.

My point is that by claiming that one can stand in the belief relation to the same proposition in different ways, I am not abandoning 1(a). 1(a) is neutral on the question of whether one stands in the relation to propositions in virtue of or partly in virtue of one's own states, and if so, whether different combinations of states and circumstances can add up to belief in the same proposition.

Another way of making the same point is to look at saying-that. According to Kaplan's theory of demonstratives, there are two ways for Lingens to say that he has a spy for a cousin. He can say "I have a spy for a cousin" or "Lingens has a spy for a cousin". The different sentences have different meanings, or characters, but in the circumstances they both express the same proposition, the singular proposition that Lingens has a spy for a cousin. It seems that we can accept this view, without giving up the view that saying is a relation between an agent and a proposition, what the agent says.

So, I don't see my view as giving up the received doctrine of belief, as identified by principles 1 and 2. I see myself as a co-defender with Stalnaker of the doctrine of belief, so defined.

§5

I think, however, that it is natural to think of the received doctrine of belief as amounting to a bit more than theses 1 and 2 entail, and to suppose that my account requires modification of the received theory, so understood. This more is that the proposition believed *captures* the elements of belief relevant to rationality. On my view, A and B might believe the same proposition, but what is rational for A to do might be different than what is rational for B to do. Of course, this can be true on the received view too because other beliefs combine with the belief in question to impose different conditions on the world, so that different actions are rational for A and B. Or A and B might have different goals and desires. But I claimed that even if we kept the auxiliary beliefs and goals and desires constant, we still had to specify *how* the propositions are believed, to understand how they can rationally motivate action.

Russell thought the proposition that Mont Blanc is more than 4000 meters high had Mont Blanc as a constituent. It is a singular proposition in David Kaplan's sense. Frege thought Russell's view made little or no sense; there were certainly no such singular propositions (1980). It seems there will always be more than one way of identifying, of thinking about, any object. So a proposition with an object in it, like a singular proposition, will be incapable of really pinning down what is going on with the belief. Russell came to think that humans couldn't believe propositions with mountains in them, but only propositions whose constituents were things they could be acquainted with, in a very special sense, which turned out to apply only to sense-data, their properties and relations, and perhaps one's own self (1912).

Now, if there are singular propositions with things like mountains and houses in them, and we do believe them, it seems that there will be certainly two different ways of believing a proposition. It is natural to think underlying such beliefs are beliefs in what Kaplan calls qualitative propositions. You and I may both believe that Mont Blanc is more than 4000 meters high, but we probably think about Mont Blanc in different ways, which would have to be brought into any rationale of what this belief leads us to do. The *augmented* received doctrine that

I am envisaging can allow for this. It holds that once we descend to the layer of qualitative propositions, or de dicto beliefs, or Fregean propositions—different pictures, perhaps, but more or less equivalent for our purposes—we will not need a further distinction between what is believed and how it is believed to capture the rational consequences of belief.

One way of reading Stalnaker, it seems, is as holding this additional doctrine, and thinking that diagonal propositions fill the bill for indexical beliefs. Another way is that he holds that once we have added diagonal propositions to our repertoire, whatever distinction needs to be made between what is believed and how it is believed will be pretty obvious and easily accommodated. In this section I'll argue against the first doctrine, that diagonal propositions obviate the need for the distinction. When Lingens believes that *he* is in a library, it may perhaps be useful for certain purposes to say that he believes the singular proposition consisting of the individual Lingens and the property of being in a library. But that doesn't get at the real underlying structure of his belief. What he really believes is a diagonal proposition. His belief consists in believing the diagonal proposition. Once we see this, on this doctrine, no further distinction between what is believed and how it is believed is necessary.

Propositions are, for Stalnaker, functions from possible worlds to truth-values. One's belief is true if the value of the proposition one believes with the actual world as argument is truth. Inquiry and rationality have to do with eliminating possibilities, and acting in ways that make sense in all of the worlds that remain consistent with what one believes.

If we adopt this inviting picture, we may have to adjust our conception of belief a bit. Ordinarily, philosophers conceive of the proposition a person believes in terms of the conditions the truth of the belief puts on its intuitive subject matter. Suppose Elwood believes that Lingens is lost in a library, having just been told this by a friend, who is going off to find him. We ordinarily think of the object of Elwood's belief, in the possible worlds framework, as the set of possible worlds in which Lingens is lost in a library, or the characteristic function of this set. But, in thinking of things in this way, we are taking a number of things about Elwood's situation as fixed, which could actually be different, so far as his conception of the world is concerned. We are describing the conditions the truth of his belief puts on the actual world *given* certain facts, which his beliefs, considered as a whole, may leave open.

Stalnaker thinks that the referentialist view is at best misleading in many cases, including identity statements and indexical belief. (Perhaps he thinks it is false, although we can give a contextual explanation of why one might report beliefs in this way.) Here's why.

Suppose that Lingens is actually Mr X, a fellow Elwood's friend introduced him to at the coffee house the night before. Elwood isn't sure who Lingens is, though. That is, he isn't sure to whom his friend refers when he says "Lingens". It might be Mr X, the fellow in the coffee house but for all Elwood knows it could as well be Mr Y, another fellow the same friend introduced him to at a party the week before. On the ordinary view of reference, we have two propositions to work with here: *that X is lost*, and *that Y is lost*. The first is the one Elwood actually believes, since actually Lingens is X. The second is the one he would have believed if "Lingens" was instead the name of Y. But neither of these propositions really gets at a proposition that reflects the conditions under which Elwood's belief is true.

Consider three little worlds, the actual world *i*, in which Lingens is X, X is lost, and Y is not lost; *j*, in which Lingens is Y, X is lost, and Y is not, and *k*, in which Lingens is Y, X is not lost, but Y is. Our two candidate propositions correspond to the following patterns of truth-values:

	i	*j*	*k*
X is lost	T	T	F
Y is lost	F	F	T

Notice that neither row gives us the pattern that corresponds to Elwood's belief being true. His belief will be true in the actual world *i*, because his belief is about X and X is lost. It would be false in *j*, for it would be about Y and Y is not lost there. It would be true in *k*, because his belief would be about Y and Y is lost there. But neither proposition gives us the pattern T, F, T. Note, however, that in the following matrix:

	i	*j*	*k*
i	*T*	*T*	*F*
j	*F*	*F*	*T*
k	*F*	*F*	*T*

the left-leaning diagonal has exactly this pattern. Here the Y-axis gives us the context, in the sense of the circumstances relevant to what is believed, i.e., whether the belief is about X or Y. The X-axis gives us the circumstances relevant to evaluating the belief as true or false. Each row tells us which proposition is expressed, commonsensically, by the belief given that it occurs in the context indicated at the head of the row. The diagonal gives us another perfectly good proposition. It tells us in which worlds the Elwood belief would be true, as opposed to the worlds in which the proposition expressed in the context is true. We might think of this as the truth-conditions of the belief, as opposed to the truth-conditions of *what is believed.*

The left-leaning diagonal is Stalnaker's candidate for the content of Elwood's belief. The diagonal gives us a function from worlds to truth-values, and so is a perfectly good proposition. This proposition is true if Elwood is thinking about X, and X is lost in the library, or if he is thinking about Y, and Y is lost in the library. It is false if the guy he is thinking about is not lost in the library. This seems to be exactly the condition that the truth of Elwood's belief puts on the world. This is not to say that this diagonal proposition is what Elwood *says* in the example, but it is the content of the belief that motivates him to say what he does.

Now let's see how this applies to an indexical belief. Consider Lingens' belief that he* is lost in the library. To simplify, assume that Lingens has narrowed the possibilities for who he might be down to two, Lingens and O'Leary. What does the truth of his belief require of the world? The appropriate diagonal proposition is that either he is Lingens, and Lingens is lost in the library, or he is O'Leary, and O'Leary is lost in the library.

Let's grant, for the time being, that Lingens believes this diagonal proposition, call it *D*. I don't see how this proposal solves the problem of indexical belief, unless we add the distinction between the belief state and the proposition believed. That is, I don't see how Stalnaker's proposal saves the strengthened received doctrine. Suppose that Lingens explains his situation to someone else he meets in the library, Julius say, the inventor of the zip, to bring in one of Gareth Evans's characters. Julius, I stipulate, is *not* lost in the library. He knows the library like he knows the back of his hand, which he knows well. Elwood tells Julius that he doesn't know whether he is Lingens or O'Leary, but whichever one he is, he is certainly lost in a library. Julius hears what Lingens says. He takes it to be sincere, the expression of a belief. He is sure that the fellow is right that he is lost in a library. The fellow's evidence that he is either Lingens or O'Leary is persuasive. So Julius believes that the fellow he is talking to is expressing a belief, that belief is true if it belongs to Lingens, and Lingens is lost in the library, or if it belongs to O'Leary, and O'Leary is lost in the library. So Julius believes *D*, the very same proposition that Lingens does. Yet Julius does not believe that he* is lost in the library. Lingens' belief in the diagonal proposition explains why *he* is asking for help, but Julius's belief in the proposition doesn't give us any reason to suppose that *he* would ask for help. So it can't be just *which* proposition is believed that does the explanatory job. Lingens believes the diagonal proposition in the way people believe diagonal propositions about their own beliefs, at least normally; Julius believes the diagonal proposition in the way one would believe a proposition about someone else's belief, normally.

Lingens believes the diagonal proposition in quite a different way than Julius does. Lingens has quite a different relation to his own belief than Julius does. Julius knows it as the belief that motivates Lingens' assertion. Any effect that Lingens' belief has on Julius's action will be filtered through this mode of presentation of Lingens' belief. Lingens knows of his belief in the way one knows of one's own beliefs. His belief has a direct effect on him, because it is part of or an aspect of his own mind. Whether or not one should say that he has a mode of presentation of his own belief, the way it enters into his thinking is quite different from the way it enters into Julius's thinking.

Perhaps Julius is a philosopher, who has adopted Stalnaker's account of indexical belief. An hour or so later he tells his class, which meets in the library, about Lingens and his belief. He puts a diagram on the board explaining just what the proposition Lingens believed was. He uses "L" for "Lingens" and "O" for O'Leary, so as to protect confidentiality. Lingens, still lost, happens on the class. He doesn't recognize Julius. But he believes Julius's account, and believes in the fellow in the library and the sincerity of what he told Julius. He believes *D*, the proposition that is identified by the left-leaning diagonal on the board. He has a great deal of sympathy for the fellow with a plight so like his own.

Lingens now believes proposition *D* in two quite different ways, corresponding to the two cognitive routes he has to the belief, on which the diagonal proposition puts conditions: as a belief that is part of him, and affects him directly, and as a belief he has been told about by Julius.

So, to review the point of this section: there is a stronger version of the received doctrine, from which my account deviates, in that this stronger version holds that there can be no explanatory gap, requiring appeal to a "way of believing," once we have located the right proposition believed. One might suppose, and perhaps Stalnaker supposes, that diagonal propositions provide a treatment of indexical belief that allows us to maintain this stronger doctrine. But it doesn't.

Without bringing in this difference in the way the diagonal proposition is apprehended, I can't see that we have any explanation about what is special about Lingens' belief.

§6

Another reading, in some ways more plausible, is that Stalnaker's view absorbs rather than rejects the distinction between what is believed and how it is believed. His view and mine would then, as I see it, both fit with the received doctrine, but deviate from the augmented received doctrine.

Let's return to Julius and Lingens. The proposition *D* characterizes one of Lingens' own beliefs, the one he expresses with "I am lost in the library." (Given Stalnaker's holism, it might be better to say that *D* partially characterizes, for certain purposes, Lingens' total belief state, but I'll just continue to talk about beliefs. It is possible, perhaps likely, that in so doing I sluff over issues I don't fully understand. See the remarks by David Israel that I quote near the end of the paper.) *D* doesn't characterize Julius's belief in the same way. *D* is *not* the diagonal proposition that corresponds to Julius's belief. Julius wouldn't express his belief in *D* by saying "I am lost in the library," but as "You are lost in the library" or "He is lost in the library," depending on the context. If we start with Julius's belief, and construct a matrix that takes account both of what the truth of his belief requires of the rest of the world, the intuitive subject matter, and what it requires of itself—how Julius's belief is connected with that subject matter—we won't get *D*, but something we can call *D′*. Julius is standing there looking at a fellow, whom we know to be Lingens. For all Julius knows, this fellow could be Lingens or could be O'Leary. That is, Julius's own belief would be true in worlds in which he is perceptually connected to Lingens or to O'Leary, not, like Lingens, in worlds in which he *is*, or the possessor of the belief *is*, Lingens or O'Leary. And then, whomever Julius's perceptions are of, the truth of his belief requires that that person have a belief that is true if that person is lost in the library. The truth of *D* doesn't impose any conditions on Julius's beliefs, but the truth of *D′* does.

On this reading of Stalnaker, his view is that the level of analysis of the contents of a person's beliefs, which is ultimately relevant to, and needed for, understanding how the belief rationally motivates the action to which it leads, is the level of the diagonal proposition that provides the truth-condition *for that very belief.* The truth of a belief, we might say, diagonally imposes conditions on itself, and on the person that has it, and on the objects that he is cognitively related to in the various ways that constitute having beliefs about. All of these facts might be reflected in other ways in his beliefs, as the example of Lingens listening to Julius's lecture shows. But only as they flow from the conditions the truth of his beliefs impose diagonally on the world will they be relevant to explaining his action. So, to explain why Lingens says, "I am lost in the library," we need to appeal to the diagonal proposition that corresponds to his beliefs. In order to explain why Julius says, "He is lost in the library," referring to Lingens, we need to appeal to the quite different diagonal proposition that explains his beliefs.

On this view, we might say that Julius really does not believe *D*, although the truth of his own belief, in *D′*, does require that the person he is talking to, who is in fact Lingens, be either Lingens and lost in the library or O'Leary and lost in the library.

§7

I want now to approach these issues from a somewhat different angle.

First let me make a point about my use of the term "essential indexical," which seems to have not been very clear to many readers. Remember the examples in the first section of this paper. It seemed that my use of the indexical "I" in expressing and reporting my belief that I had a class to teach was essential to the expression or report of belief *explaining* the fact that I was leaving for campus. I didn't think that there was some proposition that could only be expressed with the indexical "I".

My focus at the time of Perry (1977 and 1979) was the fact that we can explain certain sorts of actions by appeal to self-locating beliefs. I curl up in a ball, and would say, "There is a bear attacking *me,*" not because I believe there is a bear attacking John Perry, but because I believe there is a bear attacking *me*. My solution in terms of belief states and propositions believed allows us to distinguish two classes of believers that are similar to me in important ways. There are those that believe in the same way, all the people who think they are being attacked by a bear, who will curl up in a ball (if that's what they've been told to do) and shout "There is a bear attacking me" (if they are English speakers prone to shout explanations of their actions in stressful situations). And there are those who believe what I believe, that John Perry is being attacked by a bear. These might include people who observe the attack, and those who hear me shout, perhaps including some thousands of miles away, if I am shouting into a cell phone. I thought by locating these two groups, and these two modes of doxastic similarity, I had provided the materials necessary to explain all of the phenomena connected with essential indexicality.

It seemed to me then, and seems to me now, that the fact that "believes" seems oriented towards the second kind of doxastic similarity was an interesting fact about it. Our concept of belief doesn't seem *primarily* focused on getting at the person's internal state, but rather on what they believe about the objects the belief is about, in the ordinary sense. It differs, in this respect, from other verbs that we use to get at doxastic phenomena, like "recognizes" and "takes to be", which seem to convey more information about the state the believer is in. Belief reports are still useful in explanations, so long as facts about how the person is thinking of these objects are in the background, or can be readily inferred.

I want to try to put the lesson I take myself to have learned from thinking about indexical beliefs, the lesson I thought my articles would teach by producing a sort of "aha" insight in readers who followed the examples, more abstractly. Consider any person x who wants to achieve some goal G that requires changes in the world outside of his own thought. There will be a set of bodily movements that person can execute that will bring about results that will promote G. For most of us and most of the desires or wishes we might have, and many of our goals, the set will be empty. There is nothing I can do to prevent Hitler from invading Poland or even make it less likely that he did so. There is probably nothing I can do that will significantly make it more likely that Bush will not continue to run up the deficit with ill-considered policies. But there are things I can do to, say, help the hurricane victims in Haiti, or bring it about that this article gets finished. Let's say that the set of movements that would promote my goals are ones I would be *well advised* to make.

The set of movements I can execute to advance a given goal is surely determined by the truth of objective propositions—propositions about me, about the things I can directly effect by moving in various ways, by the more distant effects that will then ensue, and so forth. Now suppose P is an objective proposition that guarantees that executing movement M is well advised for me, that it will be a way of bringing about or at least promoting a goal G that I have. Does it follow that my *believing* P explains my executing M? Not unless everyone who believes P and desires G is well advised to execute M. Insofar as people in much different situations, in which their executing M will have much different effects, can also believe P, my belief in P cannot be the whole explanation of my executing M. They will not be well advised to execute M. So if everyone who believed P executed M, things would work out fine for me, but everyone else would be doing some that they are not well advised to do (except perhaps some lucky ones, for whom M is well advised for other reasons). If we assume that humans execute intentional movements in general more or less law-like ways, and in more or less rational ways, my belief in P can't be the whole explanation.

To slip back into the example-giving mode, suppose you and I, sitting across the table from me, each desire that I have a bit of the last bit of cake on the table halfway between us. You made the cake and want me to try it. I see it and want to eat it. I am well advised to execute movements that will bring it about that I pick up a fork, dig into the cake, and bring the fork and cake to my mouth. The same movement, if you were to make it, would bring about the result that you ate some of the cake, not me. You would be well advised to shove the plate in my direction and say, "I don't want it. It's yours. Try it!" If I were to execute the same movements, I would not promote my goal.

To return to the example of §1, how can it be that my belief that I have a class to teach together with my desire that I teach the classes I am supposed to teach, reasonably motivates me to start moving, when you could believe exactly the same propositions and desire that exactly the same desires be true, and yet not be reasonably motivated to start moving towards the classroom?

There seem to be two ways this could happen, given the objectivity of propositions. One is that not everyone can have the belief; only certain people can believe some propositions. The proposition, belief in which motivates me to do what I am well advised to do, is one that only people who are well advised to do the same thing can have. This is the solution I called "limited accessibility". I interpreted Frege and Castañeda as adopting this solution to the problem of indexical belief. Russell is another example. Eventually he thought that although the proposition with Mont Blanc in it was a perfectly good proposition, ordinary mortals couldn't believe it, because they couldn't be acquainted with Mont Blanc.[1] We can each be acquainted with propositions with our own sense-data as constituents, while we are having them. Only I can have the belief that *the person with this sense-datum* (attending to one of my sense-data) *has a class to teach.* Such a belief not only makes it well-advised to execute the movements necessary to get oneself to class, it can motivate it, since everyone who believes that proposition is well-advised to execute those movements. Others might have beliefs about my sense-data by description, but not by acquaintance, so they wouldn't believe exactly the same proposition I do; the one I believe has the sense-data as constituents, since I am acquainted with them.

I preferred the other solution. This is to suppose that what actually motivates our taking well-advised movements, when we believe the propositions that make them well advised, is a way of believing the relevant objective propositions.

It seems that we can interpret Stalnaker's view in either way. We might see him as making a distinction between diagonally believing propositions and merely

[1] At least, I think this is what Russell eventually thought.

believing them. The way of believing things that we seem to express when we use indexicals to express our beliefs or report our own beliefs, or use quasi-indexicals to report the beliefs of others, is diagonal belief. This is the first interpretation I gave above.

Alternatively, we could suppose that diagonal belief is real belief. This would be a type of limited accessibility approach. I am the only one who can believe propositions about my own beliefs that are captured by diagonal propositions. As I understand the second interpretation of Stalnaker, it imposes such limited accessibility. Only Lingens can really believe *D*, because to really believe *D*, one must possess the belief for which *D* provides the diagonal truth-conditions.

§8

In the 1970s, I introduced the concept of "acceptance" to get at ways of believing. My use of "accepts" wasn't the same as Stalnaker's and isn't a very standard use of this useful word. By "*x* accepts S" I meant that *x* was in a doxastic state that disposed him to assertively say "S". If I believe I have a class to teach in the first person way I accept "I have a class to teach." In the odd cases we looked at in §1 I don't accept "I have a class to teach." In the case of first person belief, I believe that I have a class to teach by accepting "I have a class to teach." In the other cases I believe it in virtue of accepting "John Perry has a class to teach." This sounds a lot like the sentence is an intermediary entity involved in my belief, but I didn't intend it that way. The sentence-type "I have a class to teach" wasn't an intermediate object of belief, but a way of getting at an inner state involved in belief, in virtue of an effect that state would have, in certain circumstances. Nor did I think that the inner state involved the English sentence accepted, or a translation of it into Mentalese. I wanted to claim only that the meanings or characters of the sentences gave us *better* abstract objects for classifying belief states, and getting at relevant kinds of doxastic similarity, than the propositions believed.

After writing those articles I became convinced that we need to think in terms of what I call the reflexive truth-conditions of particular utterances and beliefs, not to *replace* the distinction between what is believed and how it is believed, but to understand it better. If I'd had "learning readiness" I could have seen the desirability of this from studying "Indexical Belief." But instead I was convinced of the need to bring particular beliefs into the picture by Mark Crimmins, and the need to bring particular utterances into the picture by dealing with problems posed by Howard Wettstein, and of both by working on a theory of information content with David Israel (See Crimmins and Perry (1989); Perry, (1980); Israel and Perry (1990, 1991). See also Crimmins (1992) for a helpful characterization of a somewhat confused intermediate state I went through.)

Let *b* be the belief in question, the one that motivates me to go to class, and the one that would lead me to say, if asked, "I have a class to teach." Here I am thinking of b not as a proposition, or a sentence-type, or an abstract object of any sort, but as a cognitive particular in my head or at least in my mind. Consider the proposition $Q(b)$ that b is owned by someone who has a class to teach. This is similar to Stalnaker's diagonal proposition, but I think of it as a singular proposition with the belief b as a constituent. It's a pretty direct descendant of what Reichenbach called the token-reflexive meaning (1947).

Now suppose, as referentialists are wont to do, that the *content* of b, that is *what I believe* in virtue of having the belief b, is the singular proposition P that John Perry has a class to teach. P is what such philosophers are likely to call the *truth-conditions* of P.

Notice however that $Q(b)$ also gives truth-conditions of b in the perfectly straightforward sense that if b is true, $Q(b)$ will be true. The belief b is the sort of belief that is true if and only if its owner has a class to teach. Q is more *intimately* connected to b than P is. The singular proposition $Q(b)$ is what I call the "reflexive truth-conditions" of b, in that it gives the truth conditions of b as a condition on b *itself.* The condition Q gets at what beliefs doxastically similar to my belief b have in common with it. If you accept "I have a class to teach," your belief b' will have, as its reflexive truth-condition, $Q(b')$. The idea of a reflexive truth-condition gives us a better way of doing what "acceptance" tried to give us: a way of characterizing the sort of belief state that one has when one believes, of oneself, that one has a class to teach that distinguishes it from the sort of belief state one is in when one merely has a belief that the person one happens to be—John Perry in my case, but not yours—has a class to teach.

Reflexive truth-conditions are contrasted with subject-matter truth-conditions, the conditions the truth of the beliefs imposes on the things it is about, in the intuitive sense—the sense in which the belief Lingens expresses with "I am lost" and the belief Julius expresses with "You are lost" are both about Lingens.

The odd cases in section §1 involve beliefs with a different reflexive truth-conditions but the same subject-matter truth-conditions. Just what those conditions are depends on how we think of proper names and the thoughts expressed by them. The way I think of these things goes like this. The belief b'' has a structure, $I(n)$, where n is a notion in my mind, and I is an idea in my mind, the idea of having a class to teach. The reflexive truth-conditions of b'' are that the person n is of has a class to teach. Whom n is of is determined by the conditions in which n was introduced. In the example, n is the notion I developed to keep track of the person referred to in the time schedule by "John Perry," and so it is of the person named in the time schedule, who happens to be me.

The reflexive truth-conditions of b'' don't require that the owner of b'' has a class to teach. These truth-conditions *plus* the fact that n is in fact a notion of the owner of b'' require this, but the reflexive truth-conditions of b'' themselves do not require this. We should expect a belief to motivate actions that will serve the goals of its owner, given its reflexive truth-conditions, and the reflexive truth-conditions of his other beliefs that are involved in motivating the actions. So we can expect *b* to motivate me to get ready for class, but not *b''*. Of course, *b''* together with other beliefs, such as the belief that I am John Perry, will motivate this. My belief that I am John Perry will have the reflexive truth-conditions that its owner is the person *n* is of, and this together with the reflexive truth-conditions of *b''* should get me on the way to class. But in the odd examples of §1 I lack this latter belief.

This account does not have either of the following consequences:

- That the believer *believes* the proposition that gives the reflexive truth-conditions of his belief;
- That anyone who believes this proposition has a first-person belief that he* is so and so.

The second point is the one I made in criticizing the first interpretation of Stalnaker's view, interpreted as a defense of the augmented received view. In the case we just discussed, where I believed in the first-person way that I had a class to teach, the reflexive truth-condition of my belief *b* was that the owner of *b* has a class to teach. You could believe that very proposition without believing that you have a class to teach. You have a belief about my belief. But you do not *have* a belief with the truth-condition that its owner has a class to teach.

With regard to the first point, I am inclined to think that we do not generally believe the reflexive truth-conditions of our beliefs, or the diagonal propositions their truth requires. I am inclined to reserve "believes" for propositions about the subject-matter of a person's beliefs, objects and properties for which the person has concepts and often words. A person with no concept of beliefs, or a philosopher who knew what such things were supposed to be but didn't think there were any such things as cognitive particulars, might believe that he had a class to teach, without believing anything like the reflexive truth-conditions of his belief. Still, such people have the following relation to such propositions: they have a belief, with those reflexive truth-conditions. I think that's a different attitude from belief. On my view we have a lot of "positive doxastic attitudes" towards various kinds of propositions. For example we are attuned to propositions about the strength of gravity on earth, or the correct manner for keeping one's balance while bicycling, or the way English syntax works, that we don't believe. I reserve "belief" for the attitude that is involved in deliberating, revising, combining and expressing our attitudes; belief requires concepts and, in many cases, words.

§9

The concept of reflexive truth-conditions gives us a way of distinguishing ways of believing things, in terms of the semantics of the belief rather than the semantics of the utterance the belief might dispose one to utter. This means we can develop an account of the content relations between beliefs and the utterances that express them. The lack of such an account was one thing that bothered Stalnaker about my view (1981:148). I'll sketch such an account, squinting at various tricky issues that mostly have to do with the somewhat forensic aspect of our concept of what a person says.

Suppose my belief *b* causes my utterance *u*, in the way that beliefs cause intentional utterances of the people that have them, that, as we say, express those beliefs, or at least are attempts to do so. For example, suppose my belief *b* that Elwood has a class to teach leads to my utterance *u* of "Elwood has a class to teach." The reflexive contents of *b* and *u* will not be the same: *b* is true iff (roughly) *b*'s owner's "Elwood"-notion is about someone who has a class to teach, while *u* is true iff *u*'s speaker's use of "Elwood" refers to someone who has a class to teach. These conditions are equivalent, however, *given* that the owner of *b* is the speaker of *u*, and given (which isn't *quite* right) that names refer to the things that the relevant notions of their owner's are about.[2] You cannot perceive my belief, but you can perceive my utterance *u* and recognize what sentence type it makes use of. You will know how to interpret it, as an utterance that is true iff the person referred to by the speaker, and hence the person the speaker's belief is about, has a class to teach. You will acquire a belief, call it *b*′, in virtue of which you believe about that person, the one I refer to and have a belief about, that he has a class to teach.

My belief *b*, my utterance *u*, and your belief *b*′ each have different reflexive contents. They are equivalent, given that the person my belief is about, the person I refer to, and the person your belief is about, are the same. All three will have the same singular proposition as their subject-matter content, that Elwood has a class to teach. The whole episode has a structure of the same species as what David Israel and I call an information structure (1991). Such structures are devices for passing along information about particular things from signal to signal, in virtue of connections to those things that are extended through the causal connections between the signals. By looking at belief and communication in this way, one can understand how our object-oriented ways of describing beliefs and utterances—the facts that are the basis of the intuitions on which referentialism relies—are more than a sort of quirk of folk psychology and folk semantics, but something rather insightful. Maybe something to do with the human condition, even.

David Israel, by the way, had this to say about the version of this paper that I sent him:

[2] The qualification has to do with the following sort of case. A group of students in front of the philosophy department are discussing whether Aristotle Onassis ever read Plato. Walking up as the conversation is in progress, I take them to be talking about the philosopher, and say, "Aristotle definitely read Plato." I think I could be held responsible for saying, unintentionally, that Aristotle Onassis definitely read Plato, even though my notion of Aristotle the philosopher motivated my use of "Aristotle." I would prefer that "saying" does not have this forensic aspect, but I think it does. I'm not *sure* it does, but I think it does.

I think you get Stalnaker's view correctly FROM your perspective. What I mean by this is something that you hint at, namely what you call S's "holism". I would rather stress his "decision-theoretic" stance, in which an agent is conceived of as confronting a state space (a set of possible worlds any of which could be true, for all the agent knows—that is, they are all doxastic possibilities) in which the notions of "aboutness" and the conditions on reality imposed by a belief that you deploy don't really have a firm purchase (if any). So you don't really try to get inside BS's head or his shoes or whatever.

I'm sure Israel is right about this, at least what he says after the first sentence. Still, I like my way of looking at things. Stalnaker's theory of diagonal propositions is part of a very general and impressive theory of inquiry, rationality, semantics and pragmatics, and his theory of indexical beliefs, at least on the second interpretation, handles the phenomena. It probably tells us things about the human condition, too, maybe more interesting things. There is clearly a lot of agreement between the views. Ultimately, my reasons for preferring mine are probably essentially indexical.

REFERENCES

Castañeda, H. (1966), '"He": A Study in the Logic of Self-Consciousness', *Ratio* 8:130–57.

——(1967), 'Indicators and Quasi-Indicators', *American Philosophical Quarterly* 4:85–100.

——(1968*a*), 'On the Logic of Attributions of Self-Knowledge to Others', *Journal of Philosophy* 65:439–56.

——(1968*b*), 'On the Phenomeno-Logic of the I', in *Akten des XIX Internationalen Kongresses für Philosophie*: 260–6.

Crimmins, M. (1992), *Talk About Beliefs* (Cambridge, Mass.: Bradford Books).

——and J. Perry (1989), 'The Prince and the Phonebooth: Reporting Puzzling Beliefs', *Journal of Philosophy* 86: 685–711; reprinted in (Perry 2000, 207–32).

Frege, Gottlob (1918/1967), 'The Thought: A Logical Inquiry', in *Philosophical Logic* edited by. P. F. Strawson (Oxford: Oxford University Press: 17–38). This translation, by A. M. and M. Quinton, originally appeared in *Mind* 65 (1956): 289–311. The original, 'Der Gedanke. Eine logische Untersuchung', appeared in *Beiträge zur Philosophie des deutschen Idealismus* I (1918): 58–77.

——(1980), *Philosophical and Mathematical Correspondence*, edited by G. Gabriel *et al.*, translated by H. Kaal, abridged from the German edition by B. McGuinness (Chicago: University of Chicago Press).

Israel, D., and J. Perry (1990), 'What is Information?', in *Information, Language and Cognition*, edited by P. Hanson (Vancouver: University of British Columbia Press: 1–19).

——(1991) 'Information and Architecture'. In J. Barwise, J. Gawron, G. Plotkin and S. Tutiya (eds.), *Situation Theory and Its Applications*, vol. 2 (Stanford: CSLI Publications :147–160).

Kaplan, D. (1989), *Demonstratives*, in Joseph Almog, John Perry and Howard Wettstein (eds.) *Themes From Kaplan* (New York: Oxford University Press: 481–563).
Lewis, D. (1979), 'Attitudes *De Dicto* and *De Se*', *Philosophical Review* 88: 513–43.
Perry, J. (1977), 'Frege on Demonstratives', *Philosophical Review* 86: 474–97; reprinted in Perry (2000: 1–26).
——(1979), 'The Problem of the Essential Indexical', *Noûs* 13: 3–21; reprinted in Perry (2000: 27–44).
——(1980*a*), 'Belief and Acceptance', *Midwest Studies in Philosophy* 5: 533–42; reprinted in Perry (2000: 533–42).
——(1980*b*) 'A Problem about Continued Belief', *Pacific Philosophical Quarterly* 61: 317–22; reprinted in Perry (2000: 57–76).
——(1988) 'Cognitive Significance and NewTheories of Reference', *Noûs*, 22:1–18; reprinted in Perry (2000: 189–206).
——(2001) *Reference and Reflexivity* (Stanford: CSLI Publications).
——, D. Israel and S. Tutiya (1993) 'Executions, Motivations and Accomplishments', *The Philosophical Review* 201: 515–40.
Quine, W. (1966), 'Quantifiers and Propositional Attitudes', in *Ways of Paradox* (New York: Random House: 183–94).
Reichenbach, H. (1947) §50, 'Token-reflexive Words', in *Elements of Symbolic Logic* (New York: The Free Press), 284ff.
Russell, B. (1912), *Problems of Philosophy* (Oxford: Oxford University Press), 284ff.
Stalnaker, R. (1981), 'Indexical Belief', *Synthése* 49:129–51; reprinted in Stalnaker (1999: 130–49).
——(1984), *Inquiry* (Cambridge, Mass.: MIT/Bradford Books).

11

Understanding Assertion

Scott Soames

INTRODUCTION

In his groundbreaking 1978 article, "Assertion," Robert Stalnaker presents an elegant model of discourse designed to solve philosophical problems arising, in part, from his identification of propositions with functions from possible world-states to truth values, and his restriction of the epistemically possible to the metaphysically possible. Among these problems are those posed by Kripkean examples of the necessary aposteriori. Being necessary, all such examples are seen by Stalnaker as semantically expressing the same trivial, universally known, apriori truth. Nevertheless, assertive utterances of them often result in the assertion of propositions that are both highly informative and knowable only aposteriori. A central task of "Assertion" is to explain how this can be so.

Stalnaker's model is based on the insight that considerably more goes into determining what is said by an assertive utterance than the meaning of the sentence uttered. Additional assertion-determining factors include (i) objective features of the context of utterance, such as the speaker, audience, time, place and world-state of the context, (ii) general conversational rules, including those against asserting what is already known or presupposed to be true (or to be false), and (iii) salient beliefs and assumptions known to be shared by conversational participants. These latter encompass beliefs and assumptions about who is speaking to whom, what words are being uttered and what they mean, what is happening in and around the speech situation, the topic of conversation, what has already been established or taken for granted, and what remains on the conversational agenda. Imagine, for example, a speaker who utters *She is late* in response to the entrance of a woman who comes into a meeting after it has already begun, attracting everyone's attention. The speaker relies on the fact that everyone will recognize him to be saying, of the woman who just entered, that she is late for the meeting. In this case, obvious facts about the conversational context interact with the meaning of the sentence uttered to determine the proposition asserted.

Other cases are more indirect, and may involve reinterpretation of what the speaker might, at first, appear to have said. For example, after listening to the remarks of a well-known campus orator, Mary might turn to her companion and say "Norman really is God's fountain pen, isn't he?" knowing full well that her companion won't take her to be asserting the proposition semantically expressed by the sentence she utters—one which predicates of Norman the property of being a certain kind of artifact (used to write by depositing ink on paper) possessed by God. Since that proposition is egregiously false, Mary realizes that her conversational partner will recognize that she is not committing herself to it, and so will look for an alternative interpretation of her remark. Depending on what else is taken for granted in the conversation, Mary might be taken to have asserted (i) that God really is using Norman to communicate his thoughts and desires, (ii) that (God aside) Norman really does have the truth on the matters about which he is speaking, or (iii) that Norman is a blow-hard who takes himself to be an authority, even though he really isn't (irony). This variation is not a sign of indexicality. Although Mary's utterance would result in assertions of different propositions in different contexts, this is not a matter of semantics. It is a matter of the way in which the meaning of the sentence uttered interacts with both principles governing discourse and the background beliefs and assumptions of conversational participants.

Similar principles govern the reinterpretation of utterances of trivially obvious, literal truths. For example, a candidate awaiting the results of an election after the polls have closed might respond to an early bit of unfavorable news by saying "What will be will be." In such a case, the speaker understands that his audience will not take him to have asserted a trivial tautology, but will instead interpret him to have said something significant—typically something to the effect that since the outcome is out of his control, he is prepared to accept whatever the result proves to be.

These examples are simple and relatively uncontroversial. However, as Stalnaker correctly recognizes, they are part of a larger, more systematic picture. A central message of "Assertion" is that there is often a substantial gap between the propositions assigned to sentences by a correct semantic theory and those asserted by utterances of these sentences in different contexts. Because of this, continued progress in solving problems in semantics and philosophical logic depends on our coming to have a better understanding of the ways in which semantic and nonsemantic factors interact in determining what is asserted and conveyed by utterances, and in guiding rational discourse and inquiry. This is the central insight, and seminal contribution, of Stalnaker's inquiry—and one with which I fully agree.

Nevertheless, I am skeptical about some of the burdens taken up in the article—in particular, the attempt to render familiar examples of the necessary a posteriori compatible with the restriction of epistemically possible to metaphysically possible world-states, and the identification of the objects of assertion with functions from such states to truth values. Thus, my first task, after reconstructing Stalnaker's discourse model, will be to demonstrate why it cannot be used to reconcile the necessary aposteriori with his antecedent conceptions of possibility and of propositions. Instead, I will argue, a proper understanding of the necessary aposteriori requires the restriction of epistemic possibility to metaphysical possibility to be abandoned, and the existence of epistemically possible world-states that are metaphysically impossible to be recognized. Having reached this point, I will next isolate inherent limitations of the discourse model itself by showing that *de re* belief and the nontransparency of meaning lead to insoluble problems—even after the model has been improved by substituting the broader class of epistemically possible world-states for the narrower class of metaphysically possible world-states. Finally, I will draw lessons from these problems, distinguish aspects of Stalnaker's discourse model that need to be revised from those that should be retained, and suggest how further progress can be made in understanding the ways in which semantics and pragmatics interact in assertion.

STALNAKER'S MODEL OF DISCOURSE

According to the model, conversations take place against a set of background assumptions shared by the conversational participants which rule out certain possible world-states as not obtaining, or "being actual."[1] As the conversation proceeds, and assertions are made and accepted, new propositions are admitted into the set of shared background assumptions, and the set of world-states that remain compatible with what has been assumed or established shrinks. This set is called the *context set* (at any given point in the conversation). The aim of further discourse is to further narrow down this set of possibilities, within which the actual state of the world—the maximally complete property that the universe really instantiates—is assumed to be located. When one asserts p, the function of one's assertion is to shrink the context set by eliminating from it all world-states in which p is not true.

Stalnaker postulates three rules governing assertion.[2]

R1. A proposition asserted should always be true in some but not all of the possible world-states in the context set.

[1] Here, and throughout, I use the term *possible world-state* instead of the more familiar *possible world* to reflect my view, shared with Stalnaker, that the items under discussion are not alternate concrete universes, but ways the world could have been, i.e. maximally complete properties the universe could have instantiated. To say of such a state that it "is actual" is to say that the universe instantiates it. I also limit myself to contexts of utterance that Stalnaker calls *nondefective* (those in which the possible world-states that the speaker takes to be compatible with everything believed and assumed by conversational participants at a given point in the conversation are the same as those that the hearers take to have this property), plus contexts that he calls *close enough* (those in which differences between speakers and hearers on this point don't arise, or affect the course of the conversation).

[2] Stalnaker (1999, 88).

R2. Any assertive utterance should express a proposition, relative to each possible world-state in the context set, and that proposition should have a truth value in each possible world-state in the context set.

R3. The same proposition should be expressed relative to each possible world-state in the context set.

The rationale for R1 is that a proposition true in all world-states of the context set would be uninformative, and so would fail to perform the essential function of assertion, which is to narrow down the range of world-states that conversational participants take to be candidates for being the way the world actually is. By the same token, a proposition false in all world-states in the context set would contradict what has already been conversationally established. Since it would eliminate the entire context set, it would also fail to narrow down the range in which the actual world-state is to be located. Of course, this rule, like the others, allows for some flexibility in how it applies. If someone seems to say something that violates it, one may sometimes conclude that no violation has really taken place because the context set isn't quite what one originally thought, or because the speaker didn't really assert, or mean, what he at first seemed to assert or mean. This is not to say that violations never occur, but it is to say that common knowledge of the rule can sometimes be exploited for conversational purposes—as when a speaker deliberately says something the literal interpretation of which would violate the rule, knowing full well that he will be reinterpreted in a certain obvious way so as to be seen as conforming to it.[3]

Stalnaker's rationale for R2 is that if an utterance violates it, then for some world-state w in the context set, the assertive utterance won't determine whether it should remain in the set, or be eliminated. If the sentence uttered does not express a proposition at w, or if it does express a proposition, but one for which no truth value—truth or untruth—is defined at w, then no verdict on whether w stays or goes will, Stalnaker thinks, be forthcoming. This is to be avoided.[4]

In explaining the rationale for R3 Stalnaker employs his notion of *the propositional concept associated with an assertion.* A propositional concept is very much like one of David Kaplan's characters. For Stalnaker, it is a function from world-states, considered as possible contexts of utterance, to propositions—where propositions are taken to be nothing more than assignments of truth values to world-states, considered as circumstances of evaluation. The propositional concept associated with an utterance of a sentence S at a certain moment m in a conversation is a function that maps each world-state w in the context set at m onto a proposition—which is simply an assignment of truth values to all world-states in the context set. This assignment of truth values is (implicitly) identified with the proposition that would be expressed by S at m, if the actual context of utterance were to turn out to be w.

[3] Ibid., p. 89. [4] Ibid., pp. 89–90.

Propositional concepts can be given pictorial representations, as is indicated by Stalnaker's matrix D.

D	i	j	k
i	T	T	T
j	F	F	T
k	F	T	T

D represents the propositional concept associated with the use of S at a moment m in which the context set consists of the world-states i, j, and k. D tells us (i) that if i is the state the world is actually in at m, then the proposition (semantically) expressed by the speaker's utterance of S is the proposition that assigns truth to every world-state of the context set, (ii) that if j is the state the world is actually in at m, then the proposition (semantically) expressed assigns truth to k and falsity to i and j, and (iii) that if k is the state the world is actually in at m, then the proposition (semantically) expressed assigns falsity to i and truth to the other two world-states.[5]

Stalnaker uses D to give the following rationale for R3.

> To see why the principle must hold, look at the matrix for the propositional concept D. Suppose the context set consists of i, j, and k, and that the speaker's utterance determines D. What would he be asking his audience to do? Something like this: If we are in the world i, leave the context set the same; if we are in the world j, throw out worlds i and j, and if we are in world k, throw out just world i. *But of course the audience does not know which of those worlds we are in, and if it did the assertion would be pointless.* So the statement, made in that context, expresses an intention that is essentially ambiguous. Notice that the problem is not that the speaker's utterance has failed to determine a unique proposition. Assuming that one of the worlds i, j, or k, is in fact the actual world, then that world will fix the proposition unambiguously *The problem is that since it is unknown which proposition it is that is expressed, the expression of it cannot do the job that it is supposed to do.*[6]

The idea is that if R3 is violated, the conversational participants won't know which proposition is (semantically) expressed by the sentence uttered, because they won't know which world-state "is actual." But if the proposition asserted is always the one (semantically) expressed by the sentence uttered (in the context), then the conversational participants won't know what is asserted, and so will be at a loss as to how to update the context set and proceed with the conversation. This is the rationale for R3.

[5] Stalnaker (1978) is not fully clear about what status the propositions "expressed by" S at the world-states of the context set are supposed to have. Although they are not always the propositions that would be asserted, if those world-states were to obtain, they are often propositions semantically expressed in those eventualities—hence the parenthetical "semantically" above. However, there are exceptions to this—to which I will return—which prevent any such general identification.

[6] Ibid., pp. 90–1, my emphasis.

PRESUPPOSITIONS OF THE MODEL

Before going further it is worth pointing out certain presuppositions of the model, and dealing with obvious worries that might arise. The model presupposes that speakers have a great deal of knowledge—(i)–(iii)—about possible world-states.

(i) For every world-state w, conversational participants at a time t in the conversation know whether w is compatible with everything believed, established, or assumed in the conversation at t, and hence whether w is in the context set at t.

(ii) For any sentence S that might be uttered at t, any world-state w in the context set at t, and any proposition p, if an utterance of S at t would express p, were it to turn out that w were actual (i.e. were w to turn out to be the world-state that is actually instantiated), then conversational participants know that this is so.

(iii) For any proposition p and world-state w, conversational participants know the truth value of p in w, i.e. they know what the truth value of p would be were w to be actual.

It is natural to wonder whether speaker-hearers really have all this knowledge of world-states.

One worry concerns what might be called the *size* of world-states. Each world-state encodes a massive amount of information about the universe—far too much for our minds to encompass. But if that is so, how are we able to know anything significant about such entities? The answer is that the knowledge of world-states required by the model is not very extensive. Although each world-state encodes a massive amount of information, only a tiny fragment of it will be relevant in any given conversational setting. Because of this, we can ignore differences among world-states that are irrelevant to our conversational purposes. For example, if our conversation has been exclusively about the 2004 American League Championship Series between the Boston Red Sox and the New York Yankees, we can form equivalence classes of world-states that agree on their accounts of the series, while differing arbitrarily on extraneous matters. In representing the context set and propositional concept for an utterance u, we can then take each 'w' as standing for one of these equivalence classes. Knowing of each equivalence class (and thereby of each member in it) that it is compatible with everything assumed, established, or believed at the time of u can then be assimilated to knowing, of the account of the series on which all members of the class agree, that it is compatible with all this background information. If this is correct, then the worry about size disappears. Although nothing I have said guarantees that speaker-hearers can, in general, be relied upon to have knowledge of types (i)–(iii), presupposed by the model, the sheer quantity of information encoded in world-states is not itself an obvious barrier.

The next thing to notice is that the model presupposes systematic *de re* knowledge of world-states. For each relevant world-state w, sentence S, and proposition p, speakers in a conversation C are said to know (i) that w is (or is not) compatible with the background assumptions of C, (ii) that an utterance of S would (or would not) express p, if w were to turn out to be actual, and (iii) that p would be true (false), if w were to obtain. In each case, an occurrence of the variable 'w' appears inside the content clause of the knowledge ascription, while being bound by a quantifier outside the clause. Since this is the mark of *de re* knowledge-ascriptions, the model attributes far-reaching *de re* knowledge of world-states to conversational participants. How should we think of this?

Consider the following example: A is speaking to B at a conference on the philosophy of language. A points at a man across the room and says "He teaches at UCLA." Suppose, for whatever reason, that the following world-states are members of the context set.

w_1: A is pointing at David Kaplan, and Kaplan teaches (exclusively) at UCLA, and...

w_2: A is pointing at David Kaplan, and Kaplan teaches (exclusively) at USC, and...

Stalnaker's discourse model presupposes that conversational participants know of w_1 and w_2 that if they "are actual" (i.e. if either one obtains or is instantiated), then A's utterance will express the proposition p_k that David Kaplan teaches at UCLA. In addition, the model presupposes that conversational participants know of w_1 that if it "is actual," then p_k will be true, while knowing of w_2 that if it "is actual," then p_k will be false. Since the only relevant aspects of w_1 and w_2 are those indicated above, this *de re* knowledge of world-states amounts, essentially, to knowledge of the following propositions:

a. that if A is pointing at David Kaplan, then A's utterance expresses the proposition that David Kaplan teaches at UCLA.
b. that the proposition that David Kaplan teaches at UCLA is true, if Kaplan teaches (exclusively) at UCLA, while it is false, if Kaplan teaches (exclusively) at USC.

On this account, the *de re* knowledge of world-states presupposed by the model is pretty easy to come by. In the case of (a), it involves knowledge of David Kaplan, and of the meaning of the sentence uttered. In the case of (b), it is apriori knowledge that every speaker acquainted with Kaplan can be expected to have. Seen in this light, the presuppositions of the model may seem to be readily satisfiable.

There is, however, cause for concern. Typically, the *de re* knowledge of world-states presupposed by the model will, as in the previous example, bottom out in ordinary *de re* knowledge of individuals (natural kinds, or other constituents of the world). In our example, knowledge of the world-states w_1 and w_2, and of the proposition p_k, that the latter would be expressed if either of the former were "actual," as well as knowledge of the truth values the proposition would have in those eventualities, is really nothing more than knowledge of David Kaplan (a) that if the speaker is pointing at him, then the speaker's utterance will express p_k, and (b) that if he teaches at UCLA, then p_k will be true, whereas if he teaches at USC, p_k will be false. However, *de re* attitudes of this sort are notorious for resisting the neat logical transitions presupposed by Stalnaker's model. For example, it is well known—from the discussion of puzzling Pierre in Kripke (1979), as well as from the discussions of other examples—that one can know of one and the same individual i that he is F and that he is G, without knowing (or being in a position to know) of i that he is both F and G. Similarly, one can know of i that he is F and that if he is F, then he is G, without knowing (or being in a position to know) of i that he is G; and one can know of i that he is F, while also knowing that S expresses the proposition that he is F, without knowing (or being in a position to know) that S expresses a truth. Cases like this pose a threat to Stalnaker's model of discourse.

This threat will be examined in due course. For now, we simply note it. Before moving on, we need to understand another aspect of the *de re* knowledge of world-states presupposed by the model—*de re* knowledge of the actual world-state. According to Stalnaker, the point of rational inquiry and conversation is to reduce, as much as possible, the space of possible world-states within which the actual world-state is believed, or known, to be located. On this picture, the idealized goal of these activities is to eliminate all possible world-states but one, which can then be correctly identified as actual. Why is this desirable? Well, it is natural to think, an agent who correctly identifies the actual world-state @ is thereby in a position to know everything. This will be so if (i) coming to know of the world state @ that it is actual inevitably involves coming to know, for each genuine truth p, that p is true in @,and (ii) knowing that p is true in @ involves knowing, or being in a position to know, p. However, this reasoning is incorrect.

First consider (ii). A fundamental presupposition of the model is that for any world-state w in the context set, and any proposition p that might be asserted in the conversation, speakers and hearers know the truth value of p in w. In many conversations—those without false presuppositions or assertions—the actual world-state @ will be a member of the context set. It follows that, in these conversations, there will be many propositions p that agents know to be true in @, without knowing (or having any way of coming to know) p. Thus, knowing that p is true in the world-state @, which actually obtains, is not sufficient for knowing p.

This brings us to (i). It is tempting to think that the reason an agent can know, of @, that p is true in it without knowing p is that, in cases like this, the agent does not know, of @, that it really obtains, or is instantiated. Were the agent to know this, the thought continues, the agent could not know that p is true in

@ without thereby knowing p, as well. In fact, however, this is highly dubious. Imagine an agent who, at a certain point, says or thinks to himself "this world-state, the one I find myself in now (which I know to be such and such, and so and so) is the one that really obtains, or is instantiated." It certainly seems that such an agent demonstratively refers to the actual world-state @, with which he is acquainted, and truly says, of @, that it obtains, or is instantiated. If he is sincere, the agent should qualify as knowing, of @, that it is the actual world-state. Most likely, he and his fellow conversationalists have known this all along. But then, if identifying the actual world-state is coming to know, of that state, that it "is actual," then identifying the actual world-state cannot be the idealized goal of rational inquiry, or conversation.

This conclusion could be resisted, if it could be shown that the familiarity of ordinary agents with the way things are in the universe is inevitably too limited and fragmentary to provide them with the sort of acquaintance with @ needed to acquire *de re* knowledge of it at all. Although the idea that such knowledge is unattainable is not without force, it is also not easy to accept. For one thing, accepting it would render that staple of indexical semantics, the actuality operator, essentially useless. Since the proposition expressed by ⌜Actually S⌝ is a proposition which says, of the world-state C_w of the context, that p is true in C_w (where p is the proposition expressed by S in C), an inability to have *de re* knowledge of @ would prevent speakers from ever knowing the propositions expressed by utterances of sentences containing the actuality operator, thereby depriving such sentences of any normal use. Since these sentences do seem to have such a use, there is reason to believe that *de re* knowledge of @ is possible.

Such knowledge is also defensible on other grounds. Although *de re* knowledge of individuals normally requires some sort of contact with them, it does not require extensive or systematic knowledge of the totality of facts involving them. For example, even though my knowledge of my city, my country, my planet, my solar system, and my universe is an infinitesimal fraction of all there is to be known about these things, I am surely able to acquire some *de re* knowledge of them. If *de re* knowledge of states of individuals (including states of the universe) is similar in this respect to *de re* knowledge of individuals themselves, then it too is compatible with extreme limitations on the extent and systematicity of such knowledge. Thus, the limited and fragmentary nature of our knowledge of the actual world-state, @, presents no obvious bar to our having some *de re* knowledge of it. Finally, it should be noted that a proponent of Stalnaker's model of discourse is in no position to deny this. Since the model routinely attributes *de re* knowledge of world-states to speakers on the basis of a much slenderer acquaintance with those states than any of us have with @, the proponent of the model ought to accept the idea that conversational participants do have *de re* knowledge of @. Once this is accepted, there is, as I have argued, no plausible grounds for denying that we know of @ that it is actual, or instantiated.

This brings us back to the goal of rational inquiry and discourse presupposed by the model. We have seen that the goal cannot be that of identifying the actual world-state, in the sense of coming to know, of the actual world-state @, that it obtains, or is instantiated. What, then, should we take the goal to be? The answer that the proponent of the model ought to give is, I think, that the goal is to "identify the actual world-state" in the sense of arriving at maximally complete, descriptive knowledge of the form, *the state of the world that actually obtains, or is instantiated, is one in which p, q, r, . . .* where 'p, q, r, . . .' are filled in with a comprehensive list of the facts of @. There are two things to notice about this answer. First, it is compatible with Stalnaker's discussion of the model—since approaching the goal involves learning, or coming to accept, more and more truths, which has the effect of shrinking the set in which the world-state that actually obtains is, and is assumed to be, located. Second, on this way of understanding the goal, world-states have no priority over propositions. The goal of identifying the actual world, in the sense in which we have now come to understand it, is simply that of learning as many (relevant) truths as we can. It is hard to quarrel with that.

A LESSON ABOUT THE NECESSARY APOSTERIORI

We are almost ready to tackle Stalnaker's attempt to use his discourse model to explain Kripkean examples of the necessary aposteriori. Before we do, however, it is worth pausing to tease out an important consequence of the model regarding how the necessary aposteriori should *not* be understood. The consequence involves sentences containing the actuality operator that are often taken to be paradigmatic instances of this category of truths. Although any sentence of this sort will do, we will focus on those constructed from contingently codesignative descriptions—*the x: x is F* and *the x: x is G*—that are rigidified using *actually*. This gives us two descriptions—*the x: actually x is F* and *the x: actually x is G*—which designate the same object o in every possible world-state in which o exists, and designate nothing in any world-state in which o doesn't exist. These are used to construct (1).

(1). If [the x: actually x is F] exists, then [the x: actually x is F] = [the x: actually x is G]

(1) is necessary, since it is true by falsity of antecedent in any world-state in which o doesn't exist, and true by truth of the consequent in any world-state in which o does exist.

Is (1) knowable aposteriori? Well, one might come to know it is by first coming to know the contingent truth

(2). If [the x: x is F] exists, then [the x: x is F] = [the x: is G]

and inferring (1) from (2). Since (2) can be known only aposteriori, anyone who comes to know (1) by this route knows it aposteriori. However, since all apriori truths can also be known aposteriori, there is nothing significant about this. In order to show that (1) is a genuine instance of the necessary aposteriori, one must show that it cannot be known apriori, and so is knowable only aposteriori. However, if the lessons we have drawn from Stalnaker's model about our knowledge of possible world-states are correct, then this cannot be shown.

Consider a scenario in which we imagine a possible state of the world to ourselves, or perhaps a class of such states. We say to ourselves, *Let w be a possible world-state in which o is the unique thing which is F, and o is the unique thing which is G, and.... and....* We go on enumerating the aspects of w for awhile, and then ask *Is w a world-state with respect to which (2) is true?* We answer that, of course, it is. On Stalnaker's model this counts as knowing of w that (2) is true with respect it—which is knowing that which is expressed by (3a).

(3a). In w: if [the x: x is F exists], then [the x: x is F] = [the x: x is G]

Since (3a) is apriori-equivalent to (3b) and (3c), knowing the former, on the basis of our apriori imagining, provides a sufficient basis for coming to know the latter in the same way.

(3b). If, in w, [the x: is F] exists, then, in w, [the x: x is F] = [the x: is G]

(3c). If [the x: in w, x is F] exists, then [the x: in w, x is F] = [the x: in w, x is G]

Now let's suppose something else, namely that the state of the world w we have been imagining is, unknown to us, its actual state. In other words, the state the universe actually is in has precisely the characteristics we were imagining, even though we didn't realize this at the time. If this is so, then in knowing (3a), and hence, (3c), apriori, we knew (4) apriori as well.[7]

(4). If [the x: in @, Fx] exists, then [the x: in @, x is F] = [the x: in @, x is G]

But then, since (1) expresses the very same thing as (4), it too is knowable apriori, and so is not an instance of the necessary aposteriori. A similar conclusion holds for every purported instance of the necessary aposteriori that makes essential use of the actuality operator. This is significant, since for a number of philosophers, particularly those who attempt to explain the necessary aposteriori by appeal to so-called two-dimensionalist semantics, such sentences have provided the template for understanding necessary aposteriori truths.[8]

[7] We may, of course, have been imagining a class of world-states satisfying our stipulations, of which @ is a member, rather than imagining @ by itself. However, if, as Stalnaker's model presupposes, knowing of this class that (2) is true with respect to its members counts as knowing of each member that (2) is true with respect to it, then the argument is not affected. Note, the argument does not depend on the model's problematic identification of propositions with functions from world-states to truth values.

[8] See Davies and Humberstone (1980) and Soames (2005*b*), plus the references cited there.

STALNAKER'S ACCOUNT OF THE NECESSARY APOSTERIORI

We return to the three rules governing assertion in Stalnaker's model of discourse.

R1. A proposition asserted should always be true in some but not all of the possible world-states in the context set.

R2. Any assertive utterance should express a proposition, relative to each possible world-state in the context set, and that proposition should have a truth value in each possible world-state in the context set.

R3. The same proposition should be expressed relative to each possible world-state in the context set.

Having motivated these rules, Stalnaker uses them to explain assertive utterances of Kripkean examples of the necessary aposteriori.

As with the other principles, one may respond to apparent violations [of R3] in different ways. One could take an apparent violation as evidence that the speaker's context set was smaller than it was thought to be, and eliminate possible worlds relative to which the utterance receives a divergent interpretation. Or, one could reinterpret the utterance so that it expresses the same proposition in each possible world. Consider an example: hearing a woman talking in the next room, I tell you, *That is either Zsa Zsa Gabor or Elizabeth Anscombe.* Assuming that both demonstrative pronouns and proper names are rigid designators—terms that refer to the same individual in all possible worlds—this sentence comes out expressing either a necessary truth or a necessary falsehood, depending on whether it is one of the two mentioned women or someone else who is in the next room. Let i be the world in which it is Miss Gabor, j the world in which it is Professor Anscombe, and k a world in which it is someone else, say Tricia Nixon Cox. Now if we try to bring the initial context set into conformity with the third principle [R3] by shrinking it, say by throwing out world k, we will bring it into conflict with the first principle [R1] by making the assertion trivial. But if we look at what is actually going on in the example, if we ask what possible states of affairs the speaker would be trying to exclude from the context set if he made that statement, we can work backward to the proposition expressed. A moment's reflection shows that what the speaker is saying is that the actual world is either i or j, and not k. What he means to communicate is that the diagonal proposition of the matrix E exhibited below, the proposition expressed by ⇑E, is true.[9]

E	i	j	k
i	T	T	T
j	T	T	T
k	F	F	F

⇑E	i	j	k
i	T	T	F
j	T	T	F
k	T	T	F

[9] Stalnaker (1999, 91).

In this example, the propositional concept E associated with the sentence S uttered by the speaker tells us two things: (i) we don't know which proposition is (semantically) expressed by S in the actual context, because which proposition is expressed depends on which world-state actually obtains, and we don't know which state does obtain; (ii) none of the possible propositions expressed would serve any useful purpose in the conversation. To assert a necessary truth is to assert something which is of no use in narrowing down the location of the actual world-state within the context set; and asserting a necessary falsehood is even worse. Thus, E violates R3, and any attempt to avoid this violation by excluding one or more of the world-states will violate R1. So, if we are to avoid violation entirely, and to regard the speaker's utterance as useful and informative, we must take it as asserting some proposition other than the proposition it (semantically) expresses at i, j, or k. Which proposition?

Since whatever the actual world-state turns out to be, the speaker will be committed to the utterance of S expressing a truth in the context, that is what we should take to be asserted. The proposition asserted is the proposition that is true (false) at a world-state w (of the context set) just in case the proposition (semantically) expressed by S in w is true (false) at w—it is the assignment of truth values that arises from E by looking along the diagonal and selecting the truth value that appears in row w of column w, for each w. Stalnaker calls this the *diagonal proposition.* Since, in this example, the diagonal proposition is neither true in all world-states of the context set nor false in all those states, it can do the job that asserted propositions are supposed to do—shrink the set. Hence, he maintains, this is the proposition that is really asserted by the speaker's utterance—no matter which member of the context set turns out actually to obtain. This is what ⥉E represents, where '⥉' (pronounced DAGGER) is an operator that maps a propositional concept C1 onto the propositional concept C2 that arises from C1 by taking each of the rows of C2 to be the diagonal proposition determined by C1.

This is the prototype for Stalnaker's treatment of the necessary aposteriori, which—extrapolating and generalizing his explicit remarks—we may take as suggesting T1.

T1. Although no necessary propositions are knowable only aposteriori, a sentence S, as used in a particular conversation C, is an example of the necessary aposteriori iff the proposition (semantically) expressed by S at the world-state that really obtains in the speaker's context is necessary, but the diagonal proposition asserted by a use of S in C is contingent, and hence knowable only aposteriori.

Given T1, plus Stalnaker's discussion of the example—*That is either Zsa Zsa Gabor or Elizabeth Anscombe*—motivating it, one might get the mistaken impression that he thought that all genuine examples of the necessary aposteriori are indexical, in the sense of semantically expressing different propositions in different contexts of utterance. However, he didn't believe this. How, then, were instances of the necessary aposteriori involving names and natural kind terms to be treated? He addresses this point in the following passage.

> I suggest that a common way of bringing utterances into conformity with the third principle [R3] is to interpret them to express the diagonal proposition, or to perform on them the operation represented by the two-dimensional operator DAGGER. There are lots of examples. Consider: *Hesperus is identical with Phosphorus, it is now three o'clock, an ophthalmologist is an eye doctor.* In each case, to construct a context which conforms to the first principle [R1], a context in which the proposition expressed is neither trivial nor assumed false, one must include possible worlds in which the sentence, interpreted in the standard way, expresses different propositions. But in any plausible context in which one of these sentences might reasonably be used, it is clear that the diagonal proposition is the one that the speaker means to communicate. The two-dimensional operator DAGGER may represent a common operation used to interpret, or reinterpret, assertions and other speech acts so as to bring them into conformity with the third principle [R3] constraining acts of assertion.[10]

Let us focus on (5a) and (5b).

5a. Hesperus is identical with Phosphorus.

5b. An ophthalmologist is an eye doctor.

Since (5a,b) don't contain indexicals, their meanings, i.e. their Kaplan-style characters, will be constant functions. Each expresses the same (necessary) proposition in every context of utterance. If the propositional concepts associated with them in these conversations were simply their meanings, then the application of the dagger operation would have no effect, and Stalnaker's explanation of their informative use wouldn't get off the ground. Thus, in these cases, he must not have been taking the needed propositional concepts to be the meanings (characters) of the sentences uttered.[11]

Instead, it is natural to interpret him as taking the propositional concept associated with an utterance of S in a conversation to be that which speaker-hearers (jointly) believe the meaning of S to be. In cases in which they know all the relevant semantic facts, this will simply be the meaning of S. In cases in which they are ignorant of, or confused about, some of these facts, the propositional concept associated with S may be something less than the actual meaning of S. For example, in the case of (5a), the propositional concept may by given by the formula *x is identical with y*—with different possibilities regarding the constant

[10] Stalnaker (1999, 92).

[11] Unlike later two-dimensionalists, Stalnaker never subscribed to the general thesis that names and natural kind terms are indexical, rigidified descriptions. For discussion, see the introduction to Stalnaker (1999), esp. pp. 14–19, and also Stalnaker (2001), esp. pp. 199–200 (of Stalnaker 2003).

functions from world-states to objects which are candidates for the meanings of the names to be substituted for the 'x' and 'y' being reflected in different world-states of the context set. The case of (5b) is similar, except that the different possibilities for filling in the content of 'O' in the relevant formula—*An O is an eye doctor*—are meanings of general terms, rather than meanings of proper names. On this interpretation, the context set for an utterance of (5a) will contain some world-states in which one or both of the names *Hesperus* and *Phosphorus* stand for something other than what they both actually stand for, and the context set for an utterance of (5b) will contain some world-states in which *ophthalmologist* means something other than what it actually means. Presumably, the justification for this way of looking at things is the idea that (5a) and (5b) will be used only if (some) conversational participants are ignorant about what these words actually mean, or stand for, with the result that world-states in which the words mean, or stand for, something different from what they actually mean, or stand for, will be among the genuine possibilities left open by the conversation prior to the utterances. But then, the thought continues, different propositions will be expressed when the sentences are "interpreted in the standard way," at these world-states, considered as contexts. This, I think, is how Stalnaker intended to generalize his explanation beyond genuinely indexical sentences.[12]

At this point, however, we run into a problem. Although there may be some sentences and conversations that fit the picture, some do not. For example, it is not true that (5a) would be used only in a conversation in which (some) conversational participants are ignorant of what *Hesperus* and *Phosphorus* stand for, in the sense most relevant to Stalnaker's model. Each participant may know perfectly well that 'Hesperus' refers to this object [pointing in the evening to Venus] and that 'Phosphorus' refers to that object [pointing in the morning to Venus]. They may even have done the pointing themselves. Clearly, such speakers know of the referent of each name that it is the referent of that name. Hence the (contingent) propositions expressed by *'Hesperus' refers to x* and *'Phosphorus' refers to x* relative to an assignment of Venus to 'x' should be among those that have already been assumed or established in the conversation. But then, metaphysically possible world-states in which the names mean and refer to different things will already have been eliminated from the context set as incompatible with what has been assumed or established. Since (5a) can, nevertheless, be used in these circumstances perfectly intelligibly, Stalnaker's explanation cannot successfully be applied to this case. This is an instance of the general problem noted earlier. When *de re* attitudes are involved, speakers cannot always determine the compatibility relations presupposed by the model.

[12] On this interpretation, propositional concepts map each world-state w in the context set onto the proposition that speaker-hearers believe would be semantically expressed if w were to obtain. When they know all relevant semantic facts, these are the propositions that really would be expressed if w obtained.

FAILURE OF THE MODEL

A related problem is posed by a different example. Imagine you are sitting across from me in my office, you point to a paperweight in plain view on my desk, and ask *What is that paperweight made of?*, and I respond *It is made of wood.* Although you don't know, prior to my utterance, what the paperweight is made of, we both assume that, whatever it is made of, it is an essential property of that paperweight that it be made of that stuff. Since, in fact, the paperweight is made out of wood, my remark is an example of the necessary aposteriori. How would this conversation be represented in Stalnaker's model of discourse? Prior to the utterance there would be different possible world-states in the context set that were compatible with everything assumed or established in the conversation up to that point. We may take these to include a context/world-state i in which the thing that, in i, is the one and only one paperweight on the desk is made of wood, a context/world-state j in which the paperweight on the desk in j is made of something else, e.g. plastic, and a context/world-state k in which a paperweight in front of us in k is made out of something else again—say, metal. In short, in Stalnaker's model, the *propositional concept* PW would be associated with my utterance.

PW	i	j	k
i	T	T	T
j	F	F	F
k	F	F	F

The rules R1–R3 for assertion would then yield two conclusions: (i) that on hearing my utterance you had no way of knowing which proposition was (semantically) expressed by my sentence, because which proposition was expressed depended on which world-state—i, j, or k—actually obtained, and you didn't know, in advance of accepting my remark, which world-state did obtain; and (ii) that none of the propositions that might have been expressed would have served a useful purpose. To have asserted a necessary truth would have been to have asserted something uninformative, and of no use in narrowing down the location of the actual world-state in the context set; and to have asserted a necessary falsehood would have been a nonstarter. So, if you were to regard my utterance as successful, you had to take it as asserting some proposition other than any of the candidates for being the one it (semantically) expressed.

Which proposition might that have been? Since you knew that whatever the world-state of the context turned out to be, I would be committed to my remark being true, the proposition you must have taken me to have asserted is a proposition that is true (false) at a world-state of the context set iff the proposition expressed by my sentence at that world-state is true (false) at that world-state. This is the diagonal proposition associated with PW. Since it is neither true at all world-states in the context set, nor false at them all, asserting it does the job that assertions are intended to do. Implicitly recognizing this, we both rightly understood the diagonal proposition to be the proposition I asserted.

That is the explanation provided by Stalnaker's model. There are two things wrong with it. First, it is wrong to suppose that you had any relevant doubt about what proposition was (semantically) expressed by my utterance of *It is made of wood* in response to your question, *What* [pointing at the paperweight] *is that made of?* The proposition I expressed is one that predicates being made of wood of that very paperweight—the one we both were looking at, and saw clearly sitting on the edge of my desk. You knew that it was the object you had asked about, and about which I had given an answer. Since you also knew what wood is, you knew precisely which property was predicated of which object by my remark. Surely, then, you did know the proposition my sentence expressed. In short, there was a proposition p such that you and I both knew that my utterance expressed p, even though you didn't know, in advance of accepting my remark, whether or not p was true, and so didn't know whether or not p was necessary. Of course, given his identification of propositions with functions from metaphysically possible world-states to truth values, Stalnaker can't say this, since the fact that p is necessary would require him to say (i) that you knew p all along, and (ii) that you knew that my utterance expressed a trivial truth, simply by virtue of understanding it. Since this is absurd, he is forced to the patently counterintuitive conclusion that upon hearing my utterance, you didn't know that it expressed p (where p is the proposition it actually did express).

The second thing wrong with Stalnaker's explanation is that the world-states j and k in the context set must either be (a) ones that are not really metaphysically possible, or (b) ones that are not compatible with all the shared assumptions of the conversational participants prior to my utterance—both of which are contrary to the dictates of the model. What are the world-states i, j, and k? They are total possibilities regarding how the world might be in which one and only one paperweight is sitting on my desk, seen by us, and the subject of our discourse. The paperweight satisfying these conditions in i is made of wood, whereas the paperweights satisfying them in j and k are made of plastic in one case and metal in the other. What paperweights satisfy these conditions in j and k? If j and k are really metaphysically possible, as Stalnaker insists, then the paperweights in j and k can't be the paperweight that is really on my desk. Since that paperweight is made of wood in every genuinely possible world-state in which it exists, it is not made of plastic in j or metal in k. It follows that j and k must be world-states in which some other paperweight is between us on the desk, seen by us, and the subject of our conversation. But how can that be? Surely, one thing that was part of the shared conversational background prior to my remark was the knowledge that this very paperweight [imagine me demonstrating it again] was between us on the desk, seen by us both, and the subject of our conversation. To deny this would be tantamount to denying that we ever know, of anything we perceive or talk about, that it has one property or another. Even if we put the question of knowledge aside, surely we both believed these things about this very paperweight, which is all the model requires. But if we did have this *de re* knowledge, or these *de re* beliefs, then the discourse model's requirement that the world-states in the context set be compatible with everything assumed and established in the conversation must have eliminated all metaphysically possible world-states in which other paperweights, not made out of wood, were the one and only paperweight under discussion. But then, there is no room for the diagonalization required by Stalnaker's explanation.

This is the fundamental problem. Unless some persuasive defense can be found for excluding obvious, shared *de re* belief and knowledge from the conversational model, Stalnaker's explanation cannot succeed. I will argue that no such defense can be given. First, however, I will improve the model by liberalizing the notion of possibility it employs. Having strengthened the model so that it can accommodate cases involving essential properties of objects, like my paperweight, I will return to the problems posed by *de re* knowledge and belief, and investigate why they are intractable.

IMPROVEMENTS AND PROBLEMS

The model can be improved by dropping Stalnaker's antecedent philosophical commitment to restricting epistemic possibility to metaphysical possibility. To drop this commitment is to recognize world-states that are metaphysically impossible but epistemically possible, i.e., maximally complete properties that the universe couldn't really have had, but which we cannot know apriori that it doesn't have (on analogy with properties that ordinary objects couldn't have had, but which we cannot know apriori that they don't have). When we allow context sets to include such world-states, the propositional concept associated with my utterance about the paperweight turns out to be different from the one we earlier took it to be. On this way of looking at things, i, j, and k are different epistemic possibilities involving the very same object, o—where o is the paperweight that we actually see on my desk, are talking about, and know that we are talking about. In world-state i, o is made of wood; in j, o is made of plastic; and in k, o is made of metal. The resulting matrix is PW*.

PW*	i	j	k
i	T	F	F
j	T	F	F
k	T	F	F

Since the same proposition is expressed with respect to each epistemologically possible world-state, and since it is neither trivially true nor trivially false, no diagonalization is needed.

This improvement encourages a certain thought. Perhaps Stalnaker's model of inquiry can be divorced from the philosophically contentious motivations that partially inspired it. The idea is to give up the identification of epistemic possibility with metaphysical possibility, to give up the goal of explaining away the necessary aposteriori, and to give up the analysis of propositions as functions from metaphysically possible world-states to truth values. We retain the idea that utterances are associated with propositional concepts or matrices, plus the general model of discourse that makes use of these matrices. We also retain the idea that the point of a discourse is to narrow the set of the possibilities—now thought of as including both epistemic and metaphysical possibilities—within which the actual world-state is presumed to be located. As before, an assertion is supposed to shrink the set of possibilities compatible with everything that has previously been assumed or established in the conversation. On this new picture, the conversational rules R1–R3 remain intact.

The model can be illustrated using the following example. I say, *He is John Hawthorne* (demonstrating a man sitting at the end of the table) in a conversation in which it is common knowledge that this man—the one I am talking about—is either John Hawthorne or Ted Sider. The utterance takes place in a context in which everyone knows a few facts about John and Ted already, but not everyone knows what they look like. Perhaps everyone has talked to each of them on the phone, or read the work of each, or corresponded with each, or some combination of the three, even though many would not recognize John or Ted by sight. Let us stipulate that everyone already knows of John that his name is 'John', that he is a Rutgers professor, and that he is not Ted—similarly for everyone's antecedent knowledge of Ted. Moreover, this shared knowledge is known to be shared, and so the propositions known are part of the presupposed conversational background. In this situation I utter the sentence, *He is John Hawthorne*, demonstrating John, who is sitting at the end of the table. The sentence uttered contains a name, which, like the demonstrative *he*, is a rigid designator with respect to all possible world-states, epistemic and metaphysical alike.

What are the epistemic possibilities prior to my utterance? It might seem that the two most obvious possibilities—j and t—could be described as follows: in j there is a unique person sitting at the end of the table and that person is John, and in t there is a unique person sitting there and that person is Ted. This gives us the following matrix.

	j	t
j	T	T
t	F	F

R1–R3 dictate that we perform the diagonalization operation, which gives us an asserted proposition that is true just in case John is sitting at the end of the table, and false otherwise. That is a good result, since it, or something quite like it, would normally be regarded as having been asserted by such an utterance. If you were to report my remark by saying *Scott said that John Hawthorne was sitting there* (gesturing to the place at the end of the table), I think most people would judge what you said to be true.

Nevertheless, the way we reached this result is problematic. World-states j and t are supposed to be epistemic possibilities compatible with everything taken for granted in the conversation prior to my remark. But, as in the earlier example about the paperweight, I left out of the specifications of j and t certain things known by all conversational participants. I ignored the fact that it was known (prior to my remark) that he [imagine me pointing again at John] was sitting there and also the fact that it was known (prior to my utterance) that since there weren't two people sitting there, and since John and Ted are different people, if John was sitting there, then Ted wasn't. When these things are added to the conversational background, t becomes incompatible with what is known or assumed by conversational participants, and so is excluded from the context set.

Why? First, since it is known (prior to my utterance) that he [pointing at John] is sitting there, it follows that he, John, is an x, such that it is known that x is sitting there. This is just to say that the singular proposition p which says of John that he is sitting there is known to be true by the conversational participants, and so must be true with respect to t, if t is to be compatible with everything commonly known or assumed. Second, since it is known (prior to my remark) that if John is sitting there, Ted isn't, it again follows that John is an x such that it is known that if x is sitting there then Ted isn't. But it has already been stipulated that the proposition q that Ted is sitting there is true in t. Hence, t can be compatible with everything which is known or assumed in the conversation (prior to my utterance) only if the trio of propositions—p, q, and the conditional proposition the antecedent of which is p and the consequent of which is the negation of q—is consistent. Since this trio is inconsistent, t must be excluded from the context set, in which case our revised, Stalnaker-style explanation of what I asserted fails in a way similar to the way the original explanation of my assertion about the paperweight failed.

How, then, is it that my utterance of *He is John Hawthorne* was informative? Since I discuss this sort of issue in considerable detail in chapters 3 and 4 of Soames (2002), I will deal with it only briefly here. We know that prior to the utterance my audience already believed of John that he was John.[13] So the new belief acquired by virtue of accepting my utterance wasn't that one. What might it have been? One such belief was surely that he, the person sitting there, was John Hawthorne. Everyone in the audience could see—without any appeal to propositional concepts or diagonalization—that I was attributing the property of being John Hawthorne to the guy sitting there, at whom I was pointing. So naturally

[13] The account initially given in Soames (2002) of the relationship between the semantic content of a sentence S in a context C and what is asserted by uttering S in C is modified and extended in Soames (2005*a*).

I was committed to that being true. Moreover, if someone in the audience were to describe what I said to a third party who hadn't been present, he might say, *At first several of us didn't know who was sitting at the end of the table, but then Scott said that John Hawthorne was the one sitting there.* In ordinary life, such a report would be taken to be completely correct. If it is correct, then not only did I convey this informative proposition, I actually said (i.e., asserted) it. This is evidence that what I asserted went a little beyond the strict semantic content of the sentence I uttered in the context. In this respect, I agree with Stalnaker; in cases like this the speaker does assert a proposition which is not the proposition semantically expressed by the sentence he utters. But the mechanism by which this occurs is a rather ordinary one, and typically doesn't involve any forced two-dimensionalist diagonalization.

THE NATURE OF THE PROBLEM

If what I have just argued is correct, then Stalnaker's elegant model of discourse can be improved, but not saved, by liberalizing it to allow for epistemically possible world-states, over and above those that are metaphysically possible. The fundamental, mistaken assumption embedded in the model leading to its failure is that conversational participants can do two things: (i) identify, at the time of each utterance, precisely which possible world-states are compatible with everything previously assumed or established in the conversation; and (ii) determine which of these possible states are compatible with propositions expressed by the sentence we utter under different assumptions about which possible world-state actually obtains. In reality, we can't always do these things, no matter whether the possible world-states in question are metaphysical or epistemic. We can't do them because the relationship between sentences and the propositions they express is nontransparent in an important way. There are pairs of sentences S1 and S2, and contexts C, such that in C

(a) S1 expresses a proposition p1, S2 expresses p2, and speaker-hearers understand both sentences, while knowing that to accept S1 is to believe p1 and to accept S2 is to believe p2,

(b) p1 bears some intimate "logical" relation to p2, e.g. p1 is the negation of p2, or p1 is identical with p2, or p1 is a conditional and p2 is its antecedent,

even though

(c) speaker-hearers have no way of knowing that the relation mentioned in (b) holds between the proposition believed in virtue of accepting S1 and the proposition believed in virtue of accepting S2.

Because of this, there are cases in which speaker-hearers believe p1, and yet are in no position to recognize that in believing p2 they are believing something inconsistent with this, which, in terms of the model, rules out all epistemically possible world-states. In other cases, in which p1 and p2 are consistent, but some different relation holds between them, the fact that speaker-hearers believe both p1 and p2 may rule out some but not all possible world-states, without their being able to recognize which. Because of this non-transparency in the relationship between sentences, the propositions we believe (assert) in virtue of accepting (uttering) them, and the world-states in which these propositions are true, our beliefs and assertions cannot always interact with one another in the way the model presupposes. Because of this, the model fails.

The assumptions that lead to this result are modest. In order to reach our conclusion, one may, but need not, endorse the contentious, but I believe correct, doctrine that the semantic contents of names and indexicals (relative to contexts) are their referents, or the similarly contentious, but correct, doctrine that the semantic contents of natural kind terms are the kinds they designate. One reason these semantic assumptions are not needed is that we can generate corresponding problems for the model using pairs of synonymous expressions of other sorts—for example *catsup/ketchup* and *dwelling/abode*—where in each case a speaker can understand both expressions without realizing that they are synonymous.[14] Another reason that contentious semantic assumptions are not necessary is that what generates problems for the model are not so much semantic facts about the sentences involved, as cognitive facts about speakers who use them. When an agent looks directly at the paperweight on my desk, and sincerely utters *That* [pointing at the paperweight] *is the paperweight I am talking about*, he is correctly described as believing, of the paperweight, that he is talking about it, where the proposition believed is also expressed by *x is the paperweight I am talking about*, relative to an assignment of the object itself to 'x'. It is believing this proposition that creates trouble for the model, whether or not we identify it with the semantic content of the sentence uttered. Similar points hold for examples in which the sentence uttered contains a proper name or natural kind term. In all these cases, the propositions that prove problematic for the model are among those that conversational participants come to believe and assume at later stages of the conversation. Since these assumptions determine the context set for later utterances, the only hope of saving the model is to exclude beliefs of this sort from playing this role. However, there seems to be no reasonable way of doing this.

THE UBIQUITY OF THE *DE RE*

The fundamental reason that *de re* belief can't be excluded from the model is that the model itself presupposes such belief. As I have stressed, it presupposes *de re* knowledge (or belief) of world-states, which, we have seen, is founded in

[14] See Salmon, (1990) and Rieber (1992).

de re knowledge (or belief) of individuals or kinds. This knowledge (or belief) of world-states is of three sorts:

(i) knowledge (belief) of world-states that they are, or are not, compatible with propositions previously assumed or established in the conversation,
(ii) knowledge (belief) of world-states w in the context set, sentences S, and propositions p, that if w obtains then S expresses p, and
(iii) knowledge of the truth value of p in w, for each w in the context set.

In each of these cases, the *de re* knowledge (belief) of world-states required by the model is inextricably linked to ordinary *de re* knowledge (belief) of individuals (or kinds). Hence there is no excluding the latter.

Regarding (i), the propositions previously assumed or established in the conversation will standardly include singular propositions—knowledge of which amounts to *de re* knowledge of individuals (or kinds)—about the speaker and other conversational participants, the salient items in the context of utterance, and the various words in use plus their meanings. In many cases, the propositions previously assumed or established will also include singular propositions about the individuals (or kinds) which are topics of the conversation. It is not unusual for these singular propositions to be more readily available as commonly held assumptions of conversational participants than many of their purely descriptive counterparts. For example, there surely are cases in which conversational participants discussing an individual i each knows of i that they all know that i is the individual being talked about, and each know that it has been assumed or established that i has one or another property P—even though the descriptive information about i possessed by conversational participants varies so much from one participant to the next that there may be few, if any, (purely qualitative) descriptions D that uniquely identify i which are known by each participant to be associated by all of them with any of the terms used in the conversation. In such cases, the descriptive differences between the parties will wash away, and the most salient proposition about i known to be commonly assumed or established in the conversation may well be a singular proposition that predicates P of i. For reasons like these, there seems to be no way for a viable model of discourse to exclude singular propositions from the set of propositions commonly assumed or established in a conversation. The consequence of this for the *de re* knowledge of world-states of type (i) presupposed by Stalnaker's discourse model is easy to see. If the propositions assumed or established include a proposition that says of o that it "is F", then to know of an arbitrary world-state w whether it is compatible with what has been assumed or established in the conversation (and hence to know whether w is in the context set), one must know of w whether it is a world-state with respect to which o "is F". It is not enough to know that w is a world-state in which whatever object satisfies a certain description (in the world-state in which the conversation actually takes place) "is F"; one must know of o itself that it "is F", or that it "is not F", in w.

A similar point holds for the combination of (ii) and (iii), which generates the propositional concepts, or matrices, on which Stalnaker's model is based. To generate these matrices, speaker-hearers must know, of each sentence S and pair of world-states w1 and w2 in the context set, whether the proposition that an utterance of S would express, were w1 to obtain, would be a true, or a false, description of w2. With this in mind, suppose, as Stalnaker does in several of his own examples, that S contains a rigid designator α which rigidly designates o1 if w1 obtains, and o2 if w2 obtains. Suppose further that S says of whatever is designated by α that it "is F". Then, in order for the background knowledge and beliefs of speaker-hearers to generate the propositional concept employed by Stalnaker, speaker-hearers must know of both o1 and o2 whether they "are F" in w1 and w2. This is ordinary *de re* knowledge and belief of those objects. Since the discourse model presupposes knowledge and belief of this sort, it cannot relegate it to the sidelines.

LESSONS

For this reason, Stalnaker's model of assertion fails. However, it is important not to overreact. Although certain aspects of the model, and the uses to which it was put, must be abandoned, other features of it can be retained. Among the former are Stalnaker's revisionary account of the necessary aposteriori, his restriction of possible world-states to the metaphysically possible, and his identification of propositions with functions from such states to truth values. Among the latter are versions of his rules for assertion, R1–R3, suitably reformulated to avoid the model's problematic features.

The correct account of the necessary aposteriori is illustrated by the homely example of the paperweight on my desk, which I show you. You see it, pick it up and feel it, but can't tell what it is made of. You imagine that it might be made of plastic, or metal, or wood. In imagining this, you are imagining the very object itself having the property of being made of plastic, being made of metal, or being made of wood. In so doing, you are also imagining different properties the universe might have—the property of containing this very object being made of plastic, the property of containing this very object being made of metal, and so on. You don't know which, if any, of these properties the universe really does have. Since you can't find this out by apriori reasoning alone, these properties, or more complete versions of them, are conceivable ways the world might be that are epistemically possible. When you finally learn that the paperweight is, in fact, made out of wood, you realize that it couldn't have existed without being made out of wood, and so you realize that certain epistemically possible ways the world might be are not ways that it could genuinely have been, and so are metaphysically impossible. Since these ways are just world-states, this elementary Kripke-style example of the necessary aposteriori shows that certain epistemically possible world-states are metaphysically impossible.

This example relies on a potentially contentious metaphysical doctrine—the essentiality of constitution. However, there is nothing special about the particular essential property chosen. Other essential properties or relations (e.g. the relation of non-identity) would serve equally well. The important thing is simply that there be such properties (and relations). Given that there are, we may reason that just as there are properties (relations) that ordinary objects could possibly have had (stood in) and other properties (relations) they couldn't possibly have had (or stood in), so there are certain maximally complete properties that the universe could have had—metaphysically possible states of the world—and other maximally complete properties that the universe could not have had—metaphysically impossible states of the world. Just as some of the properties that ordinary objects couldn't have had are properties that one can coherently conceive them as having, and that one cannot know apriori that they don't have, so some maximally complete properties that the universe could not have had are properties that one can coherently conceive it as having, and that one cannot know apriori that it doesn't have. Given this, one can explain the informativeness of certain necessary truths as resulting (in part) from the fact that learning them allows one to rule out certain impossible, but nevertheless coherently conceivable, states of the world. Moreover, one can explain the function played by empirical evidence in providing the justification needed for knowledge of necessary aposteriori propositions. Empirical evidence is required to rule out certain metaphysically impossible world-states which cannot be known apriori not to be instantiated, with respect to which these propositions are false. Thus, by expanding the range of epistemically possible states of the world to include some that are metaphysically impossible, one can accommodate Kripkean examples of the necessary aposteriori. This—rather than any Stalnaker-style diagonalization—is the correct account of the necessary aposteriori.[15]

Finally, there is the matter of salvaging what we can from Stalnaker's rules governing conversation and inquiry. Stalnaker's R1 is equivalent to $R1_{NC}$.

$R1_{NC}$. A proposition asserted should never be a necessary consequence of, or necessarily inconsistent with, the set of propositions already assumed or established in the conversation. (A proposition p is a necessary consequence of a set S of propositions iff there is no metaphysically possible world-state w in which the members of S are all true, while p is not; p is necessarily inconsistent with S iff there is no metaphysically possible world-state in which p and the members of S are jointly true.)

Since necessary consequences that are not apriori consequences are not, in general, discernable to agents, $R1_{NC}$ won't do as a conversational maxim. Although $R1_{AC}$ is an improvement over $R1_{NC}$, it won't quite do, either.

[15] For more on the distinction between metaphysically and epistemically possible world-states, and its role in explaining the necessary aposteriori, see Soames (2005*b*, 198–209).

$R1_{AC}$. A proposition asserted should never be an apriori consequence of, or apriori-inconsistent with, the set of propositions already assumed or established in the conversation. (A proposition p is an apriori consequence of a set S of propositions iff there is no epistemically possible world-state w in which the members of S are all true, while p is not; p is apriori-inconsistent with S iff there is no epistemically possible world-state in which p and the members of S are jointly true.)

An important problem with $R1_{AC}$ is that some apriori consequences are highly unobvious, requiring intricate and arduous apriori reasoning to reach. For this reason, the assertion of such a consequence of propositions already assumed or established may be highly informative, and effective in furthering the purposes of the conversation. Thus, it makes sense to replace $R1_{AC}$ with $R1_{OAC}$.

$R1_{OAC}$. A proposition asserted should never be an obvious apriori consequence of, or obviously apriori-inconsistent with, the set of propositions already assumed or established in the conversation.

At this point, however, the effects of the nontransparency of belief (and assertion) must be faced. Imagine a case in which conversational participants accept a pair of utterances—*This A is B* and *That A is C* (separated by a brief span of time)—accompanied by a pair of demonstrations, each clearly indicating a given object o in full view of each of the parties, without it being recognized that the same object is demonstrated on both occasions. In such a case, the propositions assumed or established in the conversation shortly after the remarks have been accepted will include a singular proposition p that says of o that it "is both A and B", and a similar proposition q that says of o that is "is both A and C". With p and q in the conversational background, it would be a violation of the intent of $R1_{OAC}$ for any of the conversational participants to assert that nothing "is both A and B", or that nothing "is both A and C", since to do so would be to assert something obviously inconsistent with what has already been established. Similarly, it would not do to assert—without further ado—that something "is both A and B", or that something "is both A and C", since these are obvious apriori consequences of p, and q, respectively. However, it might be very informative, and not in the least inappropriate, for someone to assert of o that it "is A, B, and C", even though the proposition thereby asserted is equivalent to the conjunction of p and q, or to assert that something "is A, B, and C", even though the truth of that proposition is a trivial consequence of the conjunction of p and q. Similar remarks hold for certain assertions obviously inconsistent with the conjunction of p and q—assertions of o that it "is not A, B, and C", or that nothing "is A, B and C". Although such assertions would render the set of propositions accepted by conversational participants inconsistent, in cases in which such inconsistency is undetectable, there is no culpable violation of conversational rules.

These considerations suggest reformulating $R1_{OAC}$ so as not to focus exclusively on the proposition asserted, the propositions assumed or established, and the logical relations between them. When singular propositions are involved, the ways in which propositions are presented or entertained are as important as which propositions are asserted, believed, or accepted. This can be accommodated by reformulating $R1_{OAC}$ along the following lines.

R_O. An assertive utterance U should allow the conversational participants to correctly identify the proposition asserted, but U should never be such that it is obvious to conversational participants that the proposition asserted by U is a consequence of, or is inconsistent with, the propositions already assumed or established in the conversation.

Since a violation of R_O will occur only when the speaker's utterance presents the proposition asserted in a way that allows conversational participants to see that it is an obvious consequence of, or obviously inconsistent with, the propositions already assumed or established in the conversation, R_O accommodates the nontransparency of belief and assertion that caused problems for $R1_{OAC}$. Thus, R_O, or something like it, offers the best hope we have for salvaging what was correct about Stalnaker's rule R1.

In requiring that conversational participants be able to identify the proposition asserted, R_O also incorporates what was correct about R3. The insight embedded in the latter was, essentially, that one should be able to recognize what is asserted without settling the open questions remaining in the conversation. The reason R_O is able to accommodate this point is that the sense in which it requires the proposition asserted to be identified, or recognized, is quite weak. What is required is simply that for some p, the conversational participants be able to correctly recognize that U is an assertion of p. It is not required that U present p to conversational participants in a way that allows them to determine the truth value of p in all genuine metaphysically possible world-states compatible with propositions already assumed or established, or even to accurately assess whether the truth of all those propositions would guarantee the truth, or the falsity, of p. It is only because Stalnaker built these unreasonable requirements of cumulative, global transparency into the discourse model that R3 was made to seem to incorporate a truth more far-reaching than it really does.

Similar remarks apply to R2. Is it important that U express a proposition, and that speaker-hearers be able to recognize that it does, without answering all the questions still open in the conversation? Of course it is, but this is already implicitly incorporated into R_O. No reference to the different world-states in the context set is needed, since when the nontransparency of knowledge, belief and assertion is accommodated, and propositions are no longer identified with functions from world-states to truth values, these states fall away and the rules of discourse can best be stated directly in terms of structured propositions and assertive utterances. In many cases, there may be no set of world-states compatible with everything assumed or established, because that which has been assumed or established contains nontransparent inconsistencies. Even when there are no such inconsistencies, compatibility relations may be obscured by pockets of nontransparency among the propositions already accepted. To treat the set of world-states that are compatible with all these propositions as if it were a central component of the discourse model guiding the computations of speaker-hearers is to assume a global and cumulative transparency in our beliefs, assertions, and knowledge that is simply not a part of our cognitive or conversational lives.[16]

If I am right about all this, then Stalnaker's model of discourse must be drastically modified. Its central idea—that what is asserted in literal, nonmetaphoric speech often differs substantially from the semantic content of the sentence assertively uttered—is both true and important. However, the basic structure of the model—involving metaphysically possible world-states, propositions as functions from such states to truth values, propositional concepts, and speaker-hearer calculations involving these items—cannot accommodate many of the facts about language use for which any acceptable theory of discourse must be responsible. I have indicated that an important core of truth can be salvaged from Stalnaker's rules, R1–R3, governing assertion. However, this is only the beginning. There is much more to be said about the ways in which the propositions commonly assumed in the conversational background, together with the meaning and semantic content of the sentence uttered, contribute to the proposition, or propositions, asserted by the utterance.

One important factor which I have not been able to talk about here differs sharply from anything in the Stalnaker model. This is the phenomenon of routine pragmatic enrichment of the semantic content of the sentence uttered, explored in Soames (2002 and 2005*a*). In the Stalnaker model, the assertion of a proposition other than the one semantically expressed by the sentence S that is assertively uttered is always forced by a conflict between the conversational background, general conversational rules governing assertive utterances, and what speaker-hearers take the meaning of S (and the context) to be. By contrast, I believe that the semantic content of S can be intimately related to the proposition asserted, without there being any general but defeasible presumption that the aim of a literal, assertive utterance of S is the assertion of the proposition that S semantically expresses (in the context).

How fruitful this idea will prove to be remains to be seen. However, whatever success may be in store for us in the future will come on top of the progress made by the pioneering work done by Robert Stalnaker. No one has done more than he to open up this important field of investigation.

[16] The second clause of R2, requiring propositions asserted to have truth values, also goes by the board. Here there is no significant truth to be salvaged, since there is no conversational rule of the sort Stalnaker imagines against asserting propositions which cannot be assigned a truth value, as opposed to those that are simply false or untrue. For discussion see Soames (1989, 583–9).

REFERENCES

Davies, M., and Humberstone L. (1980), "Two Notions of Necessity," *Philosophical Studies* 38 (1980), 1–30

Kripke, S. (1979), "A Puzzle about Belief," in A. Margalit (ed.), *Meaning and Use* (Dordrecht: Reidel), 239–83

Rieber, S. (1992), "Understanding Synonyms Without Knowing That They Are Synonymous," *Analysis* 52, 224–28

Salmon, N. (1990), "A Millian Heir Rejects the Wages of *Sinn*," in C. A. Anderson and J. Owens (eds.), *Propositional Attitudes: The Role of Content in Logic, Language, and Mind* (Stanford, Calif.: CSLI), 215–47

Soames, S. (1989), "Presupposition," in D. Gabbay and F. Guenthner (eds.), *Handbook of Philosophical Logic*, Vol. 4: *Topics in the Philosophy of Language* (Dordrecht: Reidel), 553–616

—— (2002), *Beyond Rigidity* (New York: Oxford University Press)

—— (2005*a*) , "Naming and Asserting," in Z. Szabo, *Semantics vs. Pragmatics* (Oxford: Oxford University Press), 356–82

—— (2005*b*), *Reference and Description: The Case Against Two-Dimensionalism* (Princeton and Oxford: Princeton University Press)

Stalnaker, R. (1978), "Assertion," Syntax and Semantics 9: 315–32 reprinted in Stalnaker (1999: 78–95)

—— (1999), *Context and Content* (Oxford: Oxford University Press)

—— (2001), "On considering a Possible World as Actual," *Proceedings of the Aristotelian Society*, suppl. vol. LXXV: 141–74; reprinted in Stalnaker (2003: 188–200)

—— (2003), *Ways a World Might Be* (Oxford: Oxford University Press)

12

Responses

Robert Stalnaker

Steve Yablo said to me, about my task of responding to these eleven papers, "it can't have been unalloyed fun," but it almost was. Not exactly unalloyed, because my friends and colleagues have made me work, and sometimes to squirm, with their challenging arguments and constructive proposals. And it was sometimes frustrating that the inevitable finitude of time and publication space made it impossible to respond as fully as the papers deserve. But what could be more fun (it's philosophical fun we are talking about) than interacting with such a fine set of papers, all on topics of great interest to me. I learned a lot from working on them, and am deeply grateful to all of the contributors for honoring me with such excellent philosophical work.

Special thanks to the editors, Judy Thomson and Alex Byrne, for organizing this collection, for assembling such an outstanding group of contributors, and for giving me the opportunity to have the fun, and do the work, to which I now turn.[1]

PHYSICALISM AND QUALIA

Daniel Stoljar's and Sydney Shoemaker's papers are each about aspects of a materialist's attempt to account for the qualitative character of experience. All three of us want to reconcile a version of physicalism with a realistic account of qualitative experience, but we all have somewhat different ideas about how this should be done.

STOLJAR, "ACTORS AND ZOMBIES"

Daniel Stoljar focuses on an abstract form of the conceivability argument against physicalism. He spells out what he describes as a consensus view—a posteriori physicalism—and he criticizes the response that he takes this view to give to the conceivability argument.

[1] Thanks to both of the editors, also, for their helpful comments on a draft of these responses.

Let φ be a statement summarizing all of the *physical* truths (setting aside for another time legitimate concerns about just what this means) and let ψ be a statement that summarizes all of the psychological truths. The physicalist asserts that the material conditional, $(\varphi \rightarrow \psi)$ (which Stoljar calls "the psychophysical conditional"), is a necessary truth, and the a posteriori physicalist adds that it is a truth that can be known only a posteriori. The conceivability argument against physicalism infers from the premise that it is conceivable that the psychophysical conditional is false to the conclusion that this conditional is not metaphysically necessary, and so that physicalism is false. The consensus view, as Stoljar understands it, holds that the thesis that physicalism is an a posteriori doctrine gives it the resources for a distinctive response to the conceivability argument. But Stoljar does not see how the epistemological status of physicalism is relevant to the conceivability argument. On his interpretation, the consensus response just asserts the denial of the conclusion of the argument, adding an irrelevant claim that the falsity of the conclusion is known only a posteriori. "On the face of it," he writes, "the consensus view is a spectacular non-sequitur." He argues further that arguments of the same form as the conceivability argument are common in philosophy, and are generally taken to be compelling. His paradigm here is an argument given by Hilary Putnam against behaviorism, and his challenge to those who reject the conceivability argument against physicalism is to explain why they are not also committed to rejecting the parallel arguments.

I think that Stoljar's interpretation of the consensus view is an uncharitable one, but I agree that his challenge is one that needs to be met. When an argument one rejects is similar in form to arguments that seem compelling, and that one accepts, one is obliged to explain the difference. I think the challenge can be met in this case. In my view, the dialectical situation in the case of Putnam's "perfect actor" argument against behaviorism is quite different from the situation in which the conceivability argument is given.

The first step in responding to an argument for a conclusion that one rejects is to pinpoint the place where one takes the argument to go wrong: one may reject one or another of the premises, or one may claim that the argument is invalid. But for a response of this kind to be persuasive, one usually needs also to explain why a premise seemed to be true, if one claims that it is false, or why the argument seemed valid, if one claims that it is not. The most common and effective strategy in rejecting a superficially appealing philosophical argument is to diagnose an equivocation. One may, for example, claim that an argument is valid but unsound on one interpretation, and has true premises but is invalid on another interpretation. The persuasiveness of the argument (according to such a diagnosis) derives from the fact that each part of it is successful, on one of the interpretations. (A wholly successful argument of the same form as one that is criticized in this way is one whose premises are all true on the same interpretation that makes it valid.)

The conceivability argument has just one premise, but it involves the notoriously problematic notion of conceivability. On one way of understanding this notion, to conceive of something is to envision a possible situation in which it is true, which of course one cannot do unless there is a possible situation of the kind envisioned. On this construal of conceivability, the argument is obviously valid, but the physicalist, whether an a priori or a posteriori physicalist, will reject the premise. The argument, construed this way, has little dialectical force, since the premise presupposes the conclusion. If the conceivability of a certain kind of situation is to be a *reason* for believing situations of that kind to be possible, there ought to be some epistemological distance between the two notions (and the appearance that there is some distance will come from other construals of conceivability).

Alternatively, one might explain (and many have explained) conceivability in terms of conceptual coherence, where conceptual coherence is assumed to be something to which we have a priori access. It is conceivable that the psychophysical conditional is false, on this interpretation, because there is no a priori accessible conceptual incoherence in the supposition that the conditional is false. On this interpretation, the a posteriori physicalist grants that the premise is true (though the a priori physicalist will not), but on this construal of conceivability, anyone who accepts the existence of necessary a posteriori truths will reject the validity of the inference. Stoljar finds this construal of conceivability implausible, and perhaps it is. (He argues, following Kripke, that one cannot *really* conceive of water not being H_2O, and this suggests that his understanding of conceivability is closer to the first construal.) But notice that all that is at stake here, in the appeal to the second construal of conceivability, is the diagnosis of a temptation to find a bad argument persuasive. The consensus view need not rest any substantive claim on an analysis of conceivability that makes the premise of the conceivability argument true. All the proponent of this view needs is a construal of that notion that has tempted some to think that the premise is true.

Now if this is the a posteriori physicalist's diagnosis of the conceivability argument, is she committed also to rejecting Putnam's argument from the conceivability of perfect actors (who can perfectly simulate pain behavior without being in pain) to the falsity of behaviorist analyses of pain? Let me say first that the dialectical situations in the two cases seem to me very different. In the case of the issue about physicalism, one begins in a context in which the necessity of a certain claim (the psychophysical conditional) is in dispute. One side proposes a reason for believing that this claim is possibly false: that it is *conceivable* that it is false. But the Putnam argument seems to me more like a traditional counterexample argument (for example, like a Gettier counterexample to an analysis of knowledge): a previously unconsidered possibility, which conflicts with a thesis that is in dispute, is proposed. The form of a counterexample argument is not, "the following is conceivable, therefore it is possible." Rather, the point is, "a situation of the following kind (that you may not have thought of before) is possible, and therefore your thesis is false." If one does force either Putnam's argument, or Gettier's, into the conceivability-to-possibility mold, then one should construe "conceivable" in the first sense, which is the sense that makes the argument valid. The difference between Putnam's argument and the anti-physicalist's conceivability argument, on this interpretation, is that the premise is uncontroversially true in the first case, but (according to the materialist) false in the second case.

SHOEMAKER, "THE FREGE–SCHLICK VIEW"

Sydney Shoemaker's paper continues a debate about a view that denies the meaningfulness of interpersonal comparisons of qualia. Shoemaker argued against this view in his classic paper on the inverted spectrum. I defended it against his criticisms, and criticized his functionalist/physicalist account of qualia. Part of the issue turns on how one responds to what I called "Shoemaker's paradox", a challenge that he had posed for his own view, and the last part of his current paper develops and defends his response to it. I am largely in agreement about the costs and benefits of his response to the paradox; the disagreement is about how to weigh the costs, and this depends on whether my version of the Frege–Schlick view is a viable alternative that can do the work that an account of qualia needs to do. In these comments, I will try to clarify the account I advocate, and to defend its coherence. There is an issue in the background of this debate concerning the relation between the representational and qualitative content of experience. I will conclude with a brief remark about representationism—the doctrine that qualia can be explained in terms of the way experience represents things to be.

I used an analogy with a simple relational theory of space to help to characterize the conception of qualia that I want to defend. Shoemaker argues that the analogy breaks down, and he also says that he doubts that "Stalnaker intended the spatial analogy to carry a great deal of weight by itself." But I continue to think that it is a good analogy, and that it helps to answer some of the challenges Shoemaker makes to the coherence of the Frege–Schlick view. The crucial question, he suggests, is "what sort of relational properties qualia might be, on the Frege–Schlick view", but the view (or at least the version of it that I was trying to defend) is not that qualia can be identified with any kind of monadic property, intrinsic or relational. Qualia are (on this view) like spatial locations, on the relational account of space. On the kind of simple relational theory of space I had in mind, distance relations are real, but locations are conventional, so that any mapping of spatial locations onto spatial locations that preserved all distance relations between points would be an equivalent representation. Given a configuration of objects in space, (at a time) one can define a coordinate system relative to which locations can be defined, and in terms of which the structure of distance relations can be described in an economical way. To say that an object is at a certain location (relative to a certain frame of reference) is not to say that it is spatially related to any other specific objects, since the objects in terms of which the coordinate system is specified are used just to fix the reference of the spatial framework.

The view I am trying to articulate seems to fit the description Shoemaker gives of the conception of qualia defended by Mark Kalderon and David Hilbert, and by Austen Clark: "The qualitative character of color experiences is determined by their position in the subject's color experience space, i.e. by their similarities and differences from other experiences in the repertoire of the subject." Shoemaker observes that this is a view that "rules out the possibility of a symmetrical color experience space", and this may be true of the accounts that Kalderon–Hilbert and Clark develop, but it is not true of the version of the view that I want to defend. Consider again the spatial analogy (where it is clear that spatial symmetries are possible): Suppose we have just three objects in a Euclidean space, two that are five meters from the third, and six meters from each other. This is a symmetrical configuration: the first two objects are, in a sense, spatially indistinguishable, since each is five meters from one object and six meters from another which is five meters from the first. But while the two objects share all real spatial properties, it is a determinate fact that they are not in the same location, since they are six meters apart. Analogously, in a symmetrical color experience space it might be a determinate fact that one experience in the subject's repertoire is at a nonzero qualitative distance from another even if each is identically related to the other experiences in the repertoire. Just as there is a permutation of spatial locations that preserves all distance relations and that maps the location of each of the two objects onto the location of the other, so there might be a permutation of points in the color experience space that preserved all real qualitative relations while mapping the location of each of two symmetrical experiences onto the location of the other.

Shoemaker also says, about the Kalderon-Hilbert and Clark account, that "this is a conception of qualia as relational that is not available to Stalnaker" because it allows for interpersonal identity of qualia when "different subjects have identically structured color experience spaces," and so is not compatible with the Frege-Schlick view. But once we allow for the possibility of a symmetrical color experience space, we can see that having identically structured color experience spaces is not enough for the identity of qualia (just as spatially identical configurations of objects at two times, or in two possible situations, is not enough for the identity of locations). This is exactly the situation with the inverted spectrum scenario, which presupposes a symmetrical color experience space. The original problem, for a pure functionalist account of qualia, was that functional properties gave us only a relational structure, and so did not provide the resources to distinguish color qualia from their symmetrical counterparts. The spatial analogy provides a simple model of a relational structure in which one can make sense of the idea that two objects are in a sense indistinguishable with respect to a family of relational properties, but still distinguishable (in terms of their place in the relational structure) from each other.

As Shoemaker notes, I said, in my original paper on this topic, that my version of the Frege-Schlick view was not an eliminativist view, but this may have been misleading. The relational view that I am defending is eliminativist about qualia in the way that the relational theory of space is eliminativist about location. The characterization of objects in terms of their spatial location, properly understood, may be perfectly correct (according to the relational theory), so the theory is not eliminativist about location in the way that current chemical theory is eliminativist about phlogiston. Still, if locations are understood as real spatial properties, then strictly speaking, there are no locations. I take the Frege-Schlick view to say the same about qualia.

This conception of qualia does diverge from the common sense view, if there is a view that deserves this name, but I think Shoemaker's response to the paradox shows that his view does as well. I took it to be an important part of a common sense view that qualia differences are accessible to consciousness, and I argued that the paradox requires Shoemaker's account to cut the qualia up more finely, allowing that there may sometimes be a qualitative change in a person's experience when the character of the experience seems to the person to be unchanged, and even when it was impossible for the person to become aware of the difference by introspection and reflection. I suspect that the common sense view of the character of phenomena experience (perhaps infected with some philosophical theory) may involve conflicting demands, requiring both that qualia be identified with the way things appear to be, and that they be independent of the properties of the things in the world that are perceived.

Shoemaker wants an account of qualia that is compatible with a version of representationism that says that the qualitative similarities and differences in experience correspond to similarities and differences in the representational content of the experience (while recognizing, as what he calls the standard representationist account does not, that there can be qualitative differences in color experience that do not correspond to the representation of different objective properties of what is represented). He points to the color constancy phenomenon—the fact that a surface partly in direct light, partly in shadow, might look to be uniform in color even though the parts are qualitatively distinguishable—arguing that this phenomenon is a problem both for the standard version of representationism, and for the Frege–Schlick view of qualitative character. I don't see that this particular phenomenon poses any problem for standard representationism since there is a straightforward objective representational difference in the case where a surface appears to be uniform in color, but differently illuminated. If the surface didn't appear to be differently illuminated, it wouldn't appear to be uniform in color. But I agree that there are at least possible cases where things look different (to one person, at different times) even though they don't look to *be* different in any objective way. Consider the following purely fictional case: a person has two different visual systems (the X system and the Y system) operative at different times that are qualitatively quite different, but that have the capacity to carry exactly the same information about the colors of things in the environment. So experiences with clearly distinguishable quantitative character each represent exactly the same shade of color under the same environmental conditions. The only way that the subject knows which system is operative is by the way things look. This case is not like an intrapersonal inverted spectrum case, where red things look (when one system is active) the way green things look when the other system is operative. Rather the qualitative character of color experience is wholly different in the two systems. There is no illusion or misrepresentation when either system is functioning normally; it is not that, for example, dark maroon things don't look dark maroon when one or the other of the visual systems is active. Instead, there are two distinguishable ways of looking dark maroon. In this hypothetical case, the subject locates a difference in the way things look in herself, rather than in the thing being perceived. There is still a representational difference, but what is represented is a fact about oneself—which of one's visual systems is operative. I won't speculate about what the phenomenology of such a purely hypothetical visual experience would be. Maybe there would be a sense in which the difference in the way things look seems to be in the object, and not in oneself, even when one knows, and has become accustomed to the fact that the difference is not really in what one is seeing. Shoemaker's overall theory of qualia and appearance properties puts a lot of weight on the transparency, or diaphanousness intuition (very roughly, that the qualitative character of our visual experience manifests itself in properties that things in the environment appear to have) but I am not sure how to pin this elusive phenomenological intuition down, or what should be concluded from it. I am not sure exactly how representationism should be formulated, or why the Frege–Schlick view should be "less friendly" to such a view than Shoemaker's account of qualia, but I do think that a defensible version of such a thesis must make room for cases in which it is features of oneself that experience is representing.

SEMANTIC COMPETENCE

Paul Pietroski's and Richard Heck's papers are each concerned with the proper characterization of semantic competence. Each defends a kind of internalism about semantics, although not the same kind, and each can be seen as an attempt to resolve some tensions between the Chomskyan conception of linguistic competence as a highly constrained and psychologically distinctive innate capacity and the idea that language is a device for the expression and communication of thought in a social context. In the background of both papers are questions about the relation between language and thought.

PIETROSKI, "CHARACTER BEFORE CONTENT"

Pietroski's project, as I understand it, is to carve out a conception of semantics as the study of an aspect of the innate psychological language faculty that humans possess. He argues that the conception of meaning that is appropriate for a semantics of this kind should be an internal one that constrains, but does not determine, the propositional content of the sentences that are used to make assertions, and that are evaluated as true or false. (But Pietroski's project is an ecumenical one that leaves room for a separate study of truth-conditional content.) Linguistic meanings are what Pietroski calls "Begriffsplans": "blueprints, produced by the language faculty, for constructing concepts from lexicalized elements." I would have a better idea what this means if I knew what concepts are. (My attitude to the word "concept" is like Quine's toward the word "meaning": it's okay to use it in casual conversation, but only confusion will result from putting any theoretical weight on it.) As some philosophers and psychologists use the term, a concept is something like a mental word—something that *has* a meaning. As others use the term, a concept is something like a Fregean sense—something that *is* a meaning that a linguistic or mental word might have. But however one understands a notion of meaning that is detached from truth-conditional content, I am skeptical about the central argument that we need such a notion.

If I understand Pietroski's line of thought, the premise of the argument for an internal, content-independent conception of meaning is the Chomskyan thesis that "the language faculty is a largely innate cognitive system specific to human languages—as opposed to a general learning device." From this premise, Pietroski concludes that "it is prima facie implausible that natural language meaning is as tightly related to *truth*" as semantic theories usually assume. "If human I-languages are determined largely by innate aspects of human biology, then absent a beneficent deity, it seems unlikely that such languages pair sentential signals with truth conditions." I am happy to grant the premise of this argument, but I want to question the inference.

Pietroski cites with approval a methodological remark that I once made (though he also argues that I have not always followed it): "We should separate, as best we can, questions about what language is used to do from questions about the means it provides for doing it." Whether the language faculty is innate or not, it is an ability, or a cluster of abilities, and we should begin by considering what this faculty provides us with the ability to do. This question can be asked and answered on more general or more specific levels, but it seems reasonable to begin with the point that linguistic competence is the ability to use language to express and communicate information. The theoretical study of how we do this may abstract away, at some points, from the specific content of the information that competent speakers have the ability to convey, but if this is the ability we are trying to explain, then whether the explanation for the fact that we have the ability involves specialized innate structures or only a general learning device, any account of the language faculty will be an account of how we are able to produce and understand things with truth-conditional content. Consider an analogy with a simpler cognitive capacity that it seems reasonable to believe has a significant innate component: the ability to recognize faces. One characterizes what the ability is in terms of a relation between the person with the ability and a range of facts external to the person: it is the ability to match certain visual inputs with correct judgments about the person whose face one is seeing. The thesis that a specialized face-recognition capacity is encapsulated and hard wired should not lead one to conclude that "absent a beneficent deity" it seems unlikely that this faculty is able to pair visual signals with actual human faces. That we are able to do this is the fact to be explained.

The question, "what is a capacity the capacity to do?" is independent of the question, "how was the capacity acquired (was it learned or innate)?" But whatever the explanation for the fact that one has a given capacity, it may seem reasonable to say that the capacity is, in some sense, internal to the thing that has the capacity (even if it is a capacity to interact with certain environments). In theorizing about a complex ability or clusters of abilities, it may be reasonable to try to factor out internal components—aspects of the capacity that can be considered in abstraction from the external environment that plays a role in the exercise of the capacity. Perhaps this is the motive for trying to carve out a part of semantics that is independent of truth conditional content, but I am still unclear why the kind of internal semantics that Pietroski is developing is not just syntax. Most of the examples that he discusses seem to me to be examples of syntactic phenomena that can and should be explained without bringing in semantics at all, though in saying this I am making some assumptions about the relation between syntactic structure and semantic interpretation. Let's take a quick look at two of Pietroski's examples:

(1) The hiker who was lost kept walking in circles
(2) The hiker who lost was kept walking in circles
(3) Was the hiker who lost kept walking in circles?

These are two declarative and one interrogative sentences, all made from the same words. (3) is an unambiguous yes/no question with a meaning that corresponds to (2) (that is, (2) is the "yes" answer to (3)). Why can't it have a meaning corresponding to (1)? I assume that it is a syntactic fact that the structure of the interrogative sentence (3) is like that of (2), and not like that of (1) in that the "was" goes with "kept", and not with "lost". This syntactic fact, together with the general assumption that compositional semantic rules interpret complex expressions in terms of the interpretations of their syntactic components explains why the meaning of (3) cannot be a question whose "yes" answer is (1). (I am not here saying anything about how the syntax explains the syntactic fact; the point is just to say that the substance of the explanation is in the syntax.)

The second of the examples is

(17) The senator called the millionaire from Texas

This can mean that the millionaire from Texas was called by the Senator, or that the Senator made a call from Texas to the millionaire; why couldn't it also mean that the Senator from Texas called the millionaire? Again, as Pietroski notes, it seems clear that there are two syntactic structures that (17) might have, with "from Texas" forming a phrase with "millionaire," or alternatively with "called the millionaire". But the sentence cannot have the syntactic structure according to which "from Texas" combines with "Senator" to form a phrase. These syntactic facts, together with the same general assumption about the relation between syntactic structure and semantic interpretation, explain the semantic fact. (An aside: I don't agree with Pietroski that "it's hard to see how we can even start accounting for the interesting facts concerning (17) without an event analysis of some kind." Why doesn't it suffice, at least to get started, to note that "from Texas" is a predicate modifier that can operate either on nouns or verb phrases (both of which have functions from individuals to propositions as semantic values)? I suspect it is only some Davidsonian extensionality assumptions that motivate the event analysis; I also suspect that some of the tensions between semantic and ontological motivations that Pietroski discusses have their source in these assumptions.)

We do need to make a modest and very general assumption about the relation between syntactic and semantic structure in order for the syntactic theory to play the role that it seems to play in the explanation for semantic facts, but I don't see that we need a notion of meaning that is disconnected from truth-conditional content. I would agree, however, that if we focus on *compositional* semantic rules, rather than on the lexical semantics for simple expressions, we can abstract away from some of the ways in which truth-conditional content is dependent on the external environment.

One final remark, about the contingency of semantic facts: Pietroski objects to the common assumption that "semantics is fundamentally contingent/arbitrary/ conventional/learned", but I think it is important to separate these four notions, each of which is different from the others, and each of which is also subject to different interpretations. Even the deepest theoretical truths about language, those that are consequences of an innate universal grammar, are contingent truths; it is an empirical fact that human beings have the innate capacities that they in fact have, and this is a point that proponents of the Chomskyan conception of language often rightly emphasize. It is only because it is a contingent fact, not predictable from an a priori point of view, that all human languages share certain distinctive features that this fact provides empirical support for the innateness hypothesis. It might be, in a sense, a necessary truth that "John is eager to please" has the semantic character that it has, and that "dog" refers to dogs, and not cats, since it might be that semantic properties of expression types are essential to them. What is contingent is that we use expressions with those semantic properties, and that expressions with the phonetic and orthographic properties that those expressions have should have the semantic properties that they have. But the differences between deeper and more superficial semantic facts, and between what is innate and what is learned, are, I think, independent of the question of contingency.

This point connects with comments that Pietroski makes in an appendix about the two-dimensional apparatus, since as I use that framework, it is primarily a descriptive apparatus that helps to represent the ways in which facts about speech acts are contingent, and ways in which speakers and listeners might be ignorant or mistaken about the semantic properties of the expressions they produce or hear. There is surely a possible world in which a sequence of sounds resembling "John is easy to please" has the grammatical and semantic properties that make it mean that John easily pleases people, and it is conceivable, even if unlikely, given our innate linguistic abilities, that a person should misunderstand it in that way. To represent a situation in which someone did misunderstand it in this way, one could use a two-dimensional matrix with a line for a possible world in which the expression had that meaning. I agree that the two-dimensional apparatus does not contribute to an explanation of the constraints that Universal Grammar puts on semantic interpretation, but that is not its aim.

HECK, "IDIOLECTS"

Richard Heck asks which is more basic, common languages or individual idiolects. The thesis he defends is that we can explain communicative success in terms of what expressions mean *to a speaker*, without reference to a common language. The phenomena that motivate the defenders of the primacy of common languages, it is further argued, can be explained without reference to a common language.

I am sympathetic to the arguments that a conception of a common language, such as Italian or Urdu, cannot bear the explanatory burden that some philosophers want to impose on it, and to the arguments that one can explain some of the communicative norms that those philosophers appeal to without reference to languages of this kind. But I am skeptical about a presupposition of Heck's argument—that we must appeal either to common languages or to individual idiolects in order to explain communicative success. He says, in closing, that he has argued "that idiolects are necessarily involved in the explanation of successful communication", but the main burden of the argument was that common languages are *not* necessarily involved. I suspect that the reasons Heck gives for concluding that reference to a common language is "explanatorily idle" may be equally good, or better, reasons for saying the same thing about reference to idiolects. I also want to suggest that the proponents of the primacy of common languages might have a stronger argument if they make it at a different point in the overall explanatory story.

Heck's basic example is an unadorned and unproblematic communicative exchange. Richard says to Janet, "Professor Parsons is teaching" as she is about to knock on his office door. She understands and believes him, thanks him and goes away. He uttered certain words, and in doing so, said that Professor Parsons was teaching. We might also add that he did something that *meant*, in Grice's favored sense, that Professor Parsons was teaching. Why, Heck asks, did uttering those particular sounds succeed in getting Janet to believe that Professor Parsons was teaching (rather than, for example, that pigs dance in Peru)? He then argues, if I understand him, that to answer this question, we need to appeal, not to what the sentence actually means (in English), but to what Janet takes it to mean. It is neither necessary nor sufficient for communicative success, in such an exchange, that the participants each have true beliefs about the meaning of the sentence in a common language.

But when Heck writes "what explains the belief Janet forms is not what the sentence means but what she *takes* it to mean", is he saying that we must appeal to what Janet takes the sentence to mean *in English*? Since the point is that meaning in a public language plays no role in the explanation, he must mean that what explains her belief is what Janet takes the sentence to mean *in Richard's idiolect.* But what work is the reference to an idiolect doing here? I would prefer to say that what explains the belief is not what Janet takes Richard's words to mean, but what she takes *him* to mean, in using those words. To invoke the rough idea of the Gricean account of meaning, what is required for successful communication in this case is for Janet to recognize Richard's intention to get her to believe that Professor Parsons is teaching (and to recognize that he intended her to recognize that he had this intention, etc.). There remains the question why articulating those particular sounds should be a way of getting her to recognize this intention, but first, I don't see how saying that Janet believes that this is what the words mean in Richard's idiolect will be a helpful answer, and second, I don't think the way the further question is answered will be relevant to the success of the communicative act, since the act will succeed as long as Richard had this intention, and Janet recognized it. In saying this, I am not disagreeing with Heck's important point that mere concurrence of beliefs (Janet's having a true belief about what Richard intended her to take him to mean) is not sufficient for communicative success. Recognizing an intention is more than coming to have a true belief that the relevant person had the intention. Just as it might be purely accidental that certain words (in one person's mouth) meant what another person took them to mean, so it might be purely accidental that one person came to believe truly that another person was acting with a certain intention. There is, in both cases, work to be done in explaining what, beyond forming a true belief, is necessary for *recognition* (either of expression meaning, or of an intention), and so for knowledge of what has been meant. But I don't see how idiolects contribute to that further work.

Of course, as Heck observes, it is an important feature of the example that the exchange was a linguistic one in which Richard *said* (and didn't just mean) that Professor Parsons was teaching. Saying, on the Gricean account, is one way of meaning things, and the Gricean strategy is to explain saying in terms of the particular means the speaker is using to get his or her intention recognized, but exactly how the explanation should go is a difficult question that greatly preoccupied Grice, and that he never answered to his own satisfaction. This is one of the issues in the background of Heck's discussion but it is not in the foreground, since I think Heck is interested in communicative success even in cases where what is successfully communicated is meant but not said. Reference to idiolects is especially unhelpful in cases like the following:

Saul and Paul see a person in the distance whom Saul takes to be Smith, but that Paul knows is Jones. While Paul but not Saul knows who the person is, Saul but not Paul can see that he is raking leaves. So Saul says, "Smith is raking leaves," causing Paul to come to believe that the person in question (Jones) is raking leaves, which is the information Saul meant to convey. (What is in question and of interest is what the person is doing, not who he is.) Paul knew that Saul meant to refer to Jones, and so knew what he meant to communicate, but did not know which person named "Smith" he had mistakenly taken Jones to be, and so did not know what Saul actually *said*.[2]

I think it is clear that "Smith" is not a name for Jones, in Saul's idiolect, and even if it were, this would play no role in explaining the fact that in the right situation, Saul can successfully communicate information about Jones with the name "Smith". It suffices for this to be a case in which it is right to say that Paul recognized Saul's intention to refer to Jones.

I have been following Heck (and Grice) thus far in assuming that we can give an account of the aims of a communicative exchange, and of the cognitive effects that it has when successful, that is independent of the linguistic means used to accomplish those ends. Richard believes that Professor Parsons is teaching, and wants to get Janet to share his belief; language comes into it only as a means to accomplish this end. But the proponents of the primacy of common language might question this assumption, and might argue that reference to a common language, and a communicative social practice in which Richard and Janet are participants, will be required in order to explain how it is possible for Richard to have the beliefs and communicative aims that explain why he uttered the words that he uttered. I am myself unsympathetic to the idea that language is prior to thought, in general, but I think it must be granted that

[2] This example is based on one from Kripke (1979).

social communicative practices play an essential role in determining the content of many of the thoughts that we are capable of having, and I think this fact may affect some of the arguments. In particular, I think the relation between Bert and his doctor, and the differences in their beliefs about "arthritis" and arthritis, may be more complicated than Heck's discussion suggests.

The main point that Tyler Burge used the arthritis example to make was that beliefs and intentions (and not just what is said or meant) are dependent on social facts that are external to the thinker. Burge would certainly argue that the difference between Bert and his doctor is not just a difference in usage (like the example of the word "livid"), as Heck, following Chomsky, suggests, but a difference in belief about arthritis. Or at least, it is not clear how to distinguish different beliefs about arthritis from different uses of the word "arthritis". Even in the kind of case that Heck stipulates, in which Bert has a quite explicit belief that "arthritis" has, in its extension, all rheumatoid ailments (whether of the joints, bones, muscles, or wherever else one can have rheumatoid ailments), it will presumably also be true that Bert intends to refer, with his word, to *arthritis*, and the point here is not the trivial one that Bert intends to refer, with his word "arthritis," to what he means by "arthritis", but that we can correctly attribute to him the intention to refer to arthritis. He has two intentions which he thinks of as compatible—even as essentially the same intention—but which in fact conflict: the intention to refer to arthritis, and the intention to refer to the general category of rheumatoid ailments. Which of his two intentions determines the reference of his term? Heck compares Bert's situation to Kripke's Gödel–Schmidt case, and the analogy is apt: we might say about the speaker in Kripke's example that he has two conflicting intentions: to refer to the person who proved the incompleteness of arithmetic, and to refer to Gödel. If we share Kripke's judgments about the case, we will agree that it is the second of these intentions that determines reference. There is no question of a difference of idiolects here (between those who know who really proved the incompleteness theorem and those who do not). There is just a difference in factual belief. The arthritis case seems to me to be more or less like this, though reference to such things as diseases is more complicated than reference to people. But even if Bert were to stipulate that whatever others may mean, he shall mean by "arthritis" any rheumatoid ailment, of the joints or not (thus making it false that he intends to refer, with "arthritis," to arthritis), the content of his beliefs about the category of diseases he has named would still be hostage to socially determined facts. At least if Bert is as ignorant of the relevant medical facts as I am, then while he may know that arthritis (or at least one kind of arthritis) is a rheumatoid ailment, he will have almost no idea what a rheumatoid ailment is, other than that arthritis is a paradigm case of one.

I doubt that this kind of consideration will ground an argument for the essential role of a common language in the explanation of communicative success, but I think the considerations that motivate the proponents of the primacy of common language concern the role of linguistic social practices in the determination of the content of thought.

LOGIC AND METAPHYSICS

Vann McGee's and Tim Williamson's papers are both about the relation between logic and metaphysics. Each brings out, in very different ways, some of the complexities in the relationships between these two enterprises.

MCGEE, "THERE ARE MANY THINGS"

Vann McGee's paper takes us on a tour through some fascinating mathematical and metaphysical terrain. The metaphysical picture he paints, as I understand it, mixes elements I find congenial with some that I find very uncongenial. Naturally, I will concentrate on the latter, and my comments will reveal my anti-metaphysical proclivities.

McGee begins with the remark that ontology is the most general science. Its job is to give a fully comprehensive account of what there is, and he proposes a thesis that belongs to this science, a principle of plenitude: "if it is possible that there are at least so-and-so many things, then there are, in fact, at least so-and-so many things." He does not try to demonstrate that the thesis is true, and is generally modest about the possibility of establishing theses in ontology. What he does is to draw out some of the consequences that he thinks the thesis has, and to argue that it must be true in order for mathematics to have the resources to do its job. The main consequence of the thesis for modal metaphysics, he argues, is that we can do without possible worlds, replacing them with mathematical models of the possibilities. This account of modality yields a significant side benefit: a conception of logical possibility that is wider than the concept of metaphysical possibility, and that permits us to extend the explanatory range of semantic theories of modal concepts, such as conditionals.

I find it hard to take seriously the idea that there is a general science of ontology that is not just the sum of all cognitive enterprises (and I am not sure how seriously McGee takes it). Imagine someone who said that there was a science of Veritology whose job it was to give a fully comprehensive account of all truths, on any and all subjects. Who should the veritologists be? There are general problems about the concept of truth that philosophers, semanticists and logicians concern themselves with, but working on an inventory of truths is not a promising way to solve them, and neither these specialists nor anyone else is well placed to provide the inventory. I think analogous things should be said about ontology. More controversially, I find it hard to take seriously the idea that there is a plurality of absolutely all the things there are, or a meaningful question about how many things there are, altogether. The reason for my skepticism is not that I think there is a paradox that cannot be avoided on the assumption that one can quantify over absolutely everything. Even if this assumption can be rendered formally coherent, I am not sure how to play the game that makes it.[3] Some of the reasons for my puzzlement will emerge as I try to understand McGee's project.

"It is not the job of mathematics to try to tell us what the world is like; that task belongs to the various empirical sciences. The job of mathematics is to provide models of all the possibilities for what the world is like." Since an account of what the world is like will include an account of what there is, this conception of mathematics suggests that mathematics does not tell us anything about what exists. Rather, it uses the materials that are found in the actual world (that the empirical sciences tell us about) to model what there might have been. The principle of plenitude is the (empirical?) thesis that there is sufficient material to do this job. This conception of mathematics seems to imply that there are no distinctively mathematical entities. For arithmetic to be true, there need not be any special kind of object (numbers), but just enough objects of any kind that can be used to construct a model with the right structure. For set theory to be true, there need not exist a distinctive kind of thing (sets), but only enough things for a model of set theory. But McGee also says that his account "requires mathematical entities in great profusion." If we assume there are abstract objects, we can accept the principle of plenitude without accepting David Lewis's modal realism, with its ontological commitment to *concrete* objects in great profusion, including literate spiders, aquatic zebras, and such. We can make do with mathematical *abstracta*. But what are these things? (Are they, for example, *numbers* and *sets*? If so, are some of the models of elementary number theory models that use the natural numbers to model the natural numbers?)

What is the status of the principle of plenitude? Is it metaphysically necessary, if true? The profusion of mathematical abstracta are perhaps metaphysically necessary existents (at least assuming that there are such things at all), but I assume that on McGee's account, there will be *logically* possible worlds in which these things do not exist, and in which there are too few things for the principle of plenitude to be true. (Or is it that if there were fewer things, there would be fewer possibilities, and so the principle of plenitude would still be true?) Consider a logically possible world in which there are only finitely many things (though let's make it a large enough number so that no one notices). The mathematicians in this logically possible world are busy doing what, for all the world, looks like the construction of mathematical models. They talk, or at least seem to talk, about transfinite numbers and argue about the continuum hypothesis. Even the most cautious among them claim that there are infinitely many natural numbers. But these poor deluded souls are like astrologers. The world does not provide the resources to permit them

[3] See Cartwright (1994), Williamson (2003) and Glanzberg (2004) for discussion of the intelligibility of quantification over absolutely everything.

to do their job. (Though perhaps a more charitable interpretation of their activities, including of their metamathematical attempts to give a semantics for their mathematical languages, would conclude that they are not deluded.) Might the actual world be like this? McGee points to some likable features of his thesis, but does not suggest that we have reason to believe that it is true.

Even given McGee's story about mathematics and ontology, and even assuming the truth of the principle of plenitude, I am dubious about the idea that one can dispense with possible worlds (understood as ways a world might be), replacing them with mathematical models As McGee says, "a mathematical model of a flying cannonball isn't a flying cannonball," and the same goes for a possibility. (I am not sure McGee disagrees with this point, since he emphasizes that "the models we are referring to as 'metaphysically possible worlds' are not ways the world might have been, but only mathematical representations of ways the world might have been.") It would be dangerous to stop believing in flying cannonballs just because we believe in mathematical models of them, and there is also no reason to think that a profusion of models of possibilities gives a reason to dispense with the possibilities themselves.

I agree, however, with McGee that given a formal language with a well defined semantics, and a distinction between the logical and non-logical expressions, one can define a notion of logical necessity, as applied to sentences: S is logically necessary if and only if S is true in all models. And I agree further that one can construct a model for a modal language in which models play the role of possible worlds, and that one can then define a derivative notion of metaphysical possibility that applies to sentences, with the result that the metaphysically possible sentences will be a proper subset of the logically possible sentences. But I don't think this exercise really provides a unified framework in which the metaphysical possibilities can be understood as a subset of a larger set of things of the same kind, and I don't think that this kind of model yields an account of conceivability, or of epistemic possibility. "Metaphysical and logical possibility," as McGee says, "are in different lines of work." In the one case, one holds the interpretation of the language fixed and varies the possible facts, while in the other case, one varies the interpretation of the language, considering the actual truth of the sentences under the different interpretations of them.[4] A model of a nonactual metaphysically possible world (on McGee's account, as I understand it) is an interpretation of the language in which words mean something different from what they actually mean, but in which what the words *actually* mean is something metaphysically possible. This seems to me a contrived way to unify the theory of logical and metaphysical possibility. Consider a model in which "Socrates" refers to G. W. Bush, and "human" means *philosopher*. "Socrates is human" is false in this model, since that sentence is interpreted to say that GWB is a philosopher. Therefore, this "logically possible world" is metaphysically impossible, since the sentence

[4] See J. Etchemendy (1990), chs. 2 and 4, where this distinction is developed and discussed.

that, in this model, says that GWB is a philosopher, *actually* says something that cannot be false. But this representation makes no contribution to explaining the fact that what the sentence actually says is metaphysically necessary—that there is no metaphysically possible world (way things might be) in which it is false. And the fact that it is only a contingent fact that GWB is not a philosopher is just an irrelevant distraction.

I do think that to account for epistemic possibility we need to consider the way semantic ignorance and error interacts with ignorance and error about the subject matter of a statement (see my discussion of Soames for more about this). But I think the way to do this is to consider metaphysically possible worlds in which the semantic facts are different. And I don't think logical possibility has any special role here—one sometimes needs to consider possible situations in which the interpretation of the logical expressions, as well as the descriptive expressions, is different from what it actually is. One might be ignorant or mistaken, not only about whether Batman is Bruce Wayne, but also about whether a sentence of the form 'P or Q' has the same truth value as one of the form 'not(not P and not Q)'.

WILLIAMSON, "STALNAKER ON THE INTERACTION OF MODALITY WITH QUANTIFICATION AND IDENTITY"

I agree with Tim Williamson that metaphysics is difficult to disentangle from logic (and more generally from semantics), and that it would be a mistake to reason from the fact that a principle is controversial to the conclusion that it is not a logically valid principle. But I also think that it is useful to try to separate criticisms of a metaphysically controversial view on the ground that it is logically incoherent from arguments that the view, while coherent, is mistaken. The task of separating the two kinds of criticism will itself often be a contentious issue, and I agree with Williamson that there is no general procedure for finding neutral ground on which to adjudicate a dispute where logical and metaphysical principles, semantic and substantive differences, are intertwined.

An attempt to render a metaphysically controversial doctrine coherent, or to defend it against objections, will often be subject to the charge that one is changing the subject by interpreting some logical expression in a non-standard way. For example, Quine's main complaint against the Meinongian is not that he has a bloated ontology, but that he masks his bloated ontology by using the word "exist" in an idiosyncratic way, thereby creating an illusion of agreement with a proponent of a more austere ontology where there is none.[5] Closer to home, many have argued (convincingly, I think) that the metaphysical thesis that there are contingent identities is logically incoherent. There may be contingent identity

[5] See Quine (1961) for this kind of criticism of the Meinongian, and Lewis (1990) for a discussion of the difficulty of separating ontological from semantic issues.

statements where there are nonrigid singular terms, but they argue that there could not be a pair of things (of any kind) that stand in the identity relation, but only contingently so. The proponent of contingent identity responds by devising a coherent logical theory of what he calls "contingent identity", but the critic argues that what the symbol "=" is used to express in the theory is not identity, but some particularly intimate equivalence relation that may hold contingently of some pairs of (distinct) things. It just muddies the waters to call such a relation by the name "identity". I agree both with Quine's complaint about the Meinongian, and with these critics of the coherence of contingent identity, but there is not a pattern here that can be simply generalized; the issues need to be argued case by case. At this level of methodological abstraction, I believe Williamson and I are in agreement; the differences come out in the details, to which I now turn.

The specific issue that Williamson is addressing in his paper concerns the significance for metaphysics of some technical results in quantified modal logic that I sketched in a paper on the interaction of quantifiers, identity and modality. My strategy in that paper was to develop and motivate separately versions of extensional quantification theory, and of propositional modal logic—a proof theory and model theory for each—and then to combine the two theories at both the semantic and proof-theoretic levels. One upshot was that, although the separate logics were sound and complete, the combined logic was sound but incomplete: there were principles involving quantifiers and modality that were semantically valid, but not theorems. A variant semantics that validated all the theorems, but not certain sentences that were validated by the standard semantics, was used to show the independence of those sentences. It was then argued that the variant semantics (a version of counterpart theory) had some intuitive motivation, given a certain metaphysical view, and that the standard semantics could be seen as a special case of it. It was also claimed that since the variant semantics was based on the same extensional semantics and logic, it gave a standard interpretation of predication and identity. This was the basis of an argument that the principles that distinguished the two semantic theories were metaphysical principles.

Williamson is right to point out that this line of argument depended on a number of decisions that might be questioned, decisions about how to axiomatize the separate theories, and how to combine them at the semantic level. Williamson shows that I could have chosen a different sound and complete axiomatization for extensional quantification theory that would have resulted in a logic that was complete relative to the standard semantics when it was combined with the propositional modal logic. These points are well taken, but I think the decisions I made can be motivated.

Williamson is also right to question any general argument, based on this kind of consideration, to the conclusion that a principle is metaphysical rather than logical. I want to disclaim any commitment to a sharp or deep line between logical and metaphysical principles, or between logical constants and other expressions. My main concern was to try to justify the claim that the nonstandard theory of modality, predication and identity that I was discussing was not guilty of changing the subject; the idea was to make it plausible that what the theory calls "predication" and "identity" really deserve their names by arguing that the logical theory of identity and predication that I was combining with modality was a standard, orthodox one.

Let me first try to justify the way in which the two languages (extensional quantification theory and propositional modal logic) were combined. As Williamson notes, there is no general procedure for combining two languages, but the possible-worlds model theory provides a natural pattern for generalizing any extensional theory to a modal version of that theory. One can think of an extensional model as a representation of a single possible world, and of the semantics as a set of rules that define the extensions of complex expressions as a function of the extensions of their constituents. The modal generalization does for many possible worlds at once whatever the extensional theory does for a single world. The semantic values are intensions, rather than extensions, where an intension generally is a function from possible worlds to extensions. A modal model will provide for each possible world whatever the extensional model provides for the one world. (In the case of first order quantification theory, a domain of individuals, referents from the domain for the names, subsets of the domain for the one-place predicates, etc.) The compositional rules will be straightforward generalizations of the extensional rules. This is not the only way to combine modality with quantification, but it is a natural one that yields a smooth generalization of the semantic rules for identity and predication to the modal context, one that makes no nonstandard assumptions about those notions.

The only place where the generalization is not straightforward is with the semantics for variable binding. One way to see the problem is this: Tarskian semantics for quantification theory generalized the notion of truth to satisfaction, or truth relative to an assignment of values to the variables; modal semantics generalizes truth to truth relative to a possible world. Modal quantification theory has to fit the two generalizations together (truth at a world, relative to an assignment, and truth relative to an assignment, at a world). The motivation for some of the decisions made in combining the theories, at both the proof-theoretic and semantic levels, depends on certain not uncontroversial assumptions about how open sentences and variable binding should be understood.

As Williamson noted, a logic may have alternative axiomatizations that are equivalent when the theory stands alone, but differ when it is combined with another. Specifically, he observes that if the quantification theory had been axiomatized by replacing my abstraction axiom schema,

$$\vdash \forall\hat{x}(\hat{y}\varphi x \leftrightarrow \varphi^{x}/_{y}),$$

by what he calls the free abstraction principle (a rule of inference),

If B is the result of replacing $\hat{y}\varphi x$ by $(Ex \mathbin{\&} \varphi^{x}/_{y})$ in A, and $\vdash$ A, then $\vdash$ B.

then the resulting extensional logic would still be sound and complete, but the enriched modal quantification theory would also be complete, with respect to the standard semantics. He concludes that my independence result therefore just reflects my "unforced choice amongst ways of formulating the logic of quantification that are equivalent in a non-modal setting but not in a modal setting," and so shows nothing about the logical or metaphysical status of the principles that I show to be independent. I grant that the choice was unforced, but not that it is unmotivated. Williamson's free abstraction rule seems to me contrived, and less natural than my abstraction axiom, involving the replacement of a complex expression with a complex expression with a different constituent structure. It is difficult to imagine formulating the logic this way except for the purpose of validating the contested principles that involve the interaction of quantifiers with modality. But I grant that one cannot put much weight on such judgements of differences in the naturalness of an axiomatization. And I cannot, Williamson argues, appeal to the fact that the counterpart semantics invalidates his free abstraction rule as a reason to reject that way of axiomatizing quantification theory, for I would then be arguing in a circle. "For his original reason for taking the counterpart semantics seriously was precisely that it validated all the principles of first-order non-modal logic and of propositional modal logic." But I see the dialectical situation somewhat differently.

I did not mean to be giving an argument of the form: Principle X is independent in the combined logic of quantification and modality; therefore principle X is not a logical truth. I think such independence results often call for some explanation, but the explanation might be quite different from case to case, and there is no reason to take an artificial deviant semantics seriously just because it was effective for the purpose of establishing an independence result. But if the deviant semantics has some intuitive motivation, as a framework for a metaphysical theory, then the independence results can play a role in the interpretation and defense of the metaphysical framework. Specifically, such results help to identify the places where the theory agrees with orthodoxy, and to isolate the places where the deviant theory diverges from orthodoxy. By providing an argument that both the semantics and the logic (on at least one natural way of formulating the logic) of certain central concepts (in this case, identity and predication) are the same as they are in the orthodox theory, the independence results help the defender of the deviant metaphysics to rebut the charge that he is changing the subject. This kind of argument is not decisive, and in any case, it is not an argument that puts all its weight on the results about the proof theory. The fact that the *semantics* for identity and predication are the same in both of the modal theories (orthodox and counterpart) is also important. But even if this kind of consideration is not decisive, it does help to shift the focus of the argument from the question of logical coherence to the question of metaphysical plausibility.

An unforced choice of a different kind that Williamson points to is the decision, in generalizing the semantics, to permit singular terms to be nonrigid designators. If singular terms were required to be rigid (by taking their semantic values to be, in general, individuals rather than individual concepts—functions from possible worlds to individuals) then a much simpler abstraction principle for singular terms,

$$\phi Et \rightarrow (\hat{x}\varphi t \leftrightarrow \varphi^{t}/_{y})$$

which is valid in the extensional semantics, would preserve its validity in the enriched theory, and could replace the quantified abstraction axiom. To justify my choice here, I would appeal to the generality and naturalness of the generalization of extensions to intensions. It would be an artificial limitation of expressive power to restrict the language in this way.

I will conclude with a brief remark about Williamson's arguments concerning the second independence result: that in some quantified modal logics, the necessity of distinctness is not provable, even when the qualified converse Barcan formula is added. Here the result turned on the fact that the extensional logic of identity is the same as the extensional logic of indiscernibility, but (in the modal models in question) things may be discernible, but only contingently so, even though they cannot be distinct, but only contingently so. In this case, the deviant semantics does not involve a deviant metaphysics, but only the reinterpretation of a symbol: "=" is (in the deviant semantics) stipulated to mean something different from identity, so here we are explicitly changing the subject (to make a formal point). I did not want to suggest that this result is relevant to the metaphysical issues; it is a point that is entirely about the expressive resources of the language (which of course change when the language is enriched). Williamson observes that the logical difference between identity and indiscernibility is not present in an S5 modal semantics, and shows that the necessity of distinctness becomes provable in the general theory if one enriches the language by adding a global necessity operator, with its S5 logic. I agree with all of this, and I should have been clearer that my remark that "the necessity (or essentiality) of identity is more central to the logic of identity than the necessity of distinctness" was based, not on this result, but on the independent fact that the necessity of distinctness is invalid in the counterpart semantics (where "=" is interpreted as identity), a result that holds even in an S5 theory.

SUPPOSITIONS AND PRESUPPOSITIONS

LYCAN, "CONDITIONAL-ASSERTION THEORIES OF CONDITIONALS"

William Lycan's paper criticizes conditional assertion theories of (indicative) conditionals, and defends the thesis that these conditionals express propositions, though he does not here rest his case on any particular account of the truth conditions of conditionals.[6] I have, at different times, found myself on both sides of this issue, and I have decided that on both sides is where one should be. My hope is that a proper development of the contrasting accounts in a common framework will show the conflict between them to be less than is usually supposed, and will point to a unified theory that allows both for a distinctive kind of conditional speech act and for a distinctive kind of conditional proposition.[7] Lycan offers a diverse battery of arguments, some of which are congenial to such a project of reconciliation, but others of which are not. First, I will comment on his arguments against what he calls the simple illocutionary theory, and on a semanticized version of it. Second, I will dissent from some arguments that rely on judgments about the phenomena that I do not share. Third, I will look at a set of Lycan's arguments that I think together make a compelling case for the thesis that indicative conditionals at least sometimes express propositions. I will conclude with a general comment about the considerations that motivate the anti-propositionalist, and the challenge they pose.

It is not entirely clear what the simple illocutionary theory says. Lycan focuses mainly on one line in Quine's rather casual exposition of the view: that "if the antecedent turns out to have been false, our conditional affirmation is as if it had never been made," but this cannot plausibly be taken to mean that no speech act has been performed in the case where the antecedent is false, or that the speech act is intended to have an effect on the context only in case the antecedent is true. Rather (as the previous sentence in the quotation from Quine makes clear), the point is that with respect to the speaker's commitment to the truth of the consequent (and the requirement "to acknowledge error if it proves false"), the situation is as if no affirmation had been made. More generally, if one explains speech acts (questions, bets, promises, requests) in terms of specific commitments or obligations that are undertaken or created in virtue of performing the speech act, then a conditional version of that speech act will be explained in terms of a commitment or obligation of the same kind that one is held to only in case the condition turns out to be true. Such an account of conditional speech acts may not help with many of Lycan's criticisms, but it goes a long way toward answering what he calls the "Initial Implausibility" objection. Whatever the ultimate fate of the idea of conditional assertion as a distinctive speech act, I think it is worth giving a more plausible development of the idea.

Lycan considers what he calls a "semanticized" version of the conditional assertion view according to which conditionals are true when the antecedent and consequent are both true, false when the antecedent is true and the consequent false, and neither true nor false otherwise. I agree with Lycan that this is not a promising move, though my reasons are somewhat different from his. One

[6] A specific analysis is defended in his book, Lycan (1999).

[7] I discuss this prospect in Stalnaker [2005]

problem with this kind of theory is that it requires one to give up an otherwise plausible norm of assertion: that one is normally committed to the (semantic) presuppositions of one's assertions, since one is normally committed to the truth of what one says. As Michael Dummett noted a long time ago, an assertion with an improper definite description is a defective assertion, but an assertion of a conditional with an antecedent that turns out to be false is perfectly okay.[8] It seems wrong to give a semantic account that assigns the same status to the contrasting cases. In any case, it is not clear to me why this kind of theory should be thought of as a version of the conditional assertion account.

The arguments in Lycan's paper that I find least congenial are those that rest on specific assumptions about the truth conditions for conditionals, assumptions that I think even a resolute propositionalist should reject. Most notably, there is the TT argument: the semanticized version of the conditional assertion account judges a conditional to be true whenever both antecedent and consequent are true (and the unreconstructed conditional assertion theorist judges a conditional to be assertable whenever the probability of the conjunction of antecedent and consequent is sufficiently high). But Lycan judges some of such conditionals (those where the truth of the antecedent is irrelevant to the truth of the consequent) to be false. Of course many truth-conditional analyses of conditionals accept the TT inference, and I think a truth-conditional account with this property can best account for the phenomena. Lycan gives us the case of the dentist, who says to a person who in fact has perfectly healthy teeth: "if you don't undergo this treatment, you'll lose all your teeth." The person refuses the treatment, but then loses all of his teeth in an automobile accident. Lycan concludes that the dentist's statement is false in this case. We can all agree that there is something wrong with the dentist's statement, but many true statements are defective despite their truth. Suppose the dentist had made a categorical prediction, after I turned down the treatment: "Mark my words, you will lose all your teeth within the year." This prediction is just as defective as the conditional assertion, and for the same reasons, but it is hard to deny that (in the case described), it turned out, by accident, so to speak, to be right.[9]

The most direct and convincing arguments for a propositional account are the embedding arguments. If a sentential clause can be a component part of a sentence that involves connectives and operators whose meanings are rules that give the truth conditions of the whole as a function of the truth conditions of the parts, then the parts had better have truth conditions. The theorist who rejects a truth-conditional account of conditionals must either explain away

[8] Dummett (1978). Dummett suggested, in effect, that one kind of truth value gap was really a kind of truth, while the other was a kind of falsehood.

[9] Lycan also takes it to be a problem that conditional assertion accounts allow that a conditional may be assertable when its contrapositive is not, but I think a propositionalist should reject the validity of contraposition, and it must be rejected by any analysis that accepts the TT inference, but rejects the material conditional analysis.

the embedded cases, or give a non-truth-conditional account of the contexts in which they are embedded. I think some embedding can be explained away without implausibility: negations of conditional assertions are naturally read as conditional denials (conditionals with the negation on the consequent), and right-nested conditionals (if A, then if B, then, . . .) are naturally read as the accumulation of suppositions. Some others might be dismissed as unintelligible, since disjunctions of indicative conditionals, and conditionals with indicative conditional antecedents are often hard to process. But there remain cases that are hard to dismiss, such as the nice example that Lycan cites:"if John should be punished if he took the money, then Mary should be punished if she took the money." It seems intuitively clear that what is supposed in the antecedent of the main conditional is a proposition that can be assessed independently of whether its antecedent is true. (Lycan does not consider quantified conditionals, but they help to bolster the embedding argument.)[10]

Closely related to the embedding point is the propositional attitude objection, which points to the fact that conditionals occur in a wide range of propositional attitude contexts. If conditionals express propositions, then the problem of conditional belief, knowledge, intention, desire, reduce to ordinary belief, knowledge, intention, desire, etc. with a conditional proposition as content. But this consideration cuts both ways. We have an independently motivated account of conditional belief (and conditional degree of belief) that does not seem to be the same as categorical belief with a conditional content. There may be some point, independently of an account of conditional speech acts, in generalizing the idea of a conditional attitude that is distinguishable from an attitude to a conditional proposition, and there may be application for a more general account of such attitudes in a theory of reasoning. Still, I agree that there are cases that cannot be explained in this way, including the factive cases that Lycan discusses: being sad, happy embarrassed, ashamed, etc. Here it seems intuitively to be a conditional fact that one is happy, embarrassed, ashamed etc. about, and it would be a stretch to try to explain such states in terms of some conditional version of the attitudes.

Despite these considerations, there remain some arguments on the other side that are difficult to answer. Stories told by the proponents of a conditional assertion account suggest that the assertability conditions for conditionals are tied more to the epistemic situation of the speaker than to the facts about the subject matter of the propositions expressed in antecedent and consequent, and that apparently contrary conditionals may be assertable by speakers with different but compatible information.[11] Let me use a variant of one of Lycan's examples to illustrate the point. Lycan suggests that certain categorical facts, such as that

[10] See Higginbotham (2004) for a discussion of some problems about quantified conditionals.

[11] See Gibbard (1980) for the notorious story of Sly Pete, and Edgington (1986) for arguments that acceptance of a material conditional without rejection of the antecedent always suffices for acceptance of the corresponding indicative conditional.

Chapel Hill is in North Carolina, entail conditionals, such as that if Jones lives in Chapel Hill, then he lives in North Carolina. But suppose that I know for a fact that Jones either lives in Greensboro, North Carolina, or else somewhere in Virginia. Does he live in Chapel Hill? you ask. I don't know, since I don't know where Chapel Hill is, but I know it is not Greensboro, so I know that if Jones does live in Chapel Hill, it must be in Virginia. So if he lives in Chapel Hill, he does *not* live in North Carolina. I would be right to acquire the conditional belief that Jones lives in Virginia, if he lives in Chapel Hill, and it seems that I am making no mistake in forming this belief, since I am basing it entirely on things that I know to be true. (Can you rule out the possibility that Chapel Hill is in North Carolina? you ask. No, I say, but if it is, then Jones doesn't live there.) I don't reject the categorical proposition (that Chapel Hill is in North Carolina) that allegedly entails a conditional that I reject.

The underlying motivation of the conditional assertion account, as I see it, is the idea that an indicative conditional is used to express something about the speaker's epistemic situation, and does not make a claim about the objective facts that can be assessed independently of the epistemic situation (at least not a claim that goes beyond the material conditional). I think there is something right about this idea; but it may be that the best way to capture it is to interpret conditionals generally as expressing propositions that are sometimes highly context-dependent, and perhaps speaker-dependent, propositions that are in part about the speaker's epistemic situation, and in part about the subject matter of the constituent propositions. Such a theory might stand a better chance of explaining the connections between indicative and so-called subjunctive conditionals that Lycan emphasizes in his "subjunctive parallel" argument, and of explaining the interaction of epistemic and factual considerations in the interpretation of indicative conditionals.

YABLO "NON-CATASTROPHIC PRESUPPOSITION FAILURE"

The idea of Steve Yablo's constructive project is to explain how statements with false presuppositions may nevertheless sometimes succeed in making true or false claims, and to provide some apparatus for distinguishing the cases where they do from cases where they do not. One important motivation for the project is to contribute to a deflationary account of the ontology of abstract objects by explaining how statements that presuppose the existence of something may nevertheless make claims that are not committed to its existence.

I am sympathetic to the idea that what one says may be true even when what one is presupposing in saying it is false, and I am also sympathetic to what is a different point, that one may sometimes succeed in communicating something true even when what one *says* commits one to something false. But I will express some skepticism about the way these points are connected and about some of the details of the theory constructed to explain them. I will first make some general comments about the various concepts of presupposition that are involved in setting up the problem, and then look at the theoretical apparatus that Yablo develops to solve it. I will point to some formal parallels between his constructive project and the old project of analyzing dispositional predicates, parallels that underscore the daunting character of the challenges that his project faces.

Yablo begins with a *pragmatic* characterization of presupposition: "Presuppositions are propositions assumed to be true when a sentence is uttered, against the background of which the sentence is to be understood." A presupposition failure is a case where what is presupposed is false, and a failure is *catastrophic* when the result of the failure is that no evaluable claim is made: when, in Strawson's terms, "the whole assertive enterprise is wrecked." Yablo says that "the best-known theories [of presupposition] suggest that all presupposition failure ought to be catastrophic", which he takes to be implausible.. This may be true of the Frege–Strawson account of presupposition (S presupposes π iff the falsity of π entails that S is neither true nor false), but that account is not one that fits Yablo's pragmatic characterization. One of the aims of the attempt to characterize presupposition in pragmatic terms (at least one of my aims in doing this) was to separate questions of truth value and truth conditions from questions of presupposition. The fact that the assertive use of a sentence requires, for some reason, that the speaker be assuming that a certain proposition is true (and common ground) by itself says nothing about the truth value or propositional content of the sentence in situations in which the background assumption is false. Consider a simple example: I say to a colleague, "The Red Sox won again last night." In doing so, I presuppose that my addressee speaks English, has enough acquaintance with major league baseball to know what I am talking about, enough interest in it to care, and that she does not already know who won last night's game. My speech act will be infelicitous in one way or another if I am mistaken in making any of these assumptions, but no one thinks that the falsity of any of *these* presuppositions would interfere with my having made a true or false statement.[12] So such a presupposition failure would be unproblematically noncatastrophic, and no special theory is needed to explain what the assertive content is. Ordinary semantics, plus contextual facts, including facts about what is presupposed, tell us what proposition a sentence expresses in that context, and the proposition can then be evaluated as true or false, whether or not the presuppositions are satisfied. Sometimes the semantics (plus the context) will deliver only a partial proposition—one that is true or false only in some possible situations—and if the actual world is one for which the proposition is undefined, then the statement will make

[12] My statement also commits me to the presupposition that the Red Sox had won before, on some probably recent salient occasion. What happens when this presupposition is false may be more controversial.

no claim that can be evaluated in the actual world. One of the things that will generally be presupposed (sometimes falsely) when a sentence is used to make a statement is that the sentence expresses a proposition with a determinate truth value, and it is when *this* presupposition fails (one might be tempted to say) that the failure is catastrophic.

One might take this to be a solution to Yablo's problem, as stated, but I don't think that it gets at the real problem that he is concerned with, or that the real problem is so easy either to state or to solve. The rough idea, as I understand it, is something like this: the aim of a statement made against a background of presuppositions is to distinguish between the possible situations compatible with what is presupposed. One aims to add to a given body of information some additional information by excluding some of those possibilities. The informational content of the statement, on one way of understanding it, is the *increment* of information—the difference between the prior informational state defined by the presuppositions and the posterior informational state that would result from accepting the assertion. But an increment of information is not a proposition that can be assessed as true or false in possible situations that lie outside of the set of possibilities that are compatible with the presuppositions. One way to represent the increment of information is to take the content of an assertion to be a partial proposition—one that is true in possible worlds in the posterior state that would result from the acceptance of the assertion, false in the possible worlds that are to be excluded from the prior state, and neither true nor false in the possible situations outside of the prior state. Assuming that a speaker is committed to the truth of what she says, this representation implies that the speaker's statement commits her both to what is presupposed and to what is asserted. But one would like to be able to isolate the new information so that it can be assessed independently of the background information that provides the context in which it was provided. The problem is to say exactly what this means, and how it is to be done.

When the background presuppositions are required for the relevance or intelligibility of the assertion, but not for the determination of its content, as in the Red Sox example above, then this problem does not arise, since the standard semantics provides a content for the assertion that can be assessed independently of what is presupposed. But there are other cases where even though the literal interpretation of the statement commits the speaker to the truth of a presupposition (and so the presupposition failure is technically catastrophic), it nevertheless seems intuitively that the statement makes a claim that can be detached from the presupposition. The problem is to distinguish these cases, and to say what the claim is.

I am not sure that the various cases Yablo discusses should get a uniform treatment. Many of the examples seem to me cases where a speaker failed to say what he meant to say, and where it is a proposition meant, rather than what was said, whose truth value is being assessed. (Strawson's example of the US Chamber of Deputies, and Kripke's example of Smith being mistaken for Jones seem to me clear cases of this.) But I want to set aside questions about how the apparatus fits the linguistic facts and look at the structure of the formal framework, which is of interest in its own right.

Here is a quick sketch of the constructive theory, as I understand it: We begin with a sentence S that has a presupposition π. We are given a division of logical space (the space of possible worlds) into three regions: the S part of π, the not-S part of π, and the part where the presupposition π is false. We assume that S determines no truth value in the possible worlds outside of the π region, and our task is to extend, or project, S to a fully determinate proposition that discriminates between those possible worlds. The assertive content X of S (its extension or projection) should be a proposition that, when conjoined with π, entails S. More specifically, X should be a proposition that is *equivalent* to S, relative to π, One might put the problem of defining the assertive content this way: given the formula, $(\pi \rightarrow (X \leftrightarrow S))$, solve for X. There are, of course many solutions. If there were no further constraints, then X could be any determinate proposition that entails $(\pi \rightarrow S)$, and that is entailed by $(\pi \& S)$. Yablo calls the propositions meeting this condition the *extensions* of S. He assumes there will be further constraints on X, and we will consider below what he says about what those constraints should be, but he does not assume that there will be a unique solution, even given the further constraints. He defines the assertive content of S (call it '**ac**(S)') as the conjunction of the extensions of S that satisfy whatever further constraints are imposed. (This conjunction will itself be an extension of S.)

One more bit of terminology: Yablo notes that the assertive content of S (relative to presupposition π) might be compatible with the assertive content of ~S. Even though S is of course incompatible with ~S, the eligible extensions of S might be compatible with the eligible extensions of ~S, so it might happen that the **ac**(S) and **ac**(~S) were both true. When this happens, Yablo says, S *makes no claim*. More precisely, S makes a *claim* (relative to world w) if and only if at least one of **ac**(S) and **ac**(~S) is false in world w. Presupposition failure is catastrophic when the statement makes no claim.

Now I want to point to the formal parallel, and perhaps a more substantive parallel as well, between the problem of extending S beyond π and the classical problem of dispositional predicates discussed by Carnap and Goodman. The formula that defines the extensions is identical in form to a Carnapian bilateral reduction sentence used as a partial definition of a dispositional term (such as "soluble") in terms of a test condition ("being put in water") and a display predicate "(dissolves"). Goodman described the relationship between the dispositional predicate and the display predicate as the projection of a predicate defined for a limited domain onto a wider domain. The analogy might be fruitfully developed, but it should also give one pause, since the project of explaining dispositions with austere empiricist resources was a dramatic failure. Of course Yablo is not confining himself to such austere resources, and he is well aware that more resources are needed if his framework is to distinguish catastrophic from non-catastrophic presupposition failures in the way that he wants to distinguish them. But the substance of the proposal will be in the way in which one explains the distinction between the extensions that are eligible to be part of the assertive content and those that are not, and I think much of the work of making this distinction remains to be done. I will conclude with a brief remark about the notion of a truth-maker, which is one of the notions that is supposed to do some of this work.

The extensions of S beyond π are required to be π-*free*, which means, roughly, that they could have the same truth value *for the same reason* in a situation where the presupposition, π, had the opposite truth value. (This constraint points in a more substantive way to the analogy with dispositions, where the rough idea is that a dispositional property is a property that coincides with the display property when the object is subjected to the test condition, but which it has independently of whether it is subjected to the test condition.)[13] "True or false *for the same reason*" is glossed as "has the same truth-(falsity)maker". An appendix to Yablo's paper says a little more about what a truth-maker might be, but as I understand the explanation, truth-makers of propositions are themselves propositions, and I believe that it follows from Yablo's definition that every true proposition is one of its own truth-makers. But if each proposition can be its own truth- or falsity-maker, then *every* extension of S that is compatible with both π and with not-π will be π-free, since such extension will be made true (or false) by the same truth- or falsity-maker (namely itself or its negation) in a situation where π has the opposite truth-value than it has in the actual world. I don't think it is an intended consequence that propositions are always their own truth-makers: a true disjunction with unrelated disjuncts seems, intuitively, to be made true by one or the other of the disjuncts, or redundantly by both, but not by itself. In any case, Yablo is well aware that more constraints are needed to get an appropriate truth-maker (he mentions a proportionality condition toward the end of the paper). I am not sure whether there is a coherent concept of truth-maker that meets the intuitive conditions. The task of explaining such a notion, and spelling out the further constraints seems to me a challenging one, but the notion has some intuitive content, and there is plenty of philosophical work that such a notion could help us to do.

THE SECOND DIMENSION

The papers by Frank Jackson, John Perry and Scott Soames all concern my use of the two-dimensional modal semantic framework to try to clarify some issues in the philosophy of language and mind. There are some recurring themes, both

[13] Not that either Goodman or Carnap would have put it this way.

in the three papers, and in my responses, issues about how to understand knowledge and belief about who or what someone or something is, how to interpret the modifiers 'actual' and 'actually', how to account for the special character of self-locating knowledge and belief. The criticisms in these papers come from at least two contrasting directions: Jackson thinks I am not enough of a two-dimensionalist, or not the right kind, while Soames and Perry think I am too much of one. I will make a few concessions in my comments, and I recognize that there are hard issues about the nature of intentionality that underlie our disagreements, issues that I don't think any of us has yet got to the bottom of. But I continue to think that the two-dimensional framework is useful for clarifying the relations between what we think and what we say, and I also continue to be skeptical about the use of this framework to give an internalist account of intentionality.

JACKSON, "THE STORY OF 'FRED'"

The centerpiece of Frank Jackson's paper is a story about a word "fred" as it is used in a community of two speakers. The story helps to focus some questions about the content of what is said. I think I am in agreement with Jackson about what to say about this particular example, but I suspect that we will disagree about the moral of the story—about what it shows about content in less artificial cases. I will start by looking at a general distinction that Jackson discusses, and then turn to the story of 'fred', and what we should conclude from it.

Jackson starts with a distinction that I have emphasized between *semantic* questions about what the meaning or content of a linguistic expression is, and *meta-semantic* questions about how it is that a linguistic expression comes to have the meaning or content that it has. I hate to complicate things at the very beginning, but I think there are (at least) two different distinctions to be made here (which I did not clearly distinguish in the papers Jackson refers to). First, there is a distinction between two ways in which the extension of an expression (for example the truth value of an assertive utterance) may depend on the facts. Suppose that the truth of an utterance *u* depends on whether P. This might be because a correct semantics for the language in question, plus the relevant contextual facts, determines that *u* expresses a proposition that is true if P, but false if not-P. On the other hand, the dependence might hold because it is true that if P, then the semantics for the language and the relevant contextual facts would determine that *u* expresses a proposition that is true, whereas if it is not true that P, then the semantics or relevant contextual facts would be different so that a different, and false, proposition would be expressed. If the explanation for the fact that the truth of the utterance depends on the fact that P is of the first kind, it is a semantic dependence, while if the explanation is of the second kind, then it is a meta-semantic dependence.

Second, there is a different distinction between a purely descriptive claim and an explanatory claim. Consider the claims that "rot" means red (in German), that "I" (in English) is the first person singular pronoun, that "London" refers to a city in England, that "she" on a particular occasion referred to Martha Stewart, and that a particular utterance of "it is raining" expressed a proposition that is true if and only if it was raining in Pittsburgh on May 15, 1993. Each of these is a purely descriptive claim that makes no contribution to answering questions about how the expressions came to have the meanings, referents or truth conditions that they are said to have, or about what the facts are (about practices, convention, states of mind, causal relations between various things and events, or whatever) in virtue of which those expressions have those meanings, referents or contents. The descriptive claims I have mentioned are all semantic in the broad sense that they have to do with the semantic properties of expressions, and so they might be called "semantic" claims, as contrasted with the explanations for those facts, which might be called "metasemantic". But some of these descriptive claims are in part about facts that are in a sense extra-semantic—facts about the context in which expressions are used, or about the thing which the expression in question is used to refer to. Furthermore—and this is where the two distinctions threaten to interfere with each other—one might make a purely descriptive claim about facts that are metasemantic in the first sense. Consider the following descriptive claim about a familiar hypothetical case: The word 'water', as used by the inhabitants of Twin Earth (or at least a word that is phonologically and orthographically like that word) refers to a substance that is different from H_2O. Different theorists may agree with this description of what the facts would be in a certain counterfactual situation, but differ about how to explain them: On one hypothesis, the word 'water' has the same meaning, in the Twin Earth context, as it has in our context, but it is a context-dependent word, and so has a different referent. On a contrasting hypothesis, the Twin-Earthian's word is a different word, with a different meaning, since the facts in virtue of which the word has the meaning that it has are different, in relevant respects, on Twin Earth. Of course the disagreement between these two hypotheses is also a disagreement about how to describe the facts—about what the right descriptive semantics is for our word 'water'. But the disagreement turns on different hypotheses about the facts in virtue of which words have the descriptive semantics that they have.

I think it is the first of these two distinctions that is most directly relevant to the question that Jackson is posing, and I agree with him that we can address his question "without answering anything as hard as the meta-semantical question," where this means without answering deep questions about the explanation for the fact that expressions have the contents that they have. But I do think that disagreements about the nature of intentionality probably lie at the heart of any disagreements about how to understand the role of the two kinds of intensions that Jackson distinguishes.

But let me go straight to the story of 'fred,' and the dispute between the two characters, Broad and Narrow, about the content or contents of some of the sentences in the language they invent.

Our characters have introduced a word 'fred', stipulating that it should apply (rigidly) to the shape of the smallest homogeneous object (the sho). They are confident that there is such an object, but neither has any idea what it is, or what shape it has. Although B and N agree about what they have stipulated, they disagree about the content of the statements that they make, using the word they have introduced. N argues that the assertive content of "my table is fred" is a proposition that is true if and only if the table has the same shape as the sho, while B argues that, if the sho is in fact round, the assertive content is a proposition that is true if and only if the table is round, and more generally that if the sho in fact has shape X, then the statement says that the table has shape X. Jackson argues that N is right, and I agree. In fact, I would go further than Jackson in support of N, for he argues that N has a problem that requires him to qualify his answer, while I don't see the problem. But before getting to that, let me make a qualification of my own: I am assuming that the language in question is one that B and N are able to speak. (The story begins by saying the N and B enjoy making up languages "for their private use".) It is possible to give a determinate characterization of a language (or of the meaning of some expression of a language) even if one lacks the cognitive capacity to speak that language (or to use the expression to make statements). There are perfectly good languages in which the word 'fred' means exactly the same as 'is round', and B and N might specify such a language by stipulating that 'fred' shall *mean* 'round' iff the sho is round. (They won't know that they have specified this language, but that is what they will have done, given that the sho is in fact round.) On this hypothesis about what they are doing, they have given the word a determinate meaning, but they don't know what the meaning is, and so they can't *use* the word with that meaning. To underscore the point, let me use a more extreme example: Suppose B were to extend their language by introducing a certain sentence (copied from a book) as an idiom, stipulating that it shall mean, in their language, exactly what it actually means in Polish. The sentence is "Inne światy możliwe są tak samo realne jak świat aktualny." Neither B nor N knows any Polish, and they have no idea what this sentence means, though they correctly believe that it makes an assertive claim. B (being impressed with the authoritative look of the book from which the sentence was copied) writes, "I hereby assert that Inne światy możliwe są tak samo realne jak świat aktualny." Whatever he has done, he hasn't succeeded in asserting what this Polish sentence says. Now there is a perfectly good language that is basically English, with the addition of an idiom, borrowed from Polish, that means just what the Polish sentence means.(There are plenty of foreign expressions, such as "carpe diem" or "que sera sera" that have been incorporated into English.) B and N can specify the language, but they cannot speak it (or this part of it).

So since the language B and N define is (I am assuming) one they can use, it not one where 'fred' means the same as 'is round', but rather a language in which it means, as Jackson and N say, something like ' the actual shape of the sho'. It seems intuitively clear that what it takes to understand *this* expression (and to have the capacity to understand the propositions it is used to express), is to understand the role of the word 'actual' and the meanings of the other words in the description. When B says "the table is fred" (as a wild speculation, not an assertion), the assertoric content of his guess is that the table has the same shape as the sho. The difference between "the actual shape of the sho" and "the shape of the sho" is that the former contains a scope device that plays a role in the compositional semantics of the language that B and N are using. This is different from its role in the language that B and N might have defined, but do not have the capacity to use.

But Jackson sees a problem with N's answer: He argues that N must concede to B that the statement 'The table is fred' is true in all and only the possible worlds in which the table is round (given that the sho is in fact round). I find this very puzzling. I think that all that N should concede is (1) that in the language that he and B are actually using, the compositional semantics will make reference to a *description* of a proposition which in fact has these truth conditions, but that the proposition described is not expressed, and (2) that in the language that B and N might have specified, but do not have the capacity to use, the sentence in question *expresses* this proposition.

As long as we stick with the story of 'fred', I think Jackson and I are in basic agreement. I have no quarrel with the standard ways of spelling out the compositional semantics for "actually", which will be a two-dimensional semantics that yields a distinction between A- and C-intensions. But I think it is important to distinguish this language-internal use of the two-dimensional apparatus from the use that I have made of it to represent the ways in which what is said is a function of the facts.

So I agree with Jackson about 'fred', but we will part company if he wants to take 'fred', as used by N and B, as a model for the way names work in general. Ordinary names, I think, are like the way 'fred' works in the language that N and B can describe, but lack the cognitive capacity to speak. But resolving this issue will require addressing deeper questions about intentionality, and about the character of our cognitive capacities. I have suggested that B and N cannot speak a language with a proper name whose reference is fixed in the way the 'fred' is fixed because they do not know what the shape of the sho is. It seems clear, intuitively, that N and B do not know what the shape of the sho is, but it is notoriously difficult to say, in general, what one must know to know what or who something or someone is. Do I know enough to have the capacity to use the name 'Aristotle' to refer to Aristotle? What about 'Anaximander', for that presocratic philosopher, or 'Berdyetchev' for the Russian city about which I know nothing except that someone once said that Balzac was married there?[14] Can I actually *believe* singular propositions, and form singular intentions about those people and that city? Is my epistemological situation with respect to the relations between those names and the things they name better than N and B's situation with respect to the relation between 'fred' and the shape of the sho? This is a daunting problem, and I think any adequate answer will have to recognize that such knowledge claims are highly context-dependent, and that we must swallow the consequence that semantic competence is context-dependent in the same ways. But I don't think we can avoid the problem by retreating to a purely internal notion of content. I agree that there are A-intensions, in several senses, but despite the name of the character Jackson has chosen to defend them, I don't think they are narrow.

PERRY, "STALNAKER AND INDEXICAL BELIEF"

It has taken me a long time to appreciate what I now take to be John Perry's central point about self-locating belief. I tried, in the paper of mine that Perry is here discussing, to argue that the problem of indexical belief was just a special case of a more general problem that was independent of self-representation or location, and that the phenomenon of self-locating belief could be explained without deviating from a picture of a state of belief as a representation of the way the world is, in itself. I now think that this was a mistake, and I want to concede one of the main points that Perry is arguing in this paper: that my appeal to diagonalization does not, by itself, solve the problem of self-locating belief, and does not obviate the need for a modification of the received doctrine of propositions (in its strong version, in the sense Perry explains in his paper). But I don't want to concede too much; there remain disagreements about how to modify the received doctrine, and more generally about what the best theoretical framework is for representing our states of mind, and the way we describe them.

I will first make a general remark about the problem of self-locating belief, as I now see it. Second, I will consider Perry's distinction between the contents of belief and the different ways of believing them, suggesting that this is not the best way to get at the distinctive character of self-locating belief. Third, I will consider an ingenious and revealing example that he discusses, saying what I think it shows, and how it should be represented. In the end, I think Perry is right that it is what he calls my holism that is responsible for the differences that remain between us in the ways we want to account for the phenomena.

There are two different kinds of example that have been used to support and illustrate the thesis that self-locating belief is not reducible to belief about what

[14] Cf. Dennett (1978), 47.

the world is like in itself. The examples that have received the most attention are ignorance cases: cases where a person knows all the relevant impersonal, objective facts, but remains ignorant of something about his or her location in the world (who one is, or what time it is). The parallel between these examples and the familiar Frege cases of identity confusion was salient, and it motivated me to extend my favored strategy for accounting for the latter kind of case to the former. I still think the parallels are relevant to part of what is going on with such examples, but there is a second range of examples that Perry has emphasized in his many discussions of this kind of attitude (and in section 7 of the current paper), and that I think better bring out what is distinctive about self-locating belief. These are the examples in which two agents are in complete agreement in their relevant beliefs about what the objective world is like, and are also in accord in their relevant values, desires and preferences—what they want the world to be like—but are nevertheless rationally motivated to act differently because of their different perspectives on the world. (For a typical Perry example, see the case of the last piece of cake in section 7.) The differences in perspective—in where the agents locate themselves in the world as they take it to be—that explain the differences in rational behavior seem to be *cognitive* differences that should be reflected in our representation of a person's states of belief and knowledge. Given my commitment (which I share with Perry) to explaining the intentionality of mental states (in part) in terms of the role of the states in motivating rational action, it should have been no surprise that a proper account of belief must give a special role to the agent and the locus of action in the world as the agent takes it to be.

But how should self-location be represented? Perry's strategy is to exploit a distinction between the content of a belief (which is an impersonal proposition) and the way that it is believed, recognizing a distinctive first-personal way of believing propositions that might also be believed, by the same person or by someone else, in a different, non-first-personal way. There are, of course, distinctions to be drawn between what is believed and the ways it is believed, but how one draws them will depend on the way one thinks about content, and I want to build more into the content of belief and other attitudes than Perry does. Before getting to self-location, let me say what I think Perry's picture is, in general terms.

As I read him, Perry is using the distinction between what is believed and the ways something is believed to reconcile an austere and externalist conception of propositional content with Fregean views about cognitive significance by giving a role to modes of presentation, but using them to characterize, not *what* is believed, but instead the way a proposition is believed. So, for example, *what* Quine's Ralph believed was the Russellian singular proposition that Ortcutt is a spy, but to fully characterize his state of belief (those aspects of it that are relevant to rationality) we need to add that he believed this proposition in a certain way, a way corresponding to his man-in-the-brown-hat mode of presentation of Ortcutt. He does not believe this proposition in a way corresponding to the other mode in which Ortcutt is presented to him (as the man seen on the beach). But I think this way of distinguishing content of belief from manner of believing distorts the phenomena, locating an aspect of Ralph's conception of what the world is like on the wrong side of the line. Ralph's way of representing Ortcutt is essential to what he takes the world to be like, and not just to the manner in which he represents the world. Were Ralph later to come to believe that the guy he sees on the beach is *also* a spy, his conception of the world—the conditions under which his beliefs would be true—would change.

On my view, one should begin an investigation into the content of Ralph's beliefs by asking what the world is like according to Ralph. (This is the holism of my account, to which Perry refers.) Forget, for a moment, what our semantic theory tells us about how to interpret Ralph's sincere assertions and the clauses used to ascribe beliefs to him. Instead, try to say what Quine's story seems to be telling us about the kind of world that Ralph takes himself to be in. It will be easy to agree about some things about the world according to Ralph: it is a world in which there are two distinct people, one in a brown hat who is a spy, and one seen on the beach who is not. To the extent that we are clear about what the world is like according to Ralph, we can say what propositions he believes: they are those that are true in all the worlds that accord with the overall account of what the world is like according to him.

The two-dimensional story, and diagnonalization, come into this picture at the stage of explaining how certain sentences and sentential clauses are ways of expressing and referring to some of the propositions that Ralph believes. How, for example, can Ralph's statement, "that guy [pointing to Ortcutt when he is wearing the brown hat] is a spy" be a way of expressing a proposition that is true in all the possibilities compatible with Ralph's conception of the world? Our well-motivated semantics for sentences of this kind tells us that Ralph's statement expresses a singular proposition about Ortcutt—the same proposition that he would have expressed if he were instead pointing to Ortcutt when he was on the beach. The two-dimensional story is an attempt to connect the standard semantics with what such sentences seem, intuitively, to be telling us about what the world is like according to Ralph. It does this by taking account of the propositions that *would* be expressed (according to the standard semantics) if the world were the way Ralph takes it to be. The diagonal propositions determined by two-dimensional representations of the contrasting demonstrative statements will be different, with just one of them being true in the worlds that are the way Ralph takes the world to be. This is why Ralph affirms the one, but not the other.

I have conceded to Perry that the representation of a state of belief as a set of possible worlds needs to be modified in order to account for the special features of self-locating belief, but I think this can be done while keeping the basic picture intact. What needs to be done, I think, is to link the believer and the world and time at which she has the beliefs to a time and place in the worlds as she then takes them to be.[15] The resulting picture will retain the idea that our representations (self-locating and otherwise) are ways of distinguishing between possibilities, and in the amnesiac cases, the first-personal statements will make different distinctions between the possibilities than the corresponding objective statements. Let me use Perry's variation on the story of Lingens in the library to illustrate how the account I favor deals with some of the twists and turns.

Lingens has figured out that he is either Lingens or O'Leary, but does not know which. He tells his story to Julius, who thereby comes to believe that Lingens (the person he is talking to) is either Lingens or O'Leary, and is an amnesiac lost in the library. Call this proposition D. (Perry calls it D because he says it is a diagonal proposition, but diagonal propositions are not a distinctive kind of proposition, but a distinctive way of representing the relation between an expression or token thought and a proposition.) Later, O'Leary overhears Julius asserting D (by displaying the two-dimensional matrix, and pointing to its diagonal), and he accepts what Julius says. But he doesn't recognize Julius as the man to whom he earlier told his sad story, or realize that the abbreviations, L and O, used in Julius's representation of the proposition refer to Lingens and O'Leary. Perry's take on the case is that since it was proposition D that Julius was asserting, we should say that this is the proposition that Lingens came to believe when he accepted it. But he came to believe it in a different way without realizing that it is the same proposition as one that he already believed (in a different way).

As with the Ortcutt case, I think it distorts the phenomenon to say that Lingens's two beliefs (the one he expressed to Julius and the one he later acquired from Julius) are beliefs with the same content. As with Ralph, we begin by asking what the world is like according to Lingens. The answer, at stage one of the story, is something like this: there are worlds compatible with Lingens's beliefs in which Lingens is thinking a certain thought (linked to Lingens's actual thought in the actual world), and is lost in the library; there are also some worlds in which it is O'Leary who is thinking that thought, and is lost in the library. The possible worlds in which it is some third party who is thinking the thought, or in which the person thinking it is *not* lost in the library are excluded from the set of worlds compatible with Lingens's beliefs or knowledge. Now to tell the story at stage two (when Lingens overhears Julius), we need to make further distinctions between the possibilities compatible with Lingens's belief. There is a person x (the subject of Julius's assertion), a person O (who may or may not be O'Leary) and a person L (who may or may not be Lingens). Lingens (when he observes Julius making his assertion) does not know whether Julius's representation is a representation of him, so he doesn't know whether he himself is x (nor whether L is Lingens, etc.). In terms of the expanded representation of the possibilities, we could represent

[15] I gesture, very sketchily, toward the kind of account I want to give in Stalnaker (2004), and I intend to spell out these ideas more fully in work still in progress.

Lingens's original belief, and it will be different from the proposition that Lingens came to believe when he accepted what Julius told him.

The proposition that Lingens came to believe, at the second stage of the story, might be represented as a diagonal of a two-dimensional matrix for Julius's representation. Even when a representation of a proposition is given as a diagonal of a two-dimensional matrix (as in Perry's story), one still has a potential gap between the representation and the proposition it expresses. Where the addressee or onlooker is ignorant of what proposition is represented (in a sentence, the diagonal of a matrix, or whatever form the representation takes), the theorist can represent the ignorance with another two-dimensional matrix, in which the representation expresses different propositions in different possible worlds. (This is one reason it is important to emphasize that diagonals (or A-intensions, in Frank Jackson's terminology) are not a distinctive kind of proposition. It is not that a diagonal representation automatically gives one some kind of direct access to the proposition it expresses.)

Perry's story does illustrate the potential for an interesting kind of breakdown in the transmission of information, a pattern that is not essentially tied to self-location, and that might arise in simpler cases. Suppose Pierre tells me (in English) that it has been a very hot summer in London. Later, he overhears me say to one of his compatriots "il a fait très chaud à Londres cet été." I was just passing on the information that I got from Pierre, but he comes to believe something new by accepting what I said. One could describe this as a case of someone coming to believe the same proposition in a new way, but as in the cases of Ralph and Lingens, it does seem intuitively to be a distinct piece of information, and it does seem that the way the world is according to Pierre changes when he comes to believe what he hears me say.

SOAMES, "UNDERSTANDING ASSERTION"

Scott Soames argues that my model of discourse needs to be drastically modified, but I remain unpersuaded by his arguments. Before digging in my heels, I will make two mildly concessive remarks. After that, I will look at what Soames describes as some presuppositions of my account, arguing that to the extent that I make these presuppositions, they do not have the consequences he draws from them. Finally, I will discuss two of his examples which he uses to raise problems for my account.

My first concessive remark concerns an assumption that Soames correctly attributes to me, and argues that I should give up: that epistemic possibilities should be understood as a subclass of the metaphysical possibilities. (It is not that I assume, for example, that it cannot be epistemically possible that water is not H_2O, or that I assume that it *is* metaphysically possible that water is not H_2O. It is rather that I try, using the two-dimensional framework, to explain the ignorance of one who does not know that water is H_2O in terms of a metaphysical possibility in which a different proposition, roughly, the one that *would* be expressed by "water is not H_2O", if that world were realized, is true.) I continue to think that it *is* illuminating to try to understand another person's states of knowledge and belief, to the extent that we can, by describing, in our own terms, real possibilities that are the way the person takes the world to be, but I am also a proponent of Quine's maxim of shallow analysis ("where it doesn't itch, don't scratch")[16] and I am happy to grant that for some purposes it is useful to take epistemic possibilities at face value without worrying about how to construe them as metaphysically possible. Mathematical discourses, for example, are naturally described with the same tools as discourses about empirical matters, with context sets of possibilities that evolve as assertions and other speech acts take place by the same dynamic processes. One can explain some features of what is going on in such discourses without worrying about what they are ultimately about. Suppose I tell you that a certain integer, say 139129, is either prime or the square of a prime, and you take my word for it. Until you do some further computation, two epistemic possibilities remain for you, and one can think of them just as the possibilities that 139129 is prime, and that it is the square of a prime. But for other purposes, more would have to be said about just what those possibilities are, and why the computation we do is relevant to discovering which of them is realized.

Second, I want to acknowledge that there are different ways of describing the phenomena, using different theoretical resources, that may be equally legitimate. All of us must say some counterintuitive things, and idealize at one point or another, though different accounts of the phenomena put the bumps under different parts of the rug. I have some sympathy with Saul Kripke's often quoted remark about the familiar belief puzzle cases: "I am unsure that the apparatus of 'propositions' does not break down in this area."[17] There is, of course, not just one apparatus of propositions, as Kripke goes on to say, and different ways of construing propositions may break down, or face tensions, at different points. Each theoretical framework must account for the phenomena, in its own way, but one shouldn't confuse the phenomena with one's theoretical account of them. I think Soames imports some of his own theoretical assumptions into his description of the phenomena that he claims my account cannot accommodate. Specifically, all of his arguments rest on some controversial assumptions about *de re* belief (assumptions both about how it should be understood theoretically, and about what examples are cases of it), and about what it means to know what or who something or someone is. These assumptions play a role, both in his account of what he describes as the presuppositions of my model of discourse, and in the discussion of specific examples.

First, on the general presuppositions: it is said that "the model presupposes systematic *de re* knowledge of world-states," and that a "fundamental presupposition of the model is that for any world-state w in the context set,

[16] Quine (1960, 160). [17] Kripke (1980, 21).

and any proposition p that might be asserted in the conversation, speakers and hearers know the truth value of p in w." Before trying to say what I think these presuppositions come to, let me emphasize that the model of discourse (and of thought) is intended as a theorist's external representation of a discourse (and of the states of mind of the participants). It is not assumed that the subjects conceive of the subject matter of their discourse, or of the discourse itself, in terms of possible worlds, or world-states, or that they even have the concept of a possible world. The main presupposition is that discourse involves the expression of propositions, and that propositions have truth conditions. Possible worlds (or world-states) are no more than a way of representing truth conditions—the conditions under which a statement or other proposition-expressing representation would be true. They are not the subject's way of representing the truth conditions of his statements—the statements themselves do that.

Now in this context, I am not sure what it means to say that a subject has systematic de re knowledge of world states, or why I am committed to saying this. (I am actually not sure what "de re knowledge of an individual" means in general. Some philosophers have thought that there is a special kind of epistemic relation to an individual—acquaintance—that is required to grasp singular propositions about the individual, but I think this is a mistake.) I am prepared to assent to the presupposition that speakers know, for any world-state w and proposition p, whether p is true or false in w, provided that this is understood in the following innocuous way: If a speaker understands some sentence S in context (that is, knows what it says), then if the world w is described to the speaker in sufficient detail in language that she understands, she will know whether the proposition expressed by S is true or false in w. To say that this is innocuous is not to deny that all of the usual problems with the highly context-sensitive locutions "knowing who" and "knowing what" will infect the assumption that a speaker knows what some expression says. The model is not trying to defuse these problems by providing or assuming some kind of direct epistemic connection to the possible world-states in terms of which propositional content is represented.

It is clear that Soames's interpretation of this presupposition is far from innocuous. He gives the following argument: we are obviously acquainted with the actual world-state, since we can demonstrate it, and we can know of it ("this world-state, the one I find myself in now") that it obtains, or is instantiated. But then the presupposition (that for any proposition p and world w compatible with the context, one knows the truth value of p in w) implies that we know of any proposition whether it is true or false (in the actual world), which is absurd.

Let me try to defuse this argument in the same way that Kripke tried to dissolve the puzzle about identifying individuals across possible worlds, using Kripke's own homey example: "Two ordinary dice (call them die A and die B) are thrown, displaying two numbers, face up.... There are thirty-six possible states of the pair of dice, as far as the numbers shown face-up are concerned." These states "are literally thirty-six 'possible worlds', as long as we (fictively) ignore everything about the world except the two dice and what they show." Such "school" examples, using a possible worlds representation to clarify elementary probability exercises, involve the same presuppositions as the discourse model that Soames is discussing. We must assume that for any relevant proposition (for example, that the sum of the numbers face up is even), and any one of the mini-worlds (for example, A 3, B 5), the subject will know whether the proposition is true or false in the world-state. Suppose our schoolchild argues that we must reject this presupposition, since he knows which of the possibilities is actual ("it is this very one, the one we are in"), but still doesn't know how the dice landed, and so doesn't know whether certain propositions are true or false in the actual world. This point deserves the same comment as the one Kripke made about the demand for a criterion of transworld identity: "no competent schoolchild would be so perversely philosophical as to make it."[18]

Let me now turn to some of the examples that are supposed to cause trouble for my use of the diagonalization strategy for treating the puzzle cases. Again, the alleged trouble turns on assumptions about de re belief, and infects even the most familiar of the puzzle cases. Soames argues that my analysis must say that "Hesperus is Phosphorus" could be appropriately asserted only in a conversation in which one of the participants is ignorant of what 'Hesperus' or 'Phosphorus' stand for. But, he says, "each participant may know perfectly well that 'Hesperus' refers to this object [pointing in the evening to Venus] and that 'Phosphorus' refers to that object [pointing in the morning to Venus]." I am happy to grant these belief attributions, and to apply the diagonalization strategy to them, as well as to the simple identity statement. No extra assumptions are needed for this analysis; we began with the assumption that for anyone who does not know whether Hesperus is Phosphorus, there will be a possible world compatible with his beliefs in which there are two distinct planets, one called 'Hesperus' that appears in the evening and one called 'Phosphorus' that appears in the morning. (Suppose, just for illustration, that in this world-state it is Mars that appears in the evening, and Venus that appears in the morning.) The sentence 'Hesperus is Phosphorus', as used in this counterfactual possible world-state, will express a different proposition (according to the standard semantics) than it expresses in the actual world, one that is necessarily false. The diagonal will be contingent—true in the actual world and false in the possible world described. That was the original story. Exactly the same story makes it true that the sentence " 'Hesperus' refers to this object [pointing, in the evening, to Mars, the celestial body that is in the place where Venus in fact is]" will express a proposition (on the same standard semantics) that is true in the world described, but false in the actual world. But the diagonal will be true in both this counterfactual world and in the actual

[18] Kripke (1980, 17).

world, which is why we can agree that the subject in the example knows that what this statement says is true.

So while I accept Soames's demonstrative knowledge attributions, I reject the conclusion he draws from them, that "such speakers know of the referent of each name that it is the referent of that name." On my view, a de re belief attribution is correct when one can correctly and determinately describe the world according to the believer as a function of the individual. What this requires is not some intimate acquaintance relation, but only that there be a unique candidate. In the standard puzzle cases, there may be no fact of the matter about which of two distinct individuals in the world according to the believer is identical to a given individual in the actual world, or at least no fact of the matter that is independent of the context of attribution. (The idea is not that there are individuals of indeterminate identity in some possible worlds; but rather that it may be indeterminate, without contextual clues, which of two sets of possible worlds best represents the way the world is according to some believer.) Soames says that "it is well known . . . that one can know of one and the same individual i that he is F and that he is G, without knowing (or being in a position to know) of i that he is both F and G." What is perhaps well known is that certain analyses of de re belief have this consequence.[19] I would prefer to put the lesson of the puzzle cases differently: for some cases, there are contexts in which it would be correct to say that the subject believes of an individual i that he is F, other contexts in which it would be correct to say that she believes of an individual i that he is G, but no contexts in which it would be correct to say that she believes of i that he is both F and G.[20]

Finally, a quick remark about the paperweight example. This case involves ignorance of an essential property rather than identity confusion, but like Soames's other examples, it involves demonstrative identification, and what looks like a paradigm case of knowing what one is thinking about. Here I am in Soames's office, holding his paperweight in my hand, wondering what it is made of. It is in fact made of wood, but for all I know, it is made of plastic. Here, I agree, it seems, intuitively, that it is the singular proposition that this particular object is made of plastic that is compatible with my knowledge (though the judgment that the content of a belief is a singular proposition is a theoretical judgment, and not a datum). But the composition of the paperweight is essential

[19] On the analysis given in Quine (1956) and modified in David Kaplan (1968), a de re belief ascription is not the ascription of belief in a particular proposition, but a statement that the believer has an unspecified belief of a certain form. De re belief attributions involve something like existential quantification over names, or modes of presentation. Soames's favored account, as I understand it, is different in that it identifies de re belief attribution with the attribution of belief in a specific (singular) proposition, but, like John Perry, he would use something like modes of presentation to distinguish different ways of believing a proposition.

[20] Soames's example of the man at the end of the table, who is known to be either Ted Sider, or John Hawthorne, exactly parallels the Hesperus/Phosphorus case, and would be handled in the same way by the diagonalization strategy.

to it, so there is no possible world in which this specific object is made of plastic. How can what is, for me, an epistemic possibility be represented by a genuine possible world? Consider the following possible world, which I think Soames will agree is metaphysically possible: I am sitting in Soames's office, holding his plastic paperweight in my hand, wondering (just as I am in the actual world) what it is made of. This possible paperweight is a different object from his actual paperweight, though it looks and feels just like it. It does not seem unreasonable to think that a possible world of this kind is compatible with my knowledge. The two-dimensional strategy allows us to reconcile the assumption that it is with the judgment that it also is right to say that I know that it is *this* paperweight whose composition I am wondering about.

Thanks again to Daniel, Sydney, Paul, Richard, Vann, Tim, Bill, Steve, Frank, John, Scott, Alex and Judy for providing me with the chance to learn from and respond to such a rich and wide-ranging collection of ideas and arguments.

REFERENCES

Carnap, R. (1936–7) "Testability and Meaning", *Philosophy of Science*, **3**,**4**.

Cartwright, R. (1994) "Speaking of Everything", *Nous* **24**, 1–20.

Dennett, D. (1978) "Brain writing and mind reading", in Dennett, *Brainstorms* (Bradford Books), 39–50.

Dummett, M. (1978) "Truth" in Dummett, *Truth and Other Enigmas* (Cambridge, Mass.: Harvard University Press), 1–24.

Edgington, D. (1986) "Do conditionals have truth conditions?", *Critica* **18**, 3–30.

Etchemendy, J. (1990) *The Concept of Logical Consequence* (Cambridge, Mass.: Harvard University Press).

Gibbard, A. (1981) "Two recent theories of conditionals", in W. Harper, *et al.* (eds.), *Ifs* (Dordrecht: Reidel), 211–47.

Glanzberg, M. (2004) "Quantification and Realism", *Philosophy and Phenomenological Research* **69**, 541–72.

Goodman, N. (1983) *Fact Fiction and Forecast* (4th edn.) (Cambridge, Mass.: Harvard University Press).

Higginbotham, J. (2003) "Conditionals and compositionality", *Philosophical Perspectives*, **17** (*Language and Philosophical Linguistics*), 181–94.

Kaplan D. (1968) "Quantifying in", *Synthese*, **19**, 178–214.

Kripke, S. (1979) "Speaker's reference and semantic reference", in P. French *et al.* (eds.), *Contemporary Perspectives in the Philosophy of Language* (Minneapolis: University of Minnesota Press), 6–27.

Kripke, S. (1980) *Naming and Necessity* (Cambridge, Mass.: Harvard University Press).

Lewis, D. (1990) "Noneism or Allism?", *Mind* **99**, 23–31.

Lycan, W. (2001) *Real Conditionals* (New York: Oxford University Press).

Quine, W. (1956) "Quantifiers and propositional attitudes", *Journal of Philosophy* **53**, 177–87.

—— (1960) *Word and Object* (New York and London: John Wiley & Sons).
—— (1961) "On what there is", in *From a Logical Point of View* (Cambridge, Mass.: Harvard University Press), 1–19.
Stalnaker, R. (2004) "On Thomas Nagel's objective self", in Stalnaker, *Ways a World Might Be* (Oxford: Oxford University Press), 253–75.
—— (2005) "Conditional assertions and conditional propositions", in *New Work on Modality* (MIT Working Papers in Linguistics and Philosophy, **51**)
Williamson, T. (2003) "Everything", *Philosophical Perspectives*, **17** (*Language and Philosophical Linguistics*), 415–65.

Publications by Robert Stalnaker

BOOKS (INCLUDING EDITED BOOKS)

A. *Ifs: Conditionals, Belief, Decision, Chance, and Time* (edited with William Harper and Glenn Pearce). Dordrecht’: D. Reidel, 1981.
B. *Inquiry*. Cambridge, Mass.: Bradford Books, MIT Press, 1984.
C. *Context and Content: Essays on Intentionality in Speech and Thought*. Oxford: Oxford University Press, 1999.
D. *Fact and Value* (edited, with Alex Byrne and Ralph Wedgwood). Cambridge, Mass.: MIT Press, 2000. (Festschrift for Judith Thomson)
E. *Ways a World Might Be: Metaphysical and Anti-metaphysical Essays*. Oxford, Oxford University Press, 2003.

ARTICLES

1. “Events, Periods and Institutions in Historians’ Language.” *History and Theory*, 6 (1967), 159–79.
2. “A Theory of Conditionals.” N. Rescher (ed.), *Studies in Logical Theory* Oxford: Blackwell, 1968, 98–112; reprinted in [A], and in E. Sosa (ed.), *Causation and Conditionals* (London: Oxford University Press, 1975), and in Frank Jackson (ed.), *Conditionals* (Oxford and New York: Oxford University Press, 1991).
3. “Modality and Reference” (with R. H. Thomason). *Noûs*, 2 (1968), 359–72.
4. “Abstraction in First Order Modal Logic” (with R. H. Thomason). *Theoria*, 34 (1968), 203–7.
5. “Wallace on Propositional Attitudes.” *Journal of Philosophy*, 66 (1969), 803–6.
6. “Probability and Conditionals.” *Philosophy of Science*, 37 (1970), 64–80; reprinted in [A].
7. “A Semantic Analysis of Conditional Logic” (with R. H. Thomason). *Theoria*, 36 (1970). Italian trans. published in Claudio Pizzi (ed.), *Leggi di Natura, Modalita, Ipotesi* (Milan: Feltrinelli, 1978).
8. “Pragmatics.” *Synthese*, 22 (1970), 272–89; reprinted in D. Davidson and G. Harman (eds.), *Semantics of Natural Language* (Dordrecht: D. Reidel, 1972), A. P. Martinich (ed.), *The Philosophy of Language* (2nd edn.), Oxford: Oxford University Press, 1989, and in [C]. Italian trans. published in Andrea Bonomi (ed.), *La Struttura Logica Linguaggio* (Milan: Valentino Bompiani, 1973).

9. "A Semantic Theory of Adverbs" (with R. H. Thomason). *Linguistic Inquiry*, 4 (1973), 195–220.
10. "Presuppositions." *Journal of Philosophical Logic*, 2 (1973), 447–7; reprinted in D. Hockney, W. Harper and B. Freed (eds.), *Contemporary Research in Philosophical and Linguistic Semantics* (Dordrecht: D. Reidel, 1975).
11. "Pragmatic Presuppositions." Milton K. Munitz and Peter Unger (eds.), *Semantics and Philosophy* (New York: New York University Press, 1974), 197–213; reprinted in Steven Davis (ed.), *Pragmatics: A Reader* (Oxford: Oxford University Press, 1991), and in [C].
12. "Indicative Conditionals." *Philosophia*, 5 (1975), 269–86; reprinted in Asa Kasher (ed.), *Language in Focus: Foundations, Methods and Systems* (Dordrecht: D. Reidel, 1976), in [A], [C], and in Frank Jackson (ed.), *Conditionals* (Oxford: Oxford University Press, 1991).
13. "Propositions." Alfred MacKay and Daniel Merrill (eds.), *Issues in the Philosophy of Language* (New Haven: Yale University Press, 1976), 79–91.
14. "Possible Worlds." *Noûs*, 10 (1976), 65–75; reprinted in M. J. Loux (ed.), *The Possible and the Actual* (Ithaca: Cornell University Press, 1979), and in Ted Honderich and Myles Burnyeat (eds.), *Philosophy As It Is* (Penguin, 1979). Polish trans. in Tadeusza Szubki (ed.), *Metafizyka W Filozofii Aanalitycznej* (Lublin, 1995).
15. "Complex Predicates." *The Monist*, 60 (1977), 327–39.
16. "Assertion." *Syntax and Semantics*, 9 (1978), 315–32; reprinted in Steven Davis (ed.), *Pragmatics: A Reader* (New York: Oxford University Press, 1991), and in [C].
17. "Anti-Essentialism." *Midwest Studies in Philosophy*, 4 (1979), 343–55; reprinted in [E].
18. "Indexical Belief." *Synthese*, 49 (1981), 129–51; reprinted in [C].
19. "Logical Semiotic." E. Agazzio (ed.), *Modern Logic–A Survey* (Dordrecht: D. Reidel, 1981), 439–56.
20. "A Defense of Conditional Excluded Middle." [A], 87–104.
21. "Possible Worlds and Situations." *Journal of Philosophical Logic*, 15 (1986), 109–23.
22. "Replies to Schiffer and Field." *Pacific Philosophical Quarterly*, 67 (1986), 113–23.
22. "Counterparts and Identity." *Midwest Studies in Philosophy*, 11, *Studies in Essentialism* (1987), 121–40; reprinted in [E].
24. "Semantics for Belief." *Philosophical Topics*, 15 (1987), 177–90; reprinted in [C].
25. "Belief Attribution and Context." Robert Grimm and Daniel Merrill (eds.), *Contents of Thought* (Tucson: U. of Arizona Press, 1988), 140–56; reprinted in [C].

26. "Vague Identity." David Austin (ed.), *Philosophical Analysis* (Dordrecht: Kluwer, 1988), 349–60; reprinted in [E].
27. Critical notice of D. Lewis, *On the Plurality of Worlds. Mind*, 97 (1988), 117–28.
28. "On What's In the Head." *Philosophical Perspectives*, **3**: *Philosophy of Mind and Action Theory* (1989), 287–316; reprinted in David M. Rosenthal (ed.), *The Nature of Mind* (Oxford: Oxford University Press, 1991), and in [C].
29. "Mental Content and Linguistic Form." *Philosophical Studies*, 58 (1990), 129–46; reprinted in [C].
30. "Narrow Content." C. Anthony Anderson and Joseph Owens (eds.), *Propositional Attitudes: The Role of Content in Logic, Language and Mind* (Stanford: CSLI, 1990), 131–46; reprinted in [C].
31. "How To Do Semantics for the Language of Thought." Barry Loewer and Georges Rey (eds.), *Meaning and Mind: Fodor and his Critics* (Oxford: Blackwell, 1991), 229–37.
32. "The Problem of Logical Omniscience, I." *Synthese*, 89 (1991), 425–40; reprinted in [C].
33. "Notes on Conditional Semantics," Yoram Moses (ed.), *Proceedings of the Fourth Conference on Theoretical Aspects of Reasoning about Knowledge* (San Mateo, Calif.: Morgan Kaufmann Publishers, Inc., 1992), 316–27.
34. Critical notice of David Sanford, *If P, then Q: Conditionals and the Foundations of Reasoning. Notre Dame Journal of Formal Logic*, 33 (1992), 291–97.
35. "Twin Earth Revisited," *Proceedings of the Aristotelian Society* (1993), 297–311; reprinted in [C].
36. "What is the Representational Theory of Thinking? A Comment on William G. Lycan." *Mind and Language*, 8 (1993), 423–30.
37. "A Note on Nonmonotonic Modal Logic." *Artificial Intelligence*, 64 (1993), 183–96.
38. "What is a Non-monotonic Consequence Relation?" *Fundamenta Informaticae*, 21 (1994), 7–21.
39. "On the Evaluation of Solution Concepts," *Theory and Decision*, 37 (1994), 49–73; revised version in M. O. L. Bacharach, L.-A. Gérard-Varet, P. Mongin and H. S. Shin (eds.), *Epistemic Logic and the Theory of Games and Decisions* (Kluwer Academic Publisher, 1997), 345–64.
40. "Conditionals as Random Variables" (with R. C. Jeffrey) B. Skyrms and E. Eells (eds.), *Probability and Conditionals: Belief Revision and Rational Decision* (Cambridge: Cambridge University Press, 1994), 31–46.
41. "The Interaction of Modality with Quantification and Identity," W. Sinnott-Armstrong, D. Raffman, and N. Asher (eds.), *Modality, Morality and Belief: Essays in Honor of Ruth Barcan Marcus* (Cambridge: Cambridge University Press, 1994), 12–28; reprinted in [E].

42. "On What Possible Worlds Could Not Be." Adam Morton and Stephen Stich (eds.), *Benacerraf and his Critics* (Oxford: Basil Blackwell, 1996), 103–19; reprinted in [E].
43. "On a Defense of the Hegemony of Representation." Enrique Villanueva (ed.), *Perception* (Atascadero, Calif.: Ridgeview Publishing Company, 1996), 101–8.
44. "Knowledge, Belief and Counterfactual Reasoning in Games." *Economics and Philosophy*, 12 (1996), 133–62.
45. "Impossibilities." *Philosophical Topics*, 24 (1996), 193–204; reprinted in [E].
46. "Varieties of Supervenience." *Philosophical Perspectives*, 10: *Metaphysics* (1996), 221–41; reprinted in [E].
47. "On the Representation of Context," *Journal of Logic, Language and Information*, 7 (1998), 3–19; reprinted in [C].
48. "Los nombres y la referencia: semántica y metasemántica." *Theorema*, 17 (1998), 7–19. (Spanish translation of a paper based on item [iv], below.)
49. "Belief Revision in Games: Forward and Backward Induction." *Mathematical Social Sciences*, 36 (1998), 31–56.
50. "What Might Nonconceptual Content Be?" Enrique Villanueva (ed.), *Concepts* (Atascadero, Calif., Ridgeview Publishing Co., 1998), 339–52; reprinted in Y. Gunther (ed.), *Essays in Nonconceptual Content* (Cambridge, Mass.: MIT Press, 2003), 95–106.
51. "Extensive and Strategic Forms: Games and Models for Games." *Research in Economics*, 53 (1999), 293–319.
52. "Dualism, Conceptual Analysis and the Explanatory Gap" (with Ned Block), *Philosophical Review*, 108 (1999), 1–46.
53. "Logical Omniscience, II." [C], 255–73.
54. "Comparing Qualia Across Persons." *Philosophical Topics*, 26 (2000), 385–405; reprinted in [E].
55. "On Moore's Paradox." Pascal Engel (ed.), *Believing and Accepting* (Dordrecht: Kluwer, 2000), 93–100.
56. "On Considering a Possible World as Actual." *Proceedings of the Aristotelian Society*, supplementary volume 75 (2001), 141–56; reprinted in [E].
57. "Metaphysics without Conceptual Analysis." *Philosophy and Phenomenological Research*, 62 (2001), 631–6.
58. "Epistemic Consequentialism." *Proceedings of the Aristotelian Society*, supplementary vol. 76 (2002), 153–68.
59. "What Is it like to Be a Zombie?" John Hawthorne and Tamar Szabó Gendler (eds.), *Conceivability and Possibility* (New York: Oxford University Press, 2002), 385–400; reprinted in [E].
60. "Common Ground." *Linguistics and Philosophy*, 25 (2002), 701–21.
61. "Conceptual Truth and Metaphysical Necessity." [E], 201–15.
62. "On Thomas Nagel's Objective Self." [E], 253–75.

63. "Comments on 'From Contextualism to Contrastivism in Epistemology'." *Philosophical Studies*, 119 (2003), 105–17.
64. "Lewis on Intentionality." *Australasian Journal of Philosophy*, 82 (2004), 199–212; reprinted in F. Jackson and G. Priest (eds.), *Lewisian Themes: the Philosophy of David Lewis* (Oxford: Oxford University Press, 2004).
65. "Assertion Revisited: on the interpretation of two-dimensional modal semantics." *Philosophical Studies*, 118 (2004), 299–322.
66. "Conditional Assertions and Conditional Propositions." *New Work on Modality* (MIT Working Papers in Linguistics and Philosophy, 51, 2005).
67. "Saying and Meaning, Cheap Talk and Credibility." A. Benz, G. Jager and R. van Rooij (eds.), *Game Theory and Pragmatics* (New York: Palgrave MacMillan, 2005).
68. "On Logics of Knowledge and Belief." *Philosophical Studies*, 128 (2006), 169–99.

HANDBOOK AND ENCYCLOPEDIA ARTICLES

i. "Pragmatik." J. Speck (ed.), *Handbuch wissenschaftstheoretisher Begriffe*, 1975, 501–6.
ii. "Robert Stalnaker." S. Guttenplan (ed.), *A Companion to the Philosophy of Mind* (Oxford: Basil Blackwell Ltd., 1994), 561–8.
iii. "Modality and Possible Worlds." Jaegwon Kim and Ernest Sosa (eds.), *A Companion to Metaphysics* (Oxford: Basil Blackwell, 1994), 333–7.
iv. "Reference and Necessity." Crispin Wright and Bob Hale (eds.), *A Companion to the Philosophy of Language* (Oxford: Basil Blackwell, 1997), 534–54; reprinted in [E].
v. "Logical Omniscience, Problem of." R. Wilson and F. Keil (eds.), *The MIT Encyclopedia of Cognitive Science* (Cambridge, Mass.: MIT Press, 1999), 489–90.
vi. "Propositional Attitudes." R. Wilson and F. Keil (eds.), *The MIT Encyclopedia of Cognitive Science* (Cambridge, Mass.: MIT Press, 1999), 678–9.
vii. "David Lewis." David Sosa and A. P. Martinich (eds.), *A Companion to Analytic Philosophy* (Oxford: Basil Blackwell, 2001), 478–88.

List of Contributors

Richard G. Heck, Jr., Professor of Philosophy, Brown University

Frank Jackson, Distinguished Professor of Philosophy, Research School of Social Sciences, Australian National University

William G. Lycan, William Rand Kenan, Jr. Professor of Philosophy, University of North Carolina at Chapel Hill

Vann McGee, Professor of Philosophy, Massachusetts Institute of Technology

John Perry, Henry Waldgrave Stuart Professor of Philosophy, Stanford University

Paul M. Pietroski, Professor of Philosophy and Professor of Linguistics, University of Maryland at College Park

Sydney Shoemaker, Susan Linn Sage Professor of Philosophy Emeritus, Cornell University

Scott Soames, Professor of Philosophy, University of Southern California

Daniel Stoljar, Senior Fellow, Research School of Social Sciences, Australian National University

Timothy Williamson, Wykeham Professor of Logic, New College, University of Oxford

Stephen Yablo, Professor of Philosophy, Massachusetts Institute of Technology

Index

Adams, E. W. 152
Aristotle 94

Barker, S. J. 152 n. 10, 158, 159 n. 21, 160 n. 22, 161
Beaver, D. 175 n.
Belnap, N. 155–6
Beltrami, E. 105–6
Bennett, J. 161
Block, N. 10
Boolos, G. S. 99–100, 114
Braddon-Mitchell, D. 11
Broackes, J. 65
Burge, T. 71, 264

Carnap, R. 37, 279
Castañeda, H.-N. 205, 206, 215
Chalmers, D. J. 4, 109–10, 192
Chomsky, N. 38, 43, 47, 49, 52, 54, 61–2, 63, 65, 75–6, 78, 79, 88
Clark, A. 20, 255

Davidson, D. 46, 49
Davies, M. 9
Dedekind, R. 95–6
DeRose, K. 160 n. 22
Donnellan, K. 171–2, 179, 182
Dummett, M. 63, 65–6, 70, 73, 86, 89, 274

Edgington, D. 152, 158–61
Einstein, A. 94
Empedocles 109
Euclid 94, 106
Evans, G. 195

Fodor, J. 44, 54
Frege, G. 18, 48, 105, 106, 166 n. 8, 168, 181, 208, 215

George, A. 77–8, 88 n. 35
Gödel, K. 98, 110
Goodman, N. 279
Grandy, R. 160 n. 22
Grice, H. P. 68, 81, 181, 195

Heck, R. G. 61–91, 261–5
Henkin, L. 103, 110
Higginbotham, J. 50, 51, 72 n. 15
Hilbert, D. 20, 97, 106, 117, 255
Hodges, W. 101
Humberstone, L. 9
Hume, D. 102

Israel, D. 219–20

Jackson, F. 4, 5, 16, 191–202, 281–5
Jeffrey, R. 150, 154, 156

Kalderon, M. 20, 255
Kaplan, D. 172–3, 205, 207–8, 293 n. 19
Krahmer, E. 175 n.
Kripke, S. 3, 6–7, 8, 11–12, 13–14, 35 & n., 36, 108, 173, 179, 229, 264, 290, 291, 292

Langendoen, D. T. 179
Lewis, D. K. 35, 93 n., 115–16, 117, 195–6, 199
Lycan, W. G. 148–62, 272–6

McDowell, J. 82
McGee, V. 93–119, 265–8
Mautner, F. I. 102
Mill, J. S. 181
Montague, R. 48
Moore, G. E. 25

Perry, J. 16, 204–20, 285–9
Pietroski, P. M. 34–58, 258–61
Plato 94
Putnam, H. 2, 3, 12–13

Quine, W. V. 62, 129, 148, 149, 268, 290, 293 n. 19

Rayo, A. 100, 114
Reichenbach, H. 217
Rhinelander, P. 148, 149
Russell, B. 170 n. 14, 175, 176–7, 208, 215

Schein, B. 50
Schlick, F. A. M. 18
Segal, G. 87
Shoemaker, S. 2, 18–32, 254–7
Shope, R. 177

Simons, M. 168 n.
Soames, S. 54 n. 14, 222–49, 289–94
Stanley, J. 84
Stoljar, D. 1–16, 251–4
Strawson, P. 44 n., 69 n., 165 n. 4, 166, 169–72, 175, 179, 277

Tarski, A. 100–2, 103, 104–5, 116, 117
Thales 7–9

Uzquiano, G. 100

van Fraassen, B. 117–18
Veblen, O. 99, 117
von Fintel, K. 165 n. 4 & 7, 175 n.
von Wright, G. H. 153 n. 13, 157 n. 19

Wiggins, D. 65, 82
Williamson, T. 123–46, 268–72
Wittgenstein, L. 15, 16
Woods, M. 158

Yablo, S. 164–89, 276–80